Gels for Removal and Adsorption (2nd Edition)

Gels for Removal and Adsorption (2nd Edition)

Shiyang Li
Zhenxing Fang
Kaiming Peng

Basel • Beijing • Wuhan • Barcelona • Belgrade • Novi Sad • Cluj • Manchester

Shiyang Li
College of Environmental
and Chemical Engineering
Shanghai University
Shanghai
China

Zhenxing Fang
College of Science
and Technology
Ningbo University
Ningbo
China

Kaiming Peng
College of Environmental
Science and Engineering
Tongji Univerisity
Shanghai
China

Editorial Office
MDPI AG
Grosspeteranlage 5
4052 Basel, Switzerland

This is a reprint of articles from the Special Issue published online in the open access journal *Gels* (ISSN 2310-2861) (available at: www.mdpi.com/journal/gels/special_issues/3DPSX394X6).

For citation purposes, cite each article independently as indicated on the article page online and using the guide below:

Lastname, A.A.; Lastname, B.B. Article Title. *Journal Name* **Year**, *Volume Number*, Page Range.

ISBN 978-3-7258-1918-8 (Hbk)
ISBN 978-3-7258-1917-1 (PDF)
https://doi.org/10.3390/books978-3-7258-1917-1

© 2024 by the authors. Articles in this book are Open Access and distributed under the Creative Commons Attribution (CC BY) license. The book as a whole is distributed by MDPI under the terms and conditions of the Creative Commons Attribution-NonCommercial-NoDerivs (CC BY-NC-ND) license (https://creativecommons.org/licenses/by-nc-nd/4.0/).

Contents

Zhenxing Fang, Kaiming Peng and Shiyang Li
Editorial for the Special Issue "Gels for Removal and Adsorption (2nd Edition)"
Reprinted from: *Gels* 2024, *10*, 512, doi:10.3390/gels10080512 1

Juan Ye, Yanchun Yang, Li Zhang, Man Li, Yiling Wang and Yuxuan Chen et al.
Molten Alkali-Assisted Formation of Silicate Gels and Its Application for Preparing Zeolites
Reprinted from: *Gels* 2024, *10*, 392, doi:10.3390/gels10060392 5

Shuyue Wang, Yajun Wang, Xinyi Wang, Sijia Sun, Yanru Zhang and Weixiong Jiao et al.
Study on Adsorption of Cd in Solution and Soil by Modified Biochar–Calcium Alginate Hydrogel
Reprinted from: *Gels* 2024, *10*, 388, doi:10.3390/gels10060388 16

Francesco Gabriele, Cinzia Casieri and Nicoletta Spreti
Efficacy of Chitosan-Carboxylic Acid Hydrogels in Reducing and Chelating Iron for the Removal of Rust from Stone Surface
Reprinted from: *Gels* 2024, *10*, 359, doi:10.3390/gels10060359 28

Farid Hajareh Haghighi, Roya Binaymotlagh, Paula Stefana Pintilei, Laura Chronopoulou and Cleofe Palocci
Preparation of Peptide-Based Magnetogels for Removing Organic Dyes from Water
Reprinted from: *Gels* 2024, *10*, 287, doi:10.3390/gels10050287 46

Zhengdong Zhao, Yichang Jing, Yuan Shen, Yang Liu, Jiaqi Wang and Mingjian Ma et al.
Silicon-Doped Carbon Dots Crosslinked Carboxymethyl Cellulose Gel: Detection and Adsorption of Fe^{3+}
Reprinted from: *Gels* 2024, *10*, 285, doi:10.3390/gels10050285 60

Yang Liu, Mingjian Ma, Yuan Shen, Zhengdong Zhao, Xuefei Wang and Jiaqi Wang et al.
Polyhedral Oligomeric Sesquioxane Cross-Linked Chitosan-Based Multi-Effective Aerogel Preparation and Its Water-Driven Recovery Mechanism
Reprinted from: *Gels* 2024, *10*, 279, doi:10.3390/gels10040279 74

Long Quan, Xueqian Shi, Jie Zhang, Zhuju Shu and Liang Zhou
Preparation of a Novel Lignocellulose-Based Aerogel by Partially Dissolving Medulla Tetrapanacis via Ionic Liquid
Reprinted from: *Gels* 2024, *10*, 138, doi:10.3390/gels10020138 90

Ayeh Khorshidian, Niloufar Sharifi, Fatemeh Choupani Kheirabadi, Farnoushsadat Rezaei, Seyed Alireza Sheikholeslami and Ayda Ariyannejad et al.
In Vitro Release of Glycyrrhiza Glabra Extract by a Gel-Based Microneedle Patch for Psoriasis Treatment
Reprinted from: *Gels* 2024, *10*, 87, doi:10.3390/gels10020087 107

Francesca Porpora, Luigi Dei, Teresa T. Duncan, Fedora Olivadese, Shae London and Barbara H. Berrie et al.
Non-Aqueous Poly(dimethylsiloxane) Organogel Sponges for Controlled Solvent Release: Synthesis, Characterization, and Application in the Cleaning of Artworks
Reprinted from: *Gels* 2023, *9*, 985, doi:10.3390/gels9120985 124

Luis M. Araque, Roberto Fernández de Luis, Arkaitz Fidalgo-Marijuan, Antonia Infantes-Molina, Enrique Rodríguez-Castellón and Claudio J. Pérez et al.
Linear Polyethyleneimine-Based and Metal Organic Frameworks (DUT-67) Composite Hydrogels as Efficient Sorbents for the Removal of Methyl Orange, Copper Ions, and Penicillin V
Reprinted from: *Gels* **2023**, *9*, 909, doi:10.3390/gels9110909 . . . . **143**

Editorial

Editorial for the Special Issue "Gels for Removal and Adsorption (2nd Edition)"

Zhenxing Fang [1], Kaiming Peng [2] and Shiyang Li [3,*]

[1] Ningbo Key Laboratory of Agricultral Germplasm Resources Mining and Environmental Regulation, College of Science and Technology, Ningbo University, Ningbo 315300, China; fangzhenxing128@163.com
[2] College of Environmental Science and Engineering, Institution of Carbon Neutrality, Tongji University, Shanghai 200092, China; pengkaiming@tongji.edu.cn
[3] Key Laboratory of Organic Compound Pollution Control Engineering (MOE), School of Environmental and Chemical Engineering, Shanghai University, Shanghai 200444, China
* Correspondence: lishiyang@shu.edu.cn

1. Introduction

Gel materials, especially hydrogels and aerogels, have been materials of interest in adsorption technology research in recent years [1]. Hydrogels are crosslinked networks of hydrophilic polymers, which contain a large amount of water without dissolving, maintaining their three-dimensional network structure [2]. This structure endows hydrogels with excellent adsorption performance, enabling them to remove and adsorb water-soluble pollutant molecules. Aerogel is a type of material with high porosity and low density. Its unique three-dimensional network structure grants it excellent performance in the field of gas adsorption and purification [3].

The applications of hydrogel in wastewater treatment have made remarkable progress. Due to the rich groups, porous structure, and convenient controllability, hydrogels have significant advantages in the adsorption and purification of wastewater. Harmful substances such as heavy metal ions, organic pollutants, and dyes from wastewater could be efficiently removed by the functional hydrogels [4–6]. However, the hydrogel adsorbents prepared at present still face problems, such as insufficient binding groups and poor mechanical properties, which limit their practical application. Therefore, the design and preparation of novel hydrogel adsorbents with good adsorption properties and mechanical properties is a current research hotspot. The performance of hydrogels could be improved by introducing functional groups, adjusting the degree of crosslinking, and combining them with other materials [7,8]. For example, the introduction of amino, carboxyl, and other functional groups can enhance the selective adsorption capacity of hydrogels for specific pollutants. The adsorption property and mechanical strength of the hydrogel can be balanced by adjusting the crosslinking degree. The adsorption efficiency and stability of hydrogels can be further improved by combining them with other materials such as nanoparticles, fibers, etc.

The applications of aerogels in the field of gas adsorption purification are also remarkable. Inorganic oxide aerogels, such as SiO_2 aerogels, have attracted much attention due to their high adsorption efficiency, convenient desorption, and stable performance [9,10]. After amino functionalization, the selective adsorption capacity for CO_2 was significantly improved, maintaining stable performance in multiple adsorption and desorption cycles. In addition, new aerogel materials such as metal oxide aerogel and carbon aerogel gel also show unique advantages in gas adsorption and purification.

Gel materials have a broad application prospect in the field of adsorption and removal. With the continuous progress of science and technology and in-depth research, the development and application of new gel materials will continue to promote the development of this field. In the future, we expect to see more high-performance and multi-functional

Citation: Fang, Z.; Peng, K.; Li, S. Editorial for the Special Issue "Gels for Removal and Adsorption (2nd Edition)". *Gels* 2024, *10*, 512. https://doi.org/10.3390/gels10080512

Received: 30 July 2024
Accepted: 1 August 2024
Published: 3 August 2024

Copyright: © 2024 by the authors. Licensee MDPI, Basel, Switzerland. This article is an open access article distributed under the terms and conditions of the Creative Commons Attribution (CC BY) license (https://creativecommons.org/licenses/by/4.0/).

gel materials widely used in wastewater treatment, gas purification, and other fields, providing strong support for solving environmental pollution problems. In this context, this Special Issue, entitled "Gels for Removal and Adsorption (2nd Edition)", in *Gels*, has been established to shed light on gels' synthesis with various biomasses, inorganic and organic materials, crosslinking, structural characterization, and the applications for removal and adsorption. In addition to this Editorial, this Special Issue is comprised of 10 articles in total. It is encouraging that these contributions can boost the development of gels' applications, especially in the field of pollutant removal and adsorption. The following content will provide a brief overview of them.

2. Overview

2.1. Some Novel Methods to Synthesize Functional Hydrogels

Zhou et al. reported that a novel lignocellulose-based aerogel was fabricated by partially dissolving medulla tetrapanacis via ionic liquid. The prepared aerogels preserve the original honeycomb-like porous structure, and the newly formed micropores due to the partial dissolving by ionic liquid [11]. The formed micropores enhance the material's capillary force, providing efficient directional transport performance, making it well suited for applications requiring high compressive performance and selective directional transport.

Wang et al. reported that a water-driven recovery aerogel was prepared by crosslinking epoxy groups with chitosan amino groups. The main structure of epoxy groups is oligomeric silsesquioxane, which enhances its susceptibility to deformation [12]. The synthesized aerogels exhibit excellent water-driven recovery performance, regaining their original volume within a very short time (1.9 s) after strong compression ($\varepsilon > 80\%$).

Wang et al. reported that N-(3-Dimethylaminopropyl)-N'-ethylcarbodiimide hydrochloride (EDC) and N-Hydroxysuccinimide (NHS) was used to crosslink silicon-doped carbon dots and carboxymethyl cellulose. The silicon-doped carbon dots crosslinked gels show improved mechanical properties but also good biocompatibility, reactivity, and fluorescence properties [13]. The abundant crosslinking points endow the gel with excellent mechanical properties, with a compressive strength reaching 294 kPa.

2.2. Applications for Wastewater Pollutant Removal and Adsorption

Hydrogels are widely used for wastewater treatment. In this Special Issue, four articles are closely related to wastewater treatment. A brief introduction is listed as follows:

An article from Argentina, entitled "Linear Polyethyleneimine-Based and Metal Organic Frameworks (DUT-67) Composite Hydrogels as Efficient Sorbents for the Removal of Methyl Orange, Copper Ions, and Penicillin V", introduced that highly efficient composite hydrogen adsorbents could be fabricated by integrating DUT-67 metal organic frameworks into polyethyleneimine-based hydrogels. The particle size of DUT-67 was successfully modulated from 1 µm to 200 nm by varying the solvent-to-modulator ratio in a water-based synthesis [14]. The obtained hydrogel with enough mechanical strength, pore structure, and chemical affinity exhibits maximum adsorption capacities of 473 ± 21 mg L^{-1}, 86 ± 6 mg L^{-1}, and 127 ± 4 mg L^{-1}, for methyl orange, copper(II) ions, and penicillin V, respectively. This hydrogel reveals a promising application as an efficient adsorbent for environmental pollutants and pharmaceuticals.

An article from Italy, entitled "Preparation of Peptide-Based Magnetogels for Removing Organic Dyes from Water", reported that polyacrylic acid-modified γ-Fe$_2$O$_3$ nanoparticles were embedded in hydrogel through the enzymatic approach. The ability of the magnetogel to remove three organic dyes, methyl orange, methylene blue, and rhodamine 6G, was studied. The results show that the obtained peptide magnetogel could represent a valuable and environmentally friendly alternative to currently employed adsorbents [15].

An article from China, entitled "Study on Adsorption of Cd in Solution and Soil by Modified Biochar–Calcium Alginate Hydrogel", reported a fabricating composite material method by utilizing crosslinked modified biochar (prepared from pine wood) and calcium alginate hydrogels. The modified biochar and calcium alginate hydrogels exhibit a higher heavy metal adsorption capacity compared to traditional biochar and hydrogels due to their increased oxygen-containing functional groups and heavy metal adsorption sites [16]. The highest Cd^{2+} removal rate of the obtained hydrogel reached up to 85.48%. This study gives a novel approach for managing Cd-contaminated cultivated land.

Another article from China, entitled "Molten Alkali-Assisted Formation of Silicate Gels and Its Application for Preparing Zeolites", illustrated the convenient formation of silicate gels assisted by molten alkali, which was energy and time saving [17]. The silicate gels were used to fabricate zeolite successfully. The maximum adsorption capacity of obtained zeolite for ammonium can reach 49.1 mg/g. The ammonium-adsorbed zeolites might be used as an environmentally friendly ammonium fertilizer for agricultural plant growth.

2.3. Some Other Applications for Arts Remediation and Medical Care

Carretti et al. reported that a non-aqueous organogel sponge made of PDMS was used for the cleaning of artworks. The role of pore size in an elastomeric network on the ability to uptake and release organic material was studied in detail. Two different sugar templates were applied to synthesize porous organic polymers, whose porosity drops with the decrease in pore size [18]. The adsorption capacity was measured by swelling with eight solvents covering a wide range of polarities. The results demonstrate that the PDMS sponges are a potential innovative support for the controlled and selective cleaning of art surfaces.

Barati et al. reported that a chitosan-based low-cost microneedle patch was obtained by a CO_2 laser cutter [19]. The impact of Glycyrrhiza glabra extract (GgE), delivered via microneedle to the cell population on the patch, was evaluated. A lot of analysis, such as microscopic analysis, swelling, penetration, degradation, biocompatibility, and drug delivery were carried out to assess the gel-based microneedle patch's performance. In general, a GgE-loaded microneedle patch can be a good remedy for skin disorders in which cell proliferation needs to be controlled.

Gabriele reported that a series of chitosan-carboxylic acid hydrogels was prepared by coupling the reducing ability of carboxylic acids with the intrinsic chelating properties of the polysaccharide [20]. The results show that the formulation containing oxalic acid is the most effective in removing rust stains. This work developed an effective formulation for removing rust on a marble surface, which is the most challenging surface to clean.

3. Summary

In summary, this Special Issue involved a number of interesting research articles that have represented the most recent progress in various aspects of gels, including the preparation technique, the applications for pollutant removal and adsorption, the treatment of artworks, and medical care. These gels may offer new insights into the design, preparation, development, and application of biomass-based, inorganic, and organic gels. We greatly appreciate the efforts of our authors, reviewers, and editors in the disclosure of these valuable research works.

Conflicts of Interest: The authors declare no conflicts of interest.

References

1. Guo, Y.; Bae, J.; Fang, Z.; Li, P.; Zhao, F.; Yu, G. Hydrogels and Hydrogel-Derived Materials for Energy and Water Sustainability. *Chem. Rev.* **2020**, *120*, 7642–7707. [CrossRef] [PubMed]
2. Qin, S.; Niu, Y.; Zhang, Y.; Wang, W.; Zhou, J.; Bai, Y.; Ma, G. Metal Ion-Containing Hydrogels: Synthesis, Properties, and Applications in Bone Tissue Engineering. *Biomacromolecules* **2024**, *25*, 3217–3248. [CrossRef] [PubMed]
3. Rashid, A.B.; Shishir, S.I.; Mahfuz, M.A.; Hossain, M.T.; Hoque, M.E. Silica Aerogel: Synthesis, Characterization, Applications, and Recent Advancements. *Part. Part. Syst. Charact.* **2023**, *40*, 2200186. [CrossRef]
4. Zhu, H.; Chen, S.; Luo, Y. Adsorption mechanisms of hydrogels for heavy metal and organic dyes removal: A short review. *J. Agric. Food Res.* **2023**, *12*, 100552. [CrossRef]
5. Vassalini, I.; Ribaudo, G.; Gianoncelli, A.; Casula, M.F.; Alessandri, I. Plasmonic hydrogels for capture, detection and removal of organic pollutants. *Environ. Sci. Nano* **2020**, *7*, 3888–3900. [CrossRef]

6. Stanciu, M.-C.; Teacă, C.-A. Natural Polysaccharide-Based Hydrogels Used for Dye Removal. *Gels* **2024**, *10*, 243. [CrossRef] [PubMed]
7. Badsha, M.; Khan, M.; Wu, B.; Kumar, A.; Lo, I. Role of surface functional groups of hydrogels in metal adsorption: From performance to mechanism. Journal of Hazardous Materials. *J. Hazard. Mater.* **2021**, *408*, 124463. [CrossRef] [PubMed]
8. Xue, Y.; Chen, H.; Xu, C.; Yu, D.; Xu, H.; Yu, Y. Synthesis of hyaluronic acid hydrogels by crosslinking the mixture of high-molecular-weight hyaluronic acid and low-molecular-weight hyaluronic acid with 1,4-butanediol diglycidyl ether. *RSC Adv.* **2020**, *10*, 7206–7213. [CrossRef] [PubMed]
9. Zhang, Y.; Zhu, Y.; Xiong, Z.; Wu, J.; Chen, F. Bioinspired Ultralight Inorganic Aerogel for Highly Efficient Air Filtration and Oil–Water Separation. *ACS Appl. Mater. Interfaces* **2018**, *10*, 13019–13027. [CrossRef] [PubMed]
10. Feng, J.; Fan, L.; Zhang, M.; Guo, M. An efficient amine-modified silica aerogel sorbent for CO_2 capture enhancement: Facile synthesis, adsorption mechanism and kinetics. *Colloids Surf. A Physicochem. Eng. Asp.* **2023**, *656*, 130510. [CrossRef]
11. Quan, L.; Shi, X.; Zhang, J.; Shu, Z.; Zhou, L. Preparation of a Novel Lignocellulose-Based Aerogel by Partially Dissolving Medulla Tetrapanacis via Ionic Liquid. *Gels* **2024**, *10*, 138. [CrossRef] [PubMed]
12. Liu, Y.; Ma, M.; Shen, Y.; Zhao, Z.; Wang, X.; Wang, J.; Pang, J.; Wang, D.; Wang, C.; Li, J. Polyhedral Oligomeric Sesquioxane Cross-Linked Chitosan-Based Multi-Effective Aerogel Preparation and Its Water-Driven Recovery Mechanism. *Gels* **2024**, *10*, 279. [CrossRef] [PubMed]
13. Zhao, Z.; Jing, Y.; Shen, Y.; Liu, Y.; Ma, J.; Ma, M.; Pan, J.; Wang, D.; Wang, C.; Li, J. Silicon-Doped Carbon Dots Crosslinked Carboxymethyl Cellulose Gel: Detection and Adsorption of Fe^{3+}. *Gels* **2024**, *10*, 285. [CrossRef] [PubMed]
14. Araque, L.M.; Fernández de Luis, R.; Fidalgo-Marijuan, A.; Infantes-Molina, A.; Rodríguez-Castellón, E.; Pérez, C.J.; Copello, G.J.; Lázaro-Martínez, J.M. Linear Polyethyleneimine-Based and Metal Organic Frameworks (DUT-67) Composite Hydrogels as Efficient Sorbents for the Removal of Methyl Orange, Copper Ions, and Penicillin V. *Gels* **2023**, *9*, 909. [CrossRef] [PubMed]
15. Hajareh Haghighi, F.; Binaymotlagh, R.; Pintilei, P.S.; Chronopoulou, L.; Palocci, C. Preparation of Peptide-Based Magnetogels for Removing Organic Dyes from Water. *Gels* **2024**, *10*, 287. [CrossRef] [PubMed]
16. Wang, S.; Wang, Y.; Wang, X.; Sun, S.; Zhang, Y.; Jiao, W.; Lin, D. Study on Adsorption of Cd in Solution and Soil by Modified Biochar–Calcium Alginate Hydrogel. *Gels* **2024**, *10*, 388. [CrossRef] [PubMed]
17. Ye, J.; Yang, Y.; Zhang, L.; Li, M.; Wang, Y.; Chen, Y.; Ling, R.; Yan, J.; Chen, Y.; Hu, J.; et al. Molten Alkali-Assisted Formation of Silicate Gels and Its Application for Preparing Zeolites. *Gels* **2024**, *10*, 392. [CrossRef]
18. Porpora, F.; Dei, L.; Duncan, T.T.; Olivadese, F.; London, S.; Berrie, B.H.; Weiss, R.G.; Carretti, E. Non-Aqueous Poly(dimethylsiloxane) Organogel Sponges for Controlled Solvent Release: Synthesis, Characterization, and Application in the Cleaning of Artworks. *Gels* **2023**, *9*, 985. [CrossRef] [PubMed]
19. Khorshidian, A.; Sharifi, N.; Choupani Kheirabadi, F.; Rezaei, F.; Sheikholeslami, S.A.; Ariyannejad, A.; Esmaeili, J.; Basati, H.; Barati, A. In Vitro Release of Glycyrrhiza Glabra Extract by a Gel-Based Microneedle Patch for Psoriasis Treatment. *Gels* **2024**, *10*, 87. [CrossRef] [PubMed]
20. Gabriele, F.; Casieri, C.; Spreti, N. Efficacy of Chitosan-Carboxylic Acid Hydrogels in Reducing and Chelating Iron for the Removal of Rust from Stone Surface. *Gels* **2024**, *10*, 359. [CrossRef] [PubMed]

Disclaimer/Publisher's Note: The statements, opinions and data contained in all publications are solely those of the individual author(s) and contributor(s) and not of MDPI and/or the editor(s). MDPI and/or the editor(s) disclaim responsibility for any injury to people or property resulting from any ideas, methods, instructions or products referred to in the content.

Article

Molten Alkali-Assisted Formation of Silicate Gels and Its Application for Preparing Zeolites

Juan Ye [1,†], Yanchun Yang [2,†], Li Zhang [1], Man Li [1], Yiling Wang [1], Yuxuan Chen [1], Ruhui Ling [1], Jiefeng Yan [1], Yan Chen [3], Jinxing Hu [1] and Zhenxing Fang [1,*]

1. College of Science and Technology, Ningbo University, 521 Wenwei Road, Ningbo 315300, China; 13252058707@163.com (J.Y.); a13736170883@163.com (L.Z.); 18758807683@163.com (M.L.); 13003713750@163.com (Y.W.); cyx2918380445@163.com (Y.C.); 18142011326@163.com (R.L.); yanjiefeng@nbu.edu.cn (J.Y.); hujinxing@nbu.edu.cn (J.H.)
2. Ecological Environment Monitoring Station in Yanji City, Yanji 133001, China; yjjczyang@163.com
3. State Key Laboratory of Inorganic Synthesis and Preparative Chemistry, Jilin University, Changchun 130012, China; yanchen@jlu.edu.cn
* Correspondence: fangzhenxing128@163.com
† These authors contributed equally to this work.

Abstract: Fly ash was used as raw material to prepare zeolites through silicate gels, assisted by the hydrothermal method. The silicate gels could be effectively formed in a few minutes in a molten alkali environment. The zeolites could be prepared by using these silicate gels through the hydrothermal method, which realizes the transformation from useless materials to highly valuable materials. The obtained zeolites were applied to the removal of ammonium in water, achieving the highvalue utilization of fly ash. The synthesized zeolites were characterized by X-ray diffraction (XRD), scanning electron microscopy (SEM), energy dispersive spectrum (EDS), thermogravimetric (TG), and Fourier transform infrared (FTIR) spectroscopy. The study on the adsorption and removal of ammonium in water shows that the adsorption of ammonium is more in line with pseudo first-order kinetics, and the adsorption mainly occurs in the first 20 min. The adsorption can reach equilibrium in 30 min, and the maximum adsorption capacity can reach 49.1 mg/g. The adsorption capacity of ammonium has the best performance at pH = 5. Furthermore, within a certain range, an increase in temperature is beneficial for the removal of ammonium.

Keywords: fly ash; silicate gels; zeolites; ammonium adsorption; ion exchange

1. Introduction

Zeolite is widely used in pollutant control, petrochemistry, water purification, gas purification, and other fields because of its excellent catalytic ability, cation exchange ability, and adsorption ability [1–4]. The main chemical components of fly ash are silica and alumina, which are similar to zeolite [5,6]. The main difference among them lies in the different crystal structures. Therefore, fly ash is considered a potential raw material for synthesizing zeolite. However, the calcium oxide and some other oxides in fly ash can affect the crystal transformation of mullite and reduce the ion exchange capacity of zeolite. The residual carbon can also reduce the whiteness of zeolite products. Therefore, pre-treatment must be carried out to remove the impurities in fly ash before zeolite synthesis. At present, there are various methods for preparing zeolites by applying fly ash, including hydrothermal synthesis, microwave irradiation, ultrasonic-assisted aging, seed crystallization, and so on [7–9]. Among them, the hydrothermal method is the most mature and commonly used method, while other methods are still in the laboratory research stage and have not yet met the requirements of industrial production. The commonly used pre-treatment to remove magnetic iron oxide from fly ash is magnetic separation, while for the removal of iron oxide and calcium oxide it is acid treatment, and for the effective removal of residual carbon from fly ash it is burning. These traditional impurity removal processes are relatively cumbersome and energy consuming, which causes huge costs in the pre-treatment process [10,11]. In addition, some other components will inevitably be brought out during the magnetic separation of iron oxide. Alkali-assisted hydrothermal treatment could improve the utilization of fly ash, but this process is time consuming. In order to improve the utilization efficiency of silicon and aluminum sources in fly ash and the purity of zeolite, zeolite was prepared through the molten alkali-assisted hydrothermal method in this paper. Due to the strong corrosive effect of alkalis, silicon oxide and aluminum oxide in fly ash can be efficiently and quickly extracted, while other impurities such as iron oxide and calcium oxide can be retained in the form of residues. Finally, the washed and dried residues can have some other applications such as secondary combustion and raw materials for other functional materials, once again proving the feasibility of alkali melting to extract silicon and aluminum from fly ash. In other words, the silicon and aluminum sources used for preparing zeolites from fly ash can be obtained through strong alkali melting combined with simple filtration [12,13].

The unique crystal structure of zeolite has shown excellent performance in the field of catalysis. In addition to typical industrial applications such as catalytic cracking [14,15], they have also played an important role in other fields, such as environmental treatment as adsorbents for treating atmospheric and water pollution [16,17]. With the advantages of large adsorption capacity, simple operation, and high processing efficiency, zeolite has been widely used for the removal of heavy metal from wastewater [18,19]. The removal of heavy metals in water by zeolite is achieved through a combination of ion exchange and adsorption, and is related to its own properties, influenced by multiple factors such as silicon–aluminum ratio, pore size, and its surface nature [20]. In general, adsorption, ion exchange and catalytic performance is very common to see in zeolites. In real life, zeolites have been applied in dehydration drying; for ethanol dehydration with low water content, zeolite adsorption dehydration is the optimal choice. Furthermore, the adsorption of H_2S, SO_2, NO_X, and formaldehyde by zeolite can improve the air environment. In this report, the synthesized zeolites could also be used for ammonium adsorption. The ammonium-adsorbed zeolites can be used as fertilizers with slow-release ammonium for plant growth [21,22].

2. Results and Discussion

Strong alkalis, NaOH and KOH, were applied to extract silicon and aluminum from fly ash. Additional potassium silicate (sodium) or potassium aluminate (sodium) was added to adjust the target silicon–aluminum ratio (1:1, 2.8:1, and 3.5:1, respectively). The whole preparation process is depicted in Figure 1. The products obtained from crystallization at 100 °C for 24 h were analyzed by XRD, and the results are shown in Figure 2. The XRD diffraction spectrum of the obtained products matches well with the standard card JCPDS No.38-0216 and the crystal structure shows a tetragonal crystal structure. Similarly, it can be seen that there is no significant change in the diffraction spectrum (i.e., crystal configuration) of the product with an increasing silicon–aluminum ratio (from 1:1 to 2.8:1 and then to 3.5:1) [23]. It can be seen that in this reaction system, the crystal configuration of the product cannot be affected by simply changing the silicon–aluminum ratio, which may be due to the large amounts of base in the reaction system. It can be concluded that the strong base condition contributes to the fomation of a tetragonal crystal structure for zeolites.

It can be seen that the obtained zeolites prepared under any ratio of silicon to aluminum conditions have a uniform morphology. This also indicates that the zeolites can complete a good crystallization process in this reaction system, which is consistent with the experimental results of the XRD. In addition, as the silicon–aluminum ratio increases, the size of the product tends to decrease (the silicon–aluminum ratio of a, c, and e increases in sequence). This experimental result can be explained by the crystal growth theory. As the silicon–aluminum ratio increases, more $[SiO_4]^{4-}$ tetrahedrons can be formed in the reaction system to explode more crystal nuclei. In other words, it can provide more sites for crystal growth, and the size of the product will inevitably decrease without changing the concentration of other materials.

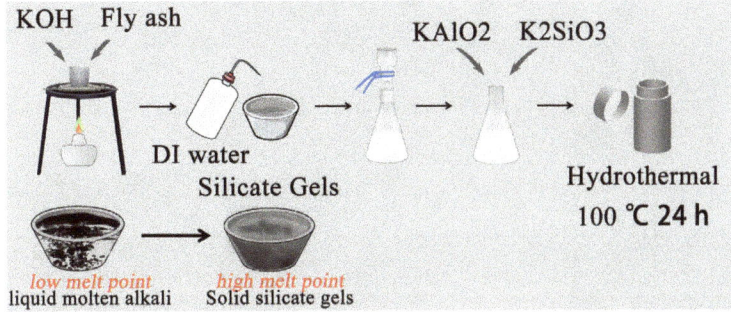

Figure 1. Schematic diagram of molten alkali-assisted preparation process.

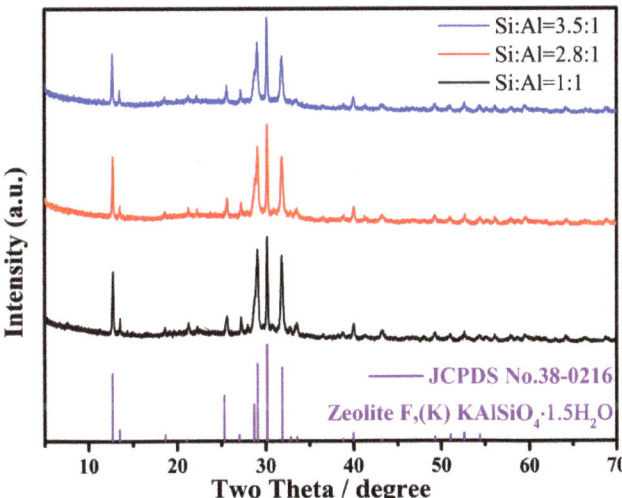

Figure 2. XRD patterns of obtained products from various ratios of Si to Al.

Pseudo color processing (Figure 3 shows) on the high magnification SEM images (coloring the repeating units with bright colors) was performed to create a more intuitive display, and it can be clearly seen that the size of the zeolite crystal gradually decreases with the increasing ratio of silicon to aluminum. The crystal growth theory can provide a reasonable explanation for the experimental results. Due to increasing the silicon–aluminum ratio, the size of the product can become smaller, exposing more adsorption activity. Subsequent characterization is mainly based on the sample with the maximum silicon–aluminum ratio.

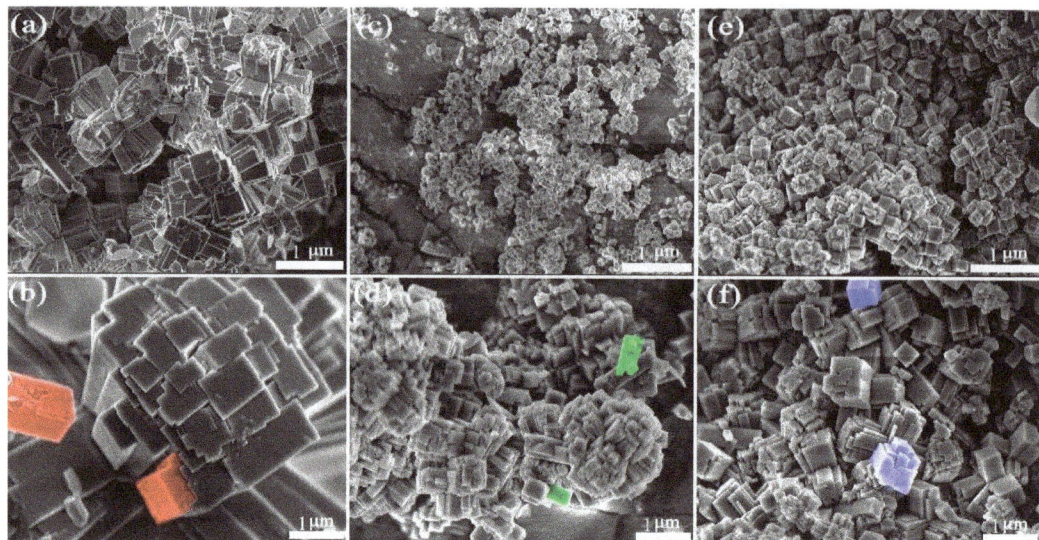

Figure 3. SEM images of zeolite-F from various ratios of Si to Al. (**a,b**) 1:1, (**c,d**) 2.8:1, and (**e,f**) 3.5:1.

As shown in Figure 4, the energy dispersion spectrum of the zeolite prepared under high silicon–aluminum ratio conditions indicates that the product contains main elements such as K, Si, Al, and O, etc. Among them, the highest content of C is mainly due to the background conductive adhesive used for testing, while Pt is the precious metal sprayed on the surface of the sample before testing to increase the conductivity of the sample. Other contents are as low as to be negligible. The experimental results demonstrate that the atomic ratio of K, Si, Al, and O in the molecular sieve we prepared is approximately 1:1:1:6. This result reveals that the obtained zeolite should belong to the type with a low ratio of silicon to aluminum [24,25].

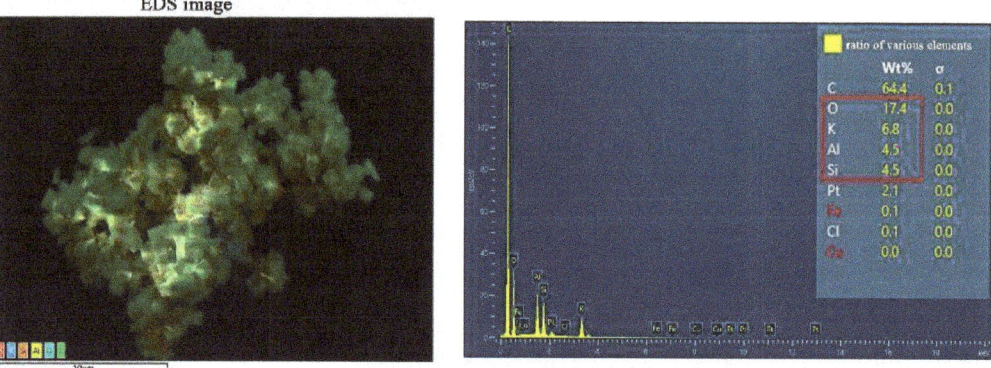

Figure 4. EDS results of zeolite-F obtained at the maximum silicon–aluminum ration.

In addition, we also conducted element distribution scanning (mapping test) on the sample, and the results are shown in Figure 5. The green background in the upper left corner is carbon, which is the background used during the test as a conductive adhesive; the red color in the lower right corner is S, indicating that the prepared zeolite contains a small amount of sulfur element, which is due to the sulfur content in the raw material fly ash. It also proves that the obtained product was prepared from fly ash as the raw material.

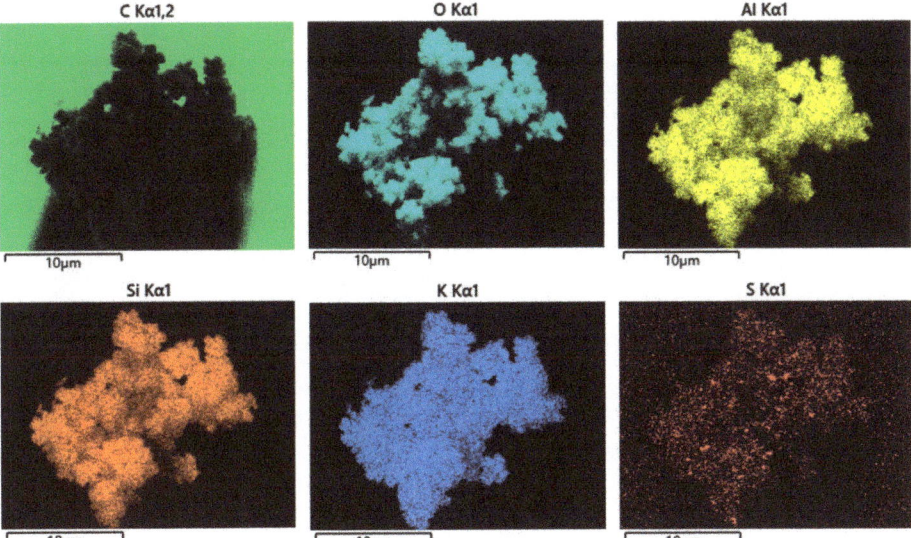

Figure 5. Mapping results of zeolite-F obtained at the maximum silicon–aluminum ration, every color represents one element ref to Figure 4.

Figure 6 shows the FTIR spectra of fly ash and the synthetic zeolites. It can be seen that fly ash exhibits a strong -OH stretching vibration and -OH bending vibration at 3442 cm^{-1} and 1600 cm^{-1}; the characteristic peak appearing at 1098 cm^{-1} is a Si-O-Si and Al-O-Si asymmetric stretching vibration. After the conversion of fly ash into zeolite-F, the infrared absorption peak undergoes a significant change, with the characteristic absorption peak at 1098 cm^{-1} shifting to 981 cm^{-1} [20,26,27]. This is because there is a large number of potassium ions in the structure of the zeolites, which inevitably leads to the existence of a Si-O-K skeleton. Compared to the Si-O-Si or Si-O-Al structures in fly ash, due to the larger atomic radius of K, the bond length of the Si-O-K structure in the moleculr is longer, resulting in a decrease in the wavenumber of the absorption peaks.

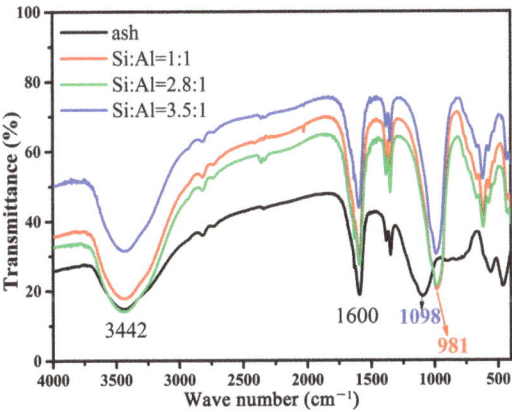

Figure 6. FTIR spectra of obtained products at various ratio of Si to Al and fly ash.

Gravimetric analysis of the synthesized zeolite was conducted to determine the content of crystalline water in the crystal structure of synthetic zeolite. The results are shown in Figure 7; there were two gradients of weight loss, which can be more clearly observed in the experimental results of first-order derivative DTG. Before 140 °C, the synthesized molecular sieve experienced a weight loss of 5.25%. The weight loss here is due to the presence of adsorbed water, including surface- and pore-adsorbed water. Continuing to increase the temperature, the synthesized molecular sieve continued to lose weight due to the detachment of crystalline water within the crystal structure, resulting in a weight loss of 5.75%. From this, it can be seen that the mass ratio of crystalline water in the synthetic zeolite reaches 6.1%. Based on the EDS spectrum data, the atomic ratio of K, Si, and Al obtained is 1:1:1, indicating that the molecular formula of the prepared molecular sieve is $KSiAlO_4 \cdot 0.5H_2O$.

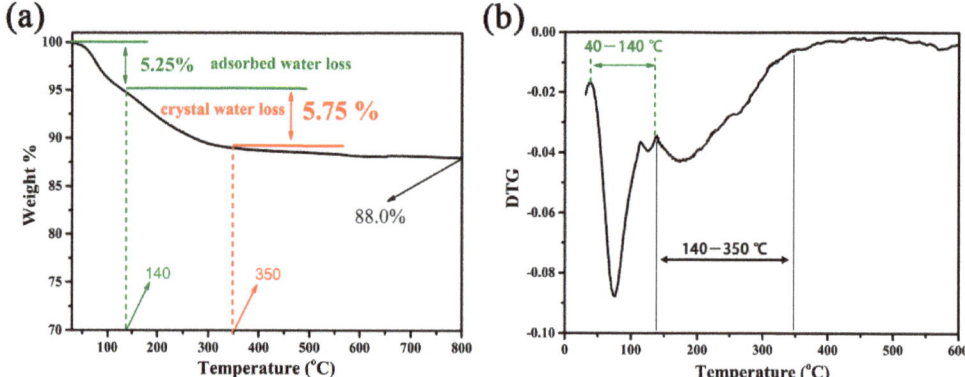

Figure 7. TG (**a**) and DTG (**b**) results of the obtained zeolite-F.

The effect of adsorption time on the adsorption performance of ammonium ions was investigated. From Figure 8a, it can be seen that adsorption mainly occurs in the first 20 min. At the beginning of adsorption, the adsorption amount of ammonium increases rapidly. Afterwards, the adsorption rate of ammonium by zeolite significantly weakens, and the adsorption curve changes to be flattened. It can also be seen that the adsorption of the ammonia–nitrogen solution by the synthesized zeolite-F can basically reach adsorption equilibrium at 30 min. pH value is an important indicator in aqueous solutions, which not only affects the state of ammonium ions in water, but also affects the surface activity and electrical properties of adsorbents. This article investigates the effect of a solution pH range between 1.0 and 12.0 on the adsorption of ammonium ions on the obtained zeolite. As shown in Figure 8b, it can be seen that the adsorption capacity of synthetic zeolite for ammonia in water first increases and then decreases with pH from 1 to 11, showing good adsorption performance at pH = 5–6. This is because under strong acidic conditions, when the pH is below the isoelectric point of the $[SiO_4]^{4-}$ surface, the surface's negative charge is weak, which affects the electrostatic adsorption of ammonium ions. When the pH value is too high, it can also affect the adsorption of ammonium ions. We believe that this may be due to the change in the form of ammonium ions under strong alkaline conditions, and the decrease in the concentration of free ammonium ions, which leads to a decrease in the removal rate of ammonium. Temperature is a crucial parameter for chemical reactions, as it not only affects the reaction rate but also the progress of the reaction. Therefore, the adsorption performance of synthetic zeolite for ammonium in water under conditions of 10, 25, 50, and 75 °C was investigated. As shown in Figure 8c, with the increase in temperature under 80 °C, the adsorption performance of the synthesized zeolite for ammonium is improved. As is well known, most adsorption reactions are exothermic, and experimental results show that heating is beneficial for adsorption. Therefore, we believe that the possible reason is that the interaction between ammonium ions and synthetic zeolite is mainly ion exchange, followed by electrostatic adsorption. Among them, the ion exchange between K^+ and NH_4^+ in synthetic zeolite is an endothermic reaction. Furthermore, the recyclability of adsorption performance was also investigated. As shown in Figure 8d, after 5 cycles of reuse (recovery at 1 M KCl solution) [28–30], the ammonium adsorption performance of the obtained zeolite reaches a stable state and the ammonium removal rate reaches 34%. Compared with the first ammonium removal rate of 52%, it can be concluded that the ammonium exchange might occur just at the surface of the zeolite.

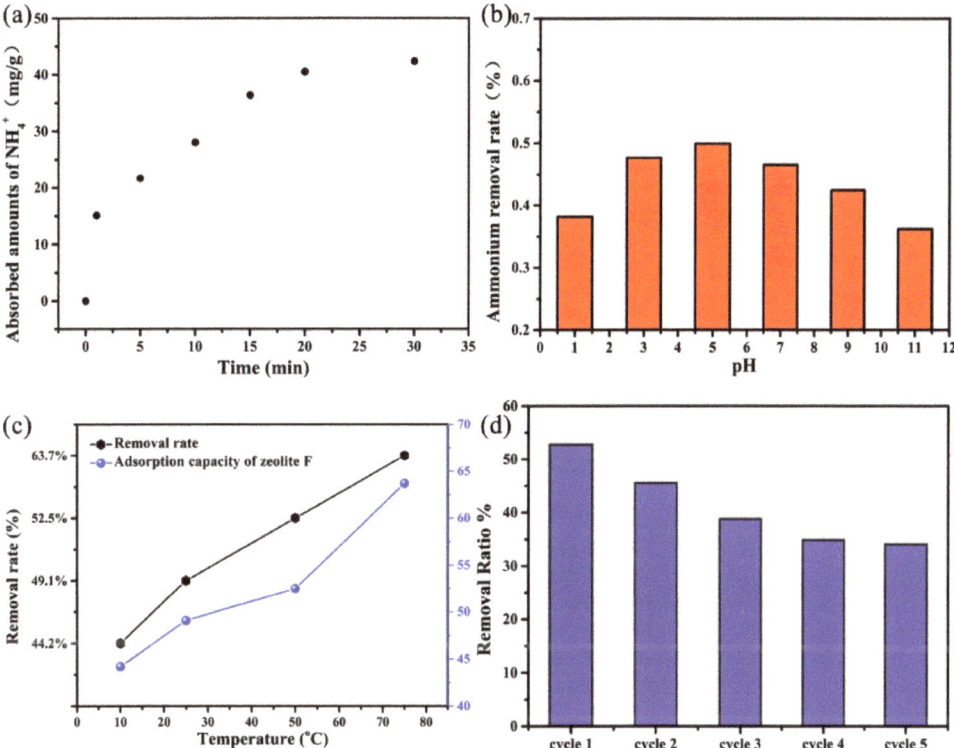

Figure 8. Influence factors time (**a**), pH (**b**), and T (**c**) on adsorption and (**d**) recylability.

As the ammonium removal is mainly caused by the ion exchange process of the obtained zeolite, the effect of the coexisting ions on the ammonium adsorption performance should be investigated. The cations with various positive charges such as Na^+, Mg^{2+}, and Al^{3+} were applied to illustrate the influencing mechanism. The result is shown in Figure 9; the ammonium removal performance reveals a decreasing trend when the number of positive charges increases. This result is consistent with the ion exchanging mechanism. Just as the schematic diagram of cations competition shows in Figure 9, the amount of surrounding ammonium decreases as the number of positive charges increases because the ion exchanging process obeys the law of charge equivalence.

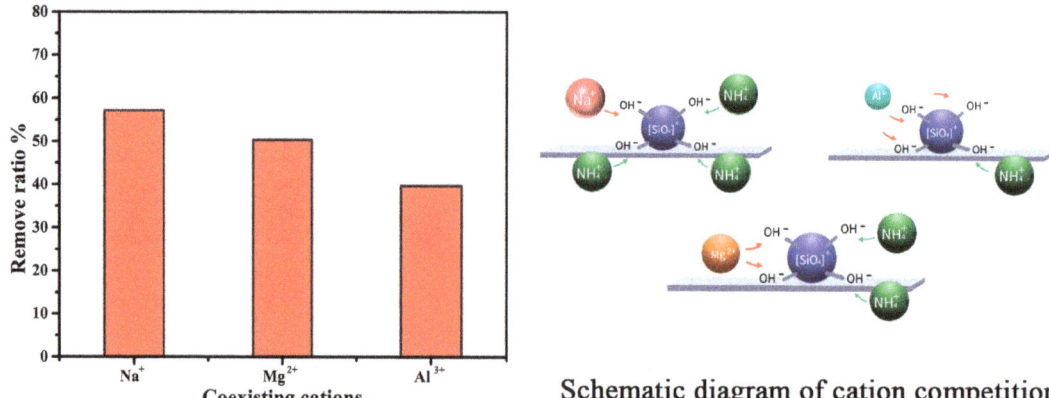

Figure 9. Influence of coexisting cations on the ammonium removal ratio.

3. Conclusions

During the pretreatment of fly ash, molten alkali was demonstrated to effectively extract Si and Al sources in fly ash. Without any adjusting of the pH of the reaction system, zeolite-F was successfully synthesized by the followed hydrothermal method, which reveals its super-convenient synthesis process. The whole process of fly ash treatment and zeolite synthesis was both energy- and time-saving, realizing the high-value usage of fly ash at the same time. The ammonium-adsorbed zeolites could also be recovered by a simple ion exchanging process. Furthermore, the ammonium-adsorbed zeolites might be used as an environmentally friendly ammonium fertilizer for agricultural plant growth.

4. Materials and Methods

4.1. Materials

The materials described in this report were all purchased from Aladdin (Shanghai, China). There was no need to purify before use. All the materials were Analytical Reagents. The purity of NaOH and KOH was larger than 95%. The fly ash was provided by Zhejiang ZhenengZhenhai Power Generation Co., Ltd (Ningbo, China). The detailed physical and chemical properties of the materials can be seen in Table 1.

Table 1. Physical and chemical properties of materials.

Materials	Melt Point/°C	Solubility/	Source
NaOH	318 (low)	strong corrosive	Aladdin
KOH	361 (low)	strong corrosive	Aladdin
Na_2SiO_3	1089 (high)	water soluble	Aladdin
$NaAlO_2$	1650 (high)	water soluble	Aladdin
K_2SiO_3	976 (high)	water soluble	Aladdin
Fly ash	/	water insoluble	SiO_2 48%, Al_2O_3 33%

4.2. Procedures of Zeolite Synthesis

4.2.1. The Formation of Silicate Gels

Calculated amounts of NaOH and KOH were added into fly ash (mass ratio was 10:1; 10 g alkali and 1 g fly ash) and the mixed powders were placed into a nickel crucible. An alcohol lamp was applied to heat the mixed powder. The NaOH and KOH soon became molten, which is beneficial to extract SiO_2 and Al_2O_3 because of its strong corrosiveness. The molten alkali reacted with SiO_2 and Al_2O_3 quickly and resulted in the production of K_2SiO_3 and $KAlO_2$, which could not be dissolved in molten alkali because of their higher molten points. Thus, some white precipitate was separated out from the transparent molten alkali system. The alkali silicate gels were formed with the extraction process and this extraction process was maintained for around another 30 min so as to fully extract the SiO_2 and Al_2O_3 from the fly ash. The rapid formation of alkali silicate gels further reveals that the extraction of SiO_2 and Al_2O_3 from fly ash by molten alkali treatment was time- and energy-saving.

4.2.2. Hydrothermal Synthesis of Zeolite-F

When the silicate gels mentioned above cooled down to room temperature, deionized water (DI water) was added to dissolve the soluble composites such as K_2SiO_3 and $KAlO_2$. A filtration process was conducted to separate the residual solid product (Fe_2O_3, CaO, and C, etc.) and the solution containing K_2SiO_3 and $KAlO_2$. Finally, a calculated Si and Al source was added to the solution to regulate the appropriate ratio of Si to Al (the ratio of Si to Al ranged from 1:1 to 3.5:1). The followed hydrothermal process was performed at 100 °C for 24 h. When cooled down to room temperature, white precipitate was formed at the bottom of teflon lining. DI water and ethanol was used to wash the surface residue chemicals by centrifugation several times. The product was finally dried at 60 °C in the vacuum oven.

4.2.3. Experiment of Ammonium Adsorption

The volume of ammonium in this experiment was 50 mL (concentration was set as 100 mg/L), which was placed in a conical flask with a capacity of 250 mL. Then, 50 mg of synthetic zeolite was added and shaken in a shaker. The temperature of the shaker was set at 25 °C and the speed was 200 rpm. An amount of 50 mg of zeolite was added to the solution, and after adsorption saturation, the supernatant was quickly centrifuged. Then, a glass fiber filter membrane with a pore size of 0.22 μm was used to filter the supernatant, and Nessler's reagent was used to color the filtration. Lasty, a spectrophotometer was used to measure the absorbance after the coloring procedure. The absorbance of the filtration was determined by the concentration of ammonium. The cyclic adsorption performance test of zeolite was conducted the same way as mentioned above.

4.3. Characterization

The XRD patterns were recorded by using PANalytical B.V. Empyrean X-ray powder diffraction (Malvern Panalytical, Enigma Business Park, Malvern, UK) with Cu Kα radiation over a range of 10–70° (2θ) with 0.02° per step. SEM images were obtained with a JSM-6700F electron microscope (JEOL, 1-2 Musashino 3-chome, Showa City, Tokyo, Japan). The thermogravimetric analysis (TGA) was performed by using a Netzch Sta 449c (Netzch Company, Selb, Germany) thermal analyzer system at a heating rate of 10 °C/min in air. The FTIR spectrum was recorded by Fourier Transform Infrared Spectrometry FTIR-650.

Author Contributions: Conceptualization, J.Y. (Juan Ye) and Z.F.; methodology, L.Z., M.L. and Y.W.; software, Y.Y., Y.C. (Yuxuan Chen) and R.L.; validation, J.Y. (Juan Ye) and Y.Y.; formal analysis, Z.F.; investigation, J.Y. (Juan Ye); resources, Y.C. (Yan Chen) and J.H.; writing—original draft preparation, J.Y. (Jiefeng Yan); writing—review and editing, Z.F. and J.H.; supervision, J.Y. (Jiefeng Yan); project administration, Y.Y.; funding acquisition, J.H. All authors have read and agreed to the published version of the manuscript.

Funding: This research received no external funding. And the APC was funded by Jinxing Hu.

Institutional Review Board Statement: This did not require ethical approval.

Informed Consent Statement: The study did not involve humans.

Data Availability Statement: The data presented in this study are openly available in article.

Acknowledgments: Thanks to the General scientific research projects of Zhejiang Provincial Department of Education (Y202352630) and the Open project of the State Key Laboratory of Inorganic Synthesis and Preparative Chemistry, Jilin University (2023-22, 2022-25). Much appreciation to The

Committee of Zhejiang College Student Chemistry Competition. The publication of this work cannot be separated from your help.

Conflicts of Interest: The authors declare no conflicts of interest.

References

1. Ding, R.; Zhu, H.; Zhou, J.; Luo, H.; Xue, K.; Yu, L.; Zhang, Y. Highly Water-Stable and Efficient Hydrogen-Producing Heterostructure Synthesized from $Mn_{0.5}Cd_{0.5}S$ and a Zeolitic Imidazolate Framework ZIF-8 via Ligand and Cation Exchange. *ACS Appl. Mater. Interfaces* **2023**, *15*, 36477–36488. [CrossRef]
2. He, J.; Deng, J.; Lan, T.; Liu, X.; Shen, Y.; Han, L.; Wang, J.; Zhang, D. Strong metal oxide-zeolite interactions during selective catalytic reduction of nitrogen oxides. *J. Hazard. Mater.* **2023**, *465*, 133164. [CrossRef]
3. Mokrzycki, J.; Franus, W.; Panek, R.; Sobczyk, M.; Rusiniak, P.; Szerement, J.; Jarosz, R.; Marcińska-Mazur, L.; Bajda, T.; Mierzwa-Hersztek, M. Zeolite Composite Materials from Fly Ash: An Assessment of Physicochemical and Adsorption Properties. *Materials* **2023**, *16*, 2142. [CrossRef]
4. Tesana, S.; Kennedy, J.V.; Yip, A.C.K.; Golovko, V.B. In Situ Incorporation of Atomically Precise Au Nanoclusters within Zeolites for Ambient Temperature CO Oxidation. *Nanomaterials* **2023**, *13*, 3120. [CrossRef]
5. Gu, J.; Liu, L.; Zhu, R.; Song, Q.; Yu, H.; Jiang, P.; Miao, C.; Du, Y.; Fu, R.; Wang, Y.; et al. Recycling Coal Fly Ash for Super-Thermal-Insulating Aerogel Fiber Preparation with Simultaneous Al_2O_3 Extraction. *Molecules* **2023**, *28*, 7978. [CrossRef]
6. Shishkin, A.; Abramovskis, V.; Zalite, I.; Singh, A.K.; Mezinskis, G.; Popov, V.; Ozolins, J. Physical, Thermal, and Chemical Properties of Fly Ash Cenospheres Obtained from Different Sources. *Materials* **2023**, *16*, 2035. [CrossRef]
7. Oliveira, M.R.; Cecilia, J.A.; Ballesteros-Plata, D.; Barroso-Martín, I.; Núñez, P.; Infantes-Molina, A.; Rodríguez-Castellón, E. Microwave-Assisted Synthesis of Zeolite A from Metakaolinite for CO_2 Adsorption. *Int. J. Mol. Sci.* **2023**, *24*, 14040. [CrossRef]
8. Zhou, Q.; Jiang, X.; Qiu, Q.; Zhao, Y.; Long, L. Synthesis of high-quality NaP1 zeolite from municipal solid waste incineration fly ash by microwave-assisted hydrothermal method and its adsorption capacity. *Sci. Total Environ.* **2023**, *855*, 158741. [CrossRef]
9. Wang, B.; Wu, J.; Yuan, Z.-Y.; Li, N.; Xiang, S. Synthesis of MCM-22 zeolite by an ultrasonic-assisted aging procedure. *Ultrason. Sonochem.* **2008**, *15*, 334–338. [CrossRef]
10. Panitchakarn, P.; Laosiripojana, N.; Viriya-Umpikul, N.; Pavasant, P. Synthesis of high-purity Na-A and Na-X zeolite from coal fly ash. *J. Air Waste Manag. Assoc.* **2014**, *64*, 586–596. [CrossRef]
11. Koshlak, H. Synthesis of Zeolites from Coal Fly Ash Using Alkaline Fusion and Its Applications in Removing Heavy Metals. *Materials* **2023**, *16*, 4837. [CrossRef]
12. Küçük, M.E.; Makarava, I.; Kinnarinen, T.; Häkkinen, A. Simultaneous adsorption of Cu(II), Zn(II), Cd(II) and Pb(II) from synthetic wastewater using NaP and LTA zeolites prepared from biomass fly ash. *Heliyon* **2023**, *9*, e20253. [CrossRef]
13. Zhou, X.; Shi, S.; Ding, B.; Jia, H.; Chen, P.; Du, T.; Wang, Y. Optimization of preparation of NaA zeolite from fly ash for CO_2 capture. *Environ. Sci. Pollut. Res.* **2023**, *30*, 102803–102817. [CrossRef]
14. Che, Q.; Yang, M.; Wang, X.; Yang, Q.; Chen, Y.; Chen, X.; Chen, W.; Hu, J.; Zeng, K.; Yang, H.; et al. Preparation of mesoporous ZSM-5 catalysts using green templates and their performance in biomass catalytic pyrolysis. *Bioresour. Technol.* **2019**, *289*, 121729. [CrossRef]
15. Tarach, K.A.; Pyra, K.; Góra-Marek, K. Opening up ZSM-5 Hierarchical Zeolite's Porosity through Sequential Treatments for Improved Low-Density Polyethylene Cracking. *Molecules* **2020**, *25*, 2878. [CrossRef]
16. Visa, M.; Enesca, A. Opportunities for Recycling PV Glass and Coal Fly Ash into Zeolite Materials Used for Removal of Heavy Metals (Cd, Cu, Pb) from Wastewater. *Materials* **2022**, *16*, 239. [CrossRef]
17. Haghjoo, S.; Lengauer, C.L.; Kazemian, H.; Roushani, M. Facile and innovative application of surfactant-modified-zeolite from Austrian fly ash for glyphosate removal from water solution. *J. Environ. Manag.* **2023**, *346*, 118976. [CrossRef]
18. Joseph, I.V.; Tosheva, L.; Miller, G.; Doyle, A.M. FAU-Type Zeolite Synthesis from Clays and Its Use for the Simultaneous Adsorption of Five Divalent Metals from Aqueous Solutions. *Materials* **2021**, *14*, 3738. [CrossRef]
19. Panek, R.; Medykowska, M.; Wiśniewska, M.; Szewczuk-Karpisz, K.; Jędruchniewicz, K.; Franus, M. Simultaneous Removal of Pb^{2+} and Zn^{2+} Heavy Metals Using Fly Ash Na-X Zeolite and Its Carbon Na-X(C) Composite. *Materials* **2021**, *14*, 2832. [CrossRef]
20. Osacký, M.; Binčík, T.; Hudcová, B.; Vítková, M.; Pálková, H.; Hudec, P.; Bačík, P.; Czímerová, A. Low-cost zeolite-based sorbents prepared from industrial perlite by-product material for Zn^{2+} and Ni^{2+} removal from aqueous solutions: Synthesis, properties and sorption efficiency. *Heliyon* **2022**, *8*, e12029. [CrossRef]
21. Fan, Y.; Huang, R.; Liu, Q.; Cao, Q.; Guo, R. Synthesis of zeolite A from fly ash and its application in the slow release of urea. *Waste Manag.* **2023**, *158*, 47–55. [CrossRef]
22. Wang, G.; Chen, C.; Li, J.; Yang, F.; Wang, L.; Lin, X.; Wu, H.; Zhang, J. A clean method for gallium recovery and the coproduction of silica-potassium compound fertilizer and zeolite F from brown corundum fly ash. *J. Hazard. Mater.* **2024**, *461*, 132625. [CrossRef]
23. Zhang, Y.; Ma, J.; Miao, J.; Yue, C.; Cheng, M.; Li, Y.; Jing, Z. Self-regulated immobilization behavior of multiple heavy metals via zeolitization towards a novel hydrothermal technology for soil remediation. *Environ. Res.* **2023**, *216 Pt 3*, 114726. [CrossRef]
24. Fan, W.; Morozumi, K.; Kimura, R.; Yokoi, T.; Okubo, T. Synthesis of nanometer-sized sodalite without adding organic additives. *Langmuir* **2008**, *24*, 6952–6958. [CrossRef]

25. Liu, H. Conversion of Harmful Fly Ash Residue to Zeolites: Innovative Processes Focusing on Maximum Activation, Extraction, and Utilization of Aluminosilicate. *ACS Omega* **2022**, *7*, 20347–20356. [CrossRef]
26. Hardin, J.L.; Oyler, N.A.; Steinle, E.D.; Meints, G.A. Spectroscopic analysis of interactions between alkylated silanes and alumina nanoporous membranes. *J. Colloid Interface Sci.* **2010**, *342*, 614–619. [CrossRef]
27. Ellerbrock, R.; Stein, M.; Schaller, J. Comparing amorphous silica, short-range-ordered silicates and silicic acid species by FTIR. *Sci. Rep.* **2022**, *12*, 11708. [CrossRef]
28. Fu, H.; Li, Y.; Yu, Z.; Shen, J.; Li, J.; Zhang, M.; Ding, T.; Xu, L.; Lee, S.S. Ammonium removal using a calcined natural zeolite modified with sodium nitrate. *J. Hazard. Mater.* **2020**, *393*, 122481. [CrossRef]
29. Zhou, C.; An, Y.; Zhang, W.; Yang, D.; Tang, J.; Ye, J.; Zhou, Z. Inhibitory effects of Ca^{2+} on ammonium exchange by zeolite in the long-term exchange and NaClO-NaCl regeneration process. *Chemosphere* **2021**, *263*, 128216. [CrossRef]
30. He, X.; Chen, W.; Sun, F.; Jiang, Z.; Li, B.; Li, X.-Y.; Lin, L. Enhanced NH_4^+ Removal and Recovery from Wastewater Using Na-Zeolite-based Flow-Electrode Capacitive Deionization: Insight from Ion Transport Flux. *Environ. Sci. Technol.* **2023**, *57*, 8828–8838. [CrossRef]

Disclaimer/Publisher's Note: The statements, opinions and data contained in all publications are solely those of the individual author(s) and contributor(s) and not of MDPI and/or the editor(s). MDPI and/or the editor(s) disclaim responsibility for any injury to people or property resulting from any ideas, methods, instructions or products referred to in the content.

Article

Study on Adsorption of Cd in Solution and Soil by Modified Biochar–Calcium Alginate Hydrogel

Shuyue Wang, Yajun Wang *, Xinyi Wang, Sijia Sun, Yanru Zhang, Weixiong Jiao and Dasong Lin *

Agro-Environmental Protection Institute, Ministry of Agriculture and Rural Affairs, Tianjin 300191, China; vivraxxvvy_w@163.com (S.W.)
* Correspondence: wangyajun@caas.cn (Y.W.); lindasong608@126.com (D.L.)

Abstract: Contamination with cadmium (Cd) is a prominent issue in agricultural non-point source pollution in China. With the deposition and activation of numerous Cd metal elements in farmland, the problem of excessive pollution of agricultural produce can no longer be disregarded. Considering the issue of Cd pollution in farmland, this study proposes the utilization of cross-linked modified biochar (prepared from pine wood) and calcium alginate hydrogels to fabricate a composite material which is called MB-CA for short. The aim is to investigate the adsorption and passivation mechanism of soil Cd by this innovative composite. The MB-CA exhibits a higher heavy metal adsorption capacity compared to traditional biochar and hydrogel due to its increased oxygen-containing functional groups and heavy metal adsorption sites. In the Cd solution adsorption experiment, the highest Cd^{2+} removal rate reached 85.48%. In addition, it was found that the material also has an excellent pH improvement effect. Through the adsorption kinetics experiment and the soil culture experiments, it was determined that MB-CA adheres to the quasi-second-order kinetic model and is capable of adsorbing 35.94% of Cd^{2+} in soil. This study validates the efficacy of MB-CA in the adsorption and passivation of Cd in soil, offering a novel approach for managing Cd-contaminated cultivated land.

Keywords: modified biochar; calcium alginate; Cd pollution; in situ passivation

Citation: Wang, S.; Wang, Y.; Wang, X.; Sun, S.; Zhang, Y.; Jiao, W.; Lin, D. Study on Adsorption of Cd in Solution and Soil by Modified Biochar–Calcium Alginate Hydrogel. *Gels* **2024**, *10*, 388. https://doi.org/10.3390/gels10060388

Academic Editor: Dirk Kuckling

Received: 10 April 2024
Revised: 3 May 2024
Accepted: 13 May 2024
Published: 6 June 2024

Copyright: © 2024 by the authors. Licensee MDPI, Basel, Switzerland. This article is an open access article distributed under the terms and conditions of the Creative Commons Attribution (CC BY) license (https://creativecommons.org/licenses/by/4.0/).

1. Introduction

Soil is the material foundation for human survival, and good soil quality is significant for maintaining the basic functions of ecosystems. Heavy metal pollution is one of the biggest environmental pollution problems worldwide [1]. Unlike other pollutants, heavy metals are prone to accumulation in organisms and do not have biodegradability, ultimately leading to disease or death [2]. Cd is a harmful element to the human body and a non-essential element for plants [3,4]. It is easy to transfer and has strong toxicity. In addition, Cd pollution is also highly harmful and the most widespread metal pollution in China [5,6]. The Cd content in Chinese paddy soil ranges from 0.01 to 5.50 $mg \cdot kg^{-1}$, with a median of 0.23 $mg \cdot kg^{-1}$. Compared with the other provinces, the Cd content of the paddy soil in Hunan (0.73 $mg \cdot kg^{-1}$), Guangxi (0.70 $mg \cdot kg^{-1}$), and Sichuan (0.46 $mg \cdot kg^{-1}$) provinces is higher. Mining, smelting, sewage farming, air pollution, and the application of Cd-containing fertilizers are the main causes of Cd pollution in many paddy fields and dry lands in China [7]. Cd pollution in soil has toxic effects on soil organisms, affecting soil microbial populations, community structure, biochemical reactions, and soil enzyme activity [8,9]. For plants, Cd interferes with the absorption of nutrients, inhibiting photosynthesis, causing oxidative stress and gene damage, and affecting plant growth metabolism [10,11]. Normally, if the Cd content in the soil exceeds 8 $mg \cdot kg^{-1}$, most crops will exhibit visible Cd toxicity symptoms.

Applying in situ passivation materials in cultivated land is an effective way to reduce the bioavailability of Cd in the soil [12,13]. Traditional passivated materials include biochar, limestone, shell powder, silicate, zeolite, phosphate rock powder, etc. [14–18]. The main principle of their remediation of Cd-contaminated soil is to adjust the soil pH or combine with Cd ions to form stable compounds. However, the conventional passivation materials often cause issues such as soil compaction and secondary pollution, while the utilization of unmodified biochar is also constrained by its functionality, leading to unsatisfactory performance in heavy metal removal. In addition, the remediation effect of traditional materials on Cd ions is not stable due to the complexity of soil and water systems in nature [19].

Modified biochar has emerged as a prominent research focus in the field of environmental adsorption materials in recent years, owing to its exceptional adsorption capacity [20–22]. It is prepared by modifying traditional biochar through ball milling, chemical reactions, and other modification methods. Compared with traditional biochar, modified biochar has a larger specific surface area and more abundant oxygen-containing functional groups, providing many heavy metal adsorption sites [23]. Li and Shi found that the specific surface area, pore volume, and microporous volume of iron and dicyandiamide co-modified walnut shell biochar were 967.1084 $m^2 \cdot g^{-1}$, 0.7425 $cm^3 \cdot g^{-1}$, and 0.4624 $cm^3 \cdot g^{-1}$, respectively, which were 3.39, 6.42, and 8.81 times that of the original biochar [24]. Wang et al. found that carboxymethyl cellulose combined with nano zero-valent zinc modification can make the surface of biochar (nZVZ-CMC-PMBC) rougher, with a larger specific surface area and more developed pore size. Simultaneously, zero-valent zinc enters spherical particles and forms a nano-metal thin plate structure which strengthens the adsorption effect of the composite material on pollutants [25]. Jin et al. found that the adsorption capacity of modified biochar for heavy metal arsenic was significantly increased after modifying with potassium hydroxide [26]. Han et al. found that the surface adsorption sites of biochar increased and the adsorption capacity for Cr^{6+} was significantly improved after modifying with $FeCl_3$ [27]. Liang et al. found that the maximum adsorption capacity of heavy metals by biochar modified with amorphous MnO_2 was higher than that of unmodified biochar [28]. Liang et al. added thiol-modified sepiolite to Cd contaminated farmland soil, resulting in a 65.4% to 77.9% decrease in Cd content in rice [29]. In actual soil remediation, modified biochar is easily able to cause heavy metal migration and desorption due to its small particle size and aging, which is worthy of further study.

In recent years, hydrogel materials [30–34] have been widely used in the treatment of heavy metal ions due to their wide source of raw materials, low cost, strong adsorption capacity for metal ions, and other characteristics [35]. Alginate saline gel has good hydrophilicity and biocompatibility, and its richness in surface functional groups (such as carboxyl and hydroxyl) can capture metal ions effectively [36]. The unique swelling property of hydrogel makes heavy metal ions adsorb not only on the surface of the hydrogel, but also on the three-dimensional network structure during the removal process [37]. Compared with the traditional methods, hydrogel materials show obvious advantages in the adsorption of heavy metal ions, such as environmental friendliness, microstructure designability, and biodegradability, etc. Chan et al. [38] prepared a DNA–chitosan hydrogel, which can effectively combine with Hg^{2+}; the maximum adsorption capacity of Hg^{2+} is 50 $mg \cdot g^{-1}$. Yetimoglu et al. [39] found that the Pb^{2+} and Cd^{2+} adsorbed on AMPSG (guanidine-modified 2-acrylamido-2-methylpropan sulfonic acid)/AAc(acrylic acid)/NVP(N-vinylpyrrolidone)/HEMA(2-Hydroxyethyl methacrylate) hydrogel can be effectively desorbed through acid leaching, and the regenerated AMPSG/AAc/NVP/HEMA hydrogel did not reduce its adsorption properties. The incorporation of organic and inorganic materials, such as carbon-modified tubes, graphene, metal and metal oxides, silica-based materials, etc., into the gel system has garnered significant attention from researchers due to its potential for enhancing hydrogel performance in terms of swelling behavior, mechanical properties, and adsorption capacity [40,41]. Li et al. [42] designed a sodium lignosulfonate–guar gum composite hydrogel, which had excellent adsorption performance for heavy metals in soil. The maximum adsorption capacity of Cu^{2+} and Co^{2+} are 709.0 $mg \cdot g^{-1}$ and 601.00 $mg \cdot g^{-1}$, respectively. Wong et al. [43] developed a nano hydroxyapatite–cellulose hydrogel composite material, which removed 70.24%, 57.74%, 48.56%, 27.33%, and 25.98% of Cu^{2+}, Pb^{2+}, Fe^{2+}, Cd^{2+} and Zn^{2+} ions from palm oil factory wastewater, respectively. Yin et al. [44] prepared a modified xanthan gum–hydroxyapatite composite hydrogel (XG-g-PAA/HAP), and more than 90% of metal ions were removed within 30 min. Zhang et al. [45] mixed a dissolved cellulose solution, a TEMPO (2,2,6,6-tetramethylpiperidinyl-1-oxide)-oxidized cellulose nanofiber (TOCN) dispersion, and an alkali lignin solution in a NaOH–urea aqueous solution to prepare a composite hydrogel based on lignocellulose. The maximum adsorption amount of Cu^{2+} on the composite hydrogel reached 541 mg·g^{-1}. In addition, the presence of TOCN and lignin made the composite hydrogel show high strength performance.

The main waste of furniture production is pine sawdust, which often has a higher lignin content and lower ash content compared to agricultural straw. Therefore, it is an ideal material for preparing modified biochar. In this research, the authors aim to cross-link modified biochar and calcium alginate hydrogel to prepare a modified biochar–calcium alginate hydrogel composite, referred to as MB-CA, whose basic structure is shown in Figure 1. This composite aims to address Cd pollution in farmland by exploring its adsorption and passivation mechanism for soil Cd. Wang et al. [36] successfully synthesized a novel composite material with significant advantages by impregnating ball-milled biochar with calcium alginate particles through an innovative approach. This material not only exhibits enhanced water and fertilizer retention capacity, which can effectively lock up water and nutrients in the soil and reduce their loss, but also has good controlled-release properties, releasing nutrients on demand and providing stable and long-lasting nutrient support for plant growth. Modification of biochar is often required to improve some of its properties. Pretreatment of biomass with phosphoric acid (H_3PO_4) for biochar production can improve carbon (C) retention, porosity structure, and the sorption ability of biochar [46]. In this paper, the plan is to validate the feasibility of MB-CA remediation of Cd-contaminated soil through verification of adsorption kinetics, soil cultivation, and other experimental methodologies.

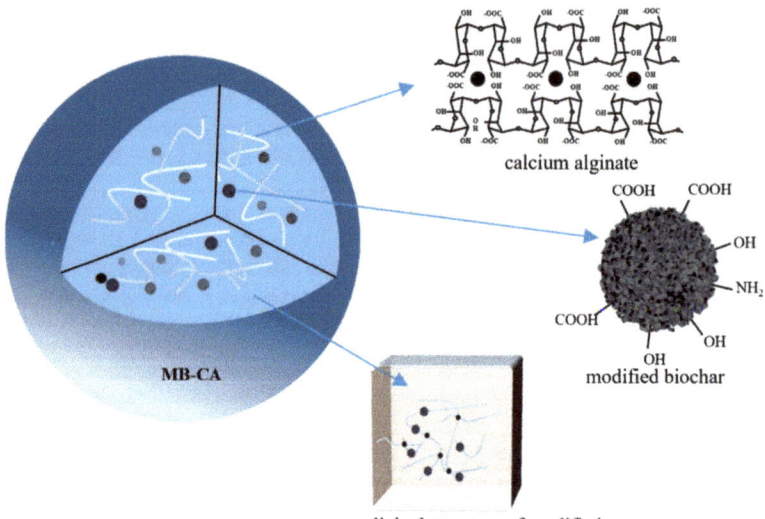

Figure 1. The fundamental structure of MB-CA.

2. Results and Discussion

2.1. Surface Morphology and Functional Groups of MB-CA

Figure 2a,b show the apparent morphology of modified biochar at different sizes, and the rough surface formed by pyrolysis and the modification of biomass can be clearly seen.

The surface microstructure of MB-CA is shown in Figure 2c,d. It can be found that the surface of the composite has deep and wide wrinkles and grooves after lyophilization. In addition, the BET surface area and average pore diameter were both measured; their values were 78.43 m$^2 \cdot$g^{-1} and 4.67 nm, respectively. The results suggest that the adsorption of Cd^{2+} by the composite is primarily attributable to complexation reactions between Cd^{2+} and the functional groups, rather than being predominantly driven by physical adsorption. To further investigate the functional groups present in the composite, FT-IR analysis was conducted. Figure 2e,f show the distribution of Cd^{2+} on the surface of the composite material after the adsorption of Cd^{2+}.

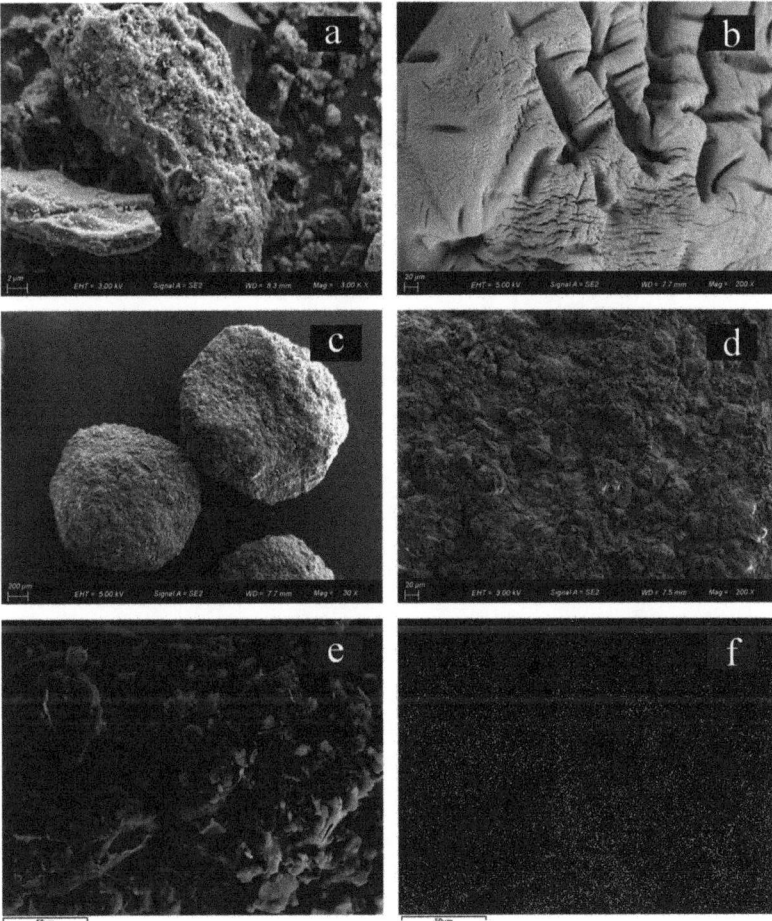

Figure 2. Surface morphology of modified biochar and MB-CA. (**a**) Modified biochar with magnification of 3 k (2 μm of scale bar); (**b**) calcium alginate hydrogel with magnification of 200 (20 μm of scale bar); (**c**) MB-CA with magnification of 30 (200 μm of scale bar); (**d**) MB-CA with magnification of 200 (20 μm of scale bar); (**e**) MB-CA after Cd^{2+} adsorption (50 μm of scale bar); and (**f**) the distribution of Cd^{2+} corresponding to Figure 2e.

The functional groups of MB-CA, calcium alginate, and the modified biochar were characterized by FT-IR spectra. As shown in Figure 3, the addition of modified biochar provides more types and quantities of functional groups in the composite. The broad peak at 3442 cm^{-1} before adsorption belongs to the free -OH (hydroxyl) in the molecule, indicating the presence of a large amount of -OH (hydroxyl) on the surface of the composite material. The peak at 2925 cm^{-1} is the vibration of aromatic C-H (hydrocarbon bond); the absorption peaks at 1589 cm^{-1} and 1436 cm^{-1} are generated by the stretching vibration of C=O (carbon oxygen double bond) and C-O (carbon oxygen single bond) in -COOH (carboxyl group), respectively, indicating the presence of a large amount of -COOH groups (carboxyl groups) on the surface of the composite material, which can provide a large number of adsorption sites and facilitate the adsorption of more Cd^{2+} by the composite material; the stretching at 1024 cm^{-1} is caused by the C-O (carbon oxygen single bond) functional group. Complexation is especially significant in wood charcoal or straw charcoal with a low inorganic mineral content. Teng et al. [47] characterized pine charcoal by FTIR, and found that there were characteristic bands such as -OH, -C-H, -C-O, -C=C, -COOH, and phenolic -OH, etc. Surface functional group complexation is the main mechanism of Cd adsorption, and the typical complexation reaction formula is: C-OH+Cd^{2+}+H$_2$O→C-OCd$^+$+H$_3$O$^+$, 2C-COOH+Cd^{2+}→(C-COO)$_2$Cd+2H$^+$ [48], etc.

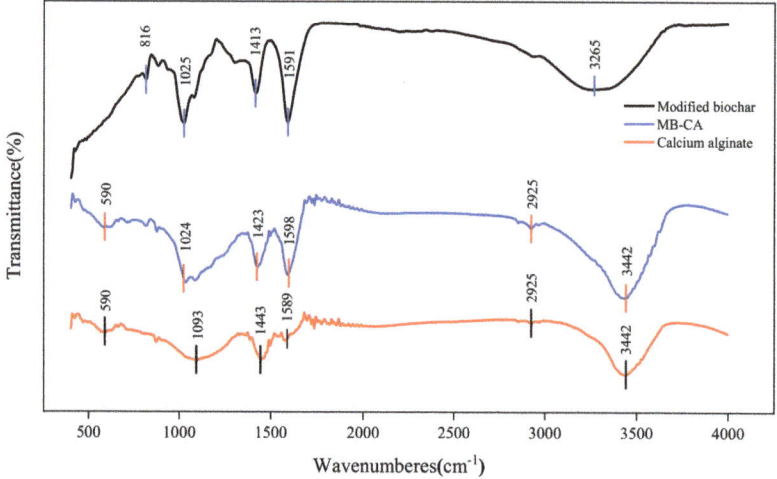

Figure 3. FTIR results of modified biochar, MB-CA, and calcium alginate.

2.2. Thermogravimetric Analysis of MB-CA

The thermal stability of the calcium alginate gel and MB-CA was verified through a thermogravimetric experiment. As shown in Figure 4b, the weightlessness curve of MB-CA exhibited three distinct weight loss phenomena at temperatures of 75.4 °C, 263.87 °C, and 717.01 °C, respectively. It is believed that these peaks correspond to the thermal degradation of water, calcium alginate, and activated carbon. At the end of the test, more than 40% of the solid residue was still left, which was considered to be mainly modified biochar residue.

2.3. The Effect of pH on the Adsorption of Cd^{2+} by MB-CA

Five 50 mL portions of Cd(NO$_3$)$_2$ solution at a concentration of 50 mg/L were prepared and the pH of each solution was adjusted to 2, 3, 4, 5 and 6, respectively. Subsequently, 0.06 g of MB-CA was added to each solution. The solutions were then magnetically agitated at room temperature (25 ± 1) °C for a duration of 24 h. Afterward, the concentrations of Cd^{2+} in the solutions as well as their respective pH values were measured. A total of 3 sets of parallel experiments were conducted and the results were averaged.

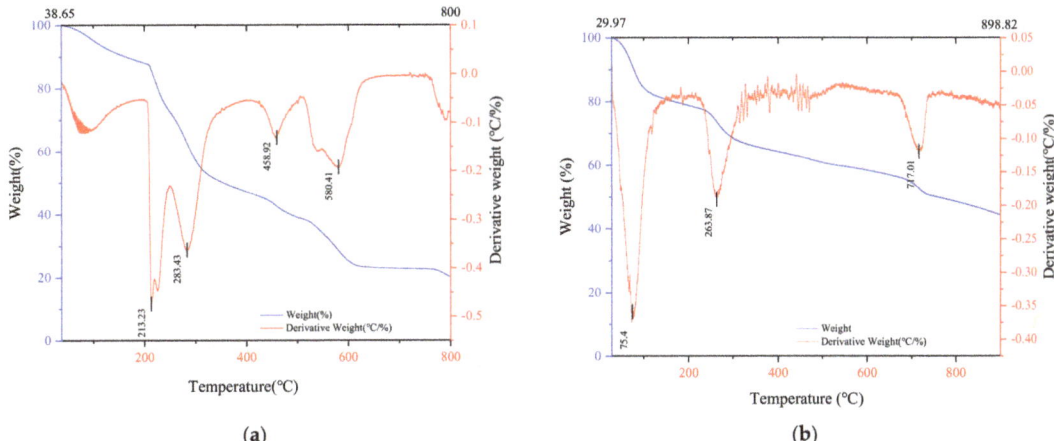

Figure 4. Weightlessness curve of calcium alginate gel and MB-CA. (**a**) Calcium alginate gel; (**b**) MB-CA.

Table 1 shows the effect of pH on the adsorption of Cd^{2+}. The adsorption capacity of MB-CA on Cd^{2+} was significantly enhanced as the pH increased from 2 to 5, leading to an increase in the concentration of Cd^{2+} within the material from 3.65 ± 1.67 mg·g^{-1} to 48.52 mg·g^{-1}. The adsorption capacity exhibited a slight decrease at pH 6. When the pH is 2, a significant abundance of positively charged H^+ ions surround the surface of MB-CA in the solution, while Cd exists as cations in the aqueous solution and competes with H^+ for adsorption sites, leading to a reduction in Cd^{2+} adsorption capacity. As the pH increases, the concentration of H^+ in the solution decreases, leading to an enhancement of MB-CA adsorption capacity which reaches its peak at pH 5 and the highest Cd^{2+} removal rate can reach 85.48%. Analyzing the reason for the above phenomenon, it may be that under strong acidic conditions, a large amount of H^+ in the solution will occupy limited binding sites and compete with Cd^{2+} for adsorption, reducing the Cd^{2+} removal rate [49]; with the increase in pH, the amount of H^+ decreases, which exposes a large number of binding sites on the surface of the material, and the adsorption capacity is also increased [50]. Interestingly, this experiment also revealed that the composite material effectively raised the pH of the solution beyond 6 when immersed in a solution with a pH ranging from 3 to 6; this observation suggests that the composite material possesses an alkalizing effect on solutions, thereby offering a novel perspective on Cd passivation mechanisms in soil. It is necessary to verify the passivation effect of the MB-CA on Cd and the pH-raising effects in the soil.

Table 1. pH changes before and after adsorption of Cd^{2+} by MB-CA.

Initial pH	pH = 2	pH = 3	pH = 4	pH = 5	pH = 6
Final pH	2.32 ± 0.06	6.04 ± 0.15	6.63 ± 0.02	6.71 ± 0.03	6.73 ± 0.01
Qe (mg·g^{-1})	3.65 ± 1.67	43.87 ± 3.15	46.65 ± 0.99	48.52 ± 4.36	44.97 ± 4.45

2.4. Dynamic Adsorption of MB-CA

The data presented in Figure 5 demonstrate that the composite material exhibits a substantial initial adsorption capacity for Cd^{2+}, followed by a gradual attainment of adsorption equilibrium. This experiment studied the kinetic behavior of MB-CA adsorption of Cd^{2+} by fitting dynamic adsorption data. Quasi-first-order kinetic models and quasi-second-order kinetic models were applied to fit the data, and the equations are shown in Equations (1) and (2)

$$lg(Q_e - Q_t) = lgQ_e - k_1/2.303t \quad (1)$$

$$t/Q_t = 1/k_2q^2e + 1/Q_e \quad (2)$$

In the formula, Q_t (mg·g^{-1}) and Q_e (mg·g^{-1}) are the adsorption capacities at time t (min) and equilibrium time of the adsorbent, respectively. k_1 (min^{-1}) and k_2 [g/(mg·min)] are the rate constants of the quasi-first-order and quasi-second-order kinetic equations, and t (min) is the adsorption time [51]. The kinetic experimental data of MB-CA adsorption is shown in Figure 5.

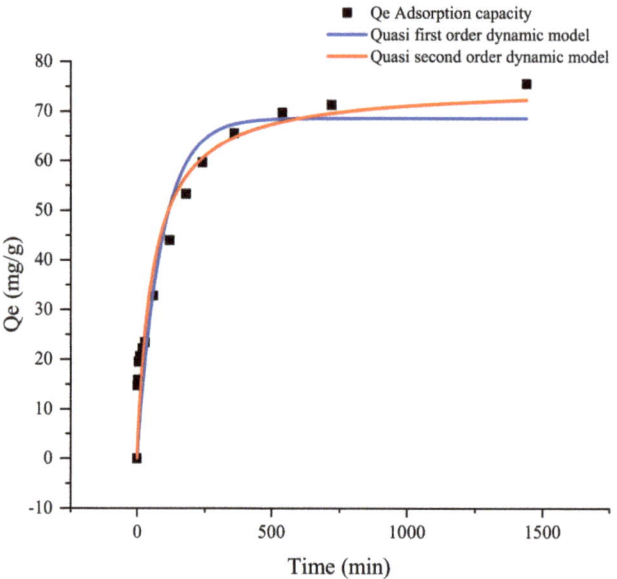

Figure 5. Kinetic adsorption diagram of MB-CA.

The initial adsorption rate of Cd^{2+} by MB-CA is rapid, primarily attributed to the outer layer of the gel absorbing Cd^{2+}, followed by a deceleration in the adsorption rate. At this stage, Cd^{2+} is predominantly adsorbed by the modified biochar and calcium alginate inside the composite. The dynamic fitting results are presented in Table 2.

Table 2. Kinetic fitting parameters of MB-CA adsorption of Cd^{2+}.

Sample Name	Quasi-First-Order Dynamic Model			Quasi-Second-Order Dynamic Model			Adsorption Capacity at Equilibrium
	$Q_{e, cal}$/mg·g^{-1}	$K_1 \times 10^{-3}$/min^{-1}	R^2	$Q_{e, cal}$/mg·g^{-1}	$K_1 \times 10^{-3}$/min^{-1}	R^2	Q_e/mg·g^{-1}
MB-CA	68.657	0.111	0.868	75.254	2.273	0.915	75.583

According to Table 2, the correlation coefficient of the quasi-second-order kinetic model for MB-CA is 0.915, which is higher than the correlation coefficient of the quasi-first-order dynamic model (0.868). Additionally, the equilibrium adsorption amount fitted by the quasi-second-order kinetic model closely approximates the actual value. According to the liquid–solid adsorption theory, diffusion is a rate control step.

2.5. The Results of Soil Culture Experiment

The results of our 30-day soil culture experiment indicate that the application of the composite material led to a gradual increase in soil pH, resulting in a rise from 6.7 to 7.1. Compared to the blank control group (MB-CA not applied), as shown in Figure 6, the application of MB-CA resulted in a reduction in Cd^{2+} concentration in contaminated soil from 0.73 mg·kg^{-1} to 0.40 mg·kg^{-1} after conducting four samplings. Additionally, the composite material also influenced the distribution of Cd in the soil, with an increase in the residual state from 58% to 67%, and a decrease in the exchangeable state from 25% to 21%. These results indicate that MB-CA effectively mitigates the total content and bioavailability of Cd in soil. This indicates that MB-CA can effectively reduce the toxicity and bioavailability of Cd in soil.

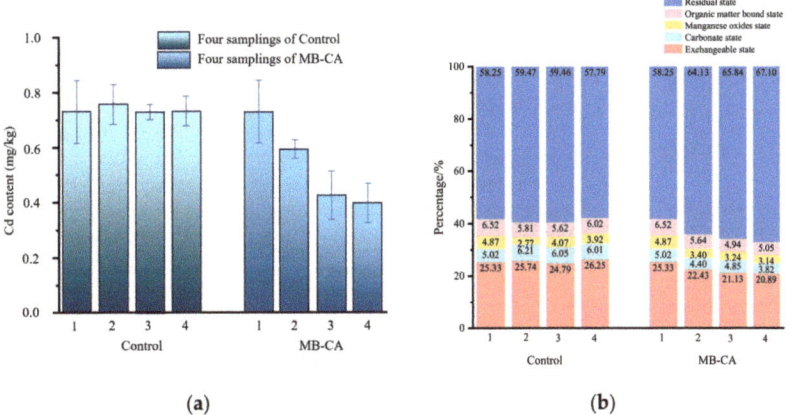

Figure 6. The influence of MB-CA on the concentration and speciation of Cd in soil. (**a**) The Cd^{2+} content; (**b**) chemical species of Cd.

3. Conclusions

MB-CA exhibits a higher abundance of oxygen-containing functional groups and heavy metal adsorption sites compared to calcium alginate. A pH ranging from 3 to 6 exhibits a favorable adsorption effect. The adsorption of Cd^{2+} by the composite is more consistent with the quasi-second-order kinetic model. The Cd^{2+} concentration and bioavailability of soil were significantly reduced by the application of MB-CA, demonstrating its significant potential for reducing the risk of agricultural products that exceed regulatory standards. This study proposes the cross-linking preparation of modified biochar with calcium alginate to form MB-CA composites. This composite material combines the adsorption properties of biochar with the gel properties of calcium alginate, which makes it show unique advantages in the field of heavy metal pollution remediation. This innovative material combination provides new possibilities for soil heavy metal pollution remediation. MB-CA composites were applied in the remediation and treatment of soil Cd pollution. Through multi-level experiments, the adsorption and passivation mechanisms of MB-CA composites on soil Cd were investigated in depth, revealing their practical effects in soil pollution remediation. This application innovation not only expands the application scope of biochar and hydrogel composites, but also provides a new technical means for soil heavy metal pollution remediation.

Future research on heavy metal adsorption by hydrogels can focus on the following aspects: (1) improving the cross-linking technology and preparation process to reduce the cost of hydrogel preparation; (2) selecting and developing new cross-linking agents to enhance the mechanical strength and adsorption capacity of hydrogels; (3) strengthening the research and development of hydrogels with selective adsorption, specific sensitivity to the adsorption environment, and high sensitivity to heavy metal ions, as well as other specific functions of hydrogels; (4) developing detoxification materials, which can make the adsorbed materials with heavy metal ions precipitated from the soil. Overall, the MB-CA composites have excellent adsorption effects and broad application prospects, which are worthy of more in-depth research.

4. Materials and Methods

4.1. Preparation of Modified Biochar/Modified Biochar Calcium Alginate Hydrogel Composite

Pulverized pine sawdust, with uniform particle size (passing through a 60-mesh sieve), was impregnated with a solution of phosphoric acid. Subsequently, it was dried and loaded into a pyrolysis furnace for slow pyrolysis under a nitrogen atmosphere at final temperatures of 450 °C, 500 °C, 550 °C, or 600 °C. After cooling down, the modified biochar was prepared and stored for later use. The modified biochar was incorporated into a 2% calcium alginate solution and subjected to ultrasound stirring for 60 min, resulting in the formation of a mixed solution containing modified biochar and calcium alginate. The above mixed solution was slowly added into a 0.2 mol·L^{-1} CaCl$_2$ solution while stirring, and the mixture was stirred for 60 min until spherical hydrogels precipitated at the bottom of the beaker to obtain the composite material MB-CA. The composite material was subsequently washed with hydrochloric acid and clean water, followed by cooling and drying. The optimal process parameters were selected based on the single indicator of Cd^{2+} adsorption strength per unit mass in a solution of Cd (NO$_3$)$_2$.

4.2. Characterization of Physical and Chemical Properties of Composite Materials MB-CA

By utilizing advanced techniques such as SEM-EDS (scanning electron microscopy–energy spectrum analysis, ZEISS Gemini Sigma 300, Oberkochen, Germany, incident electron beam: 3 and 5 kV), FT-IR (infrared spectroscopy analysis, Thermo Scientific Nicolet iS5 FT-IR, Spectral range: 400–4000 cm^{-1}, Waltham, MA, USA), TG-DTG (thermogravimetric differential thermal analysis, Swiss Mettler Toledo TGA/DSC 1/1600, Greifensee, Switzerland, temperature range: 25–800 °C, heating rate: 10 °C/min, atmosphere: nitrogen), ICP-MS (inductively coupled plasma–mass spectrometry iCAP Q, Waltham, MA, USA), etc., the physical and chemical properties of composite materials were comprehensively analyzed. This included examining their apparent morphology, functional groups, pore size distribution, and thermal stability.

4.3. Test on the Adsorption Characteristics of Composite Material MB-CA for Cd in Solution

Optimal adsorption conditions for Cd^{2+} in aqueous solutions were determined through batch intermittent adsorption experiments, employing a HCl (1 mol·L^{-1} and 0.1 mol·L^{-1}) and NaOH solution (1 mol·L^{-1} and 0.1 mol·L^{-1}), respectively, to adjust the initial pH values of various aqueous solutions within the range of 2–6. Subsequently, 0.6 g of freeze-dried composite material was added to a 30 mL solution of Cd(NO$_3$)$_2$ with a concentration ranging from 20 to 500 mg·L^{-1}, and the beaker was placed in a constant temperature shaker at 20–40 °C for continuous agitation. When the adsorption of the reaction system solution reached equilibrium, a water sample was collected approximately 2 cm below the liquid level. The sample was then filtered and diluted to determine the removal efficiency of Cd^{2+} by the appropriate material.

4.4. Adsorption Kinetics

The kinetic adsorption experiment of MB-CA was conducted at room temperature (25 ± 1) °C. Under the conditions of initial Cd^{2+} concentration of 50 mg·L^{-1}, adsorption time of 1440 min, and system pH of 5 (pH adjusted with sodium hydroxide and hydrochloric acid), 500 mL of Cd(NO$_3$)$_2$ solution was added to a 1 L beaker. Then, 0.6 g of MB-CA was added to the 500 mL solution of Cd(NO$_3$)$_2$ and vigorously stirred using magnetic force. At specific time intervals (ranging from 0 min to 1440 min), a sample of 5 mL Cd(NO$_3$)$_2$ solution was taken and diluted with a 0.24 mol·L^{-1} HNO$_3$ solution. Subsequently, the concentration of Cd^{2+} was measured using an atomic absorption analyzer. The samples were collected at 1, 3, 5, 10, 20, 30, 60, 120, 180, 240, 360, 540, 720, and 1440 min (in triplicate), diluted to a concentration of 1 ppm, and then filtered through the membrane for subsequent use. This experiment accurately demonstrates the dynamic adsorption behavior of composite materials and investigates the influence of composite materials on adsorption effectiveness.

4.5. Soil Cd Bioavailability in Soil Culture Experiment

Quantitative measurement of Cd-contaminated soil was conducted, and MB-CA (2% of the soil weight) was quantitatively applied. After mixing, the mixture was filled into pots, and soil samples were collected every 7 days. ICP-MS technology was used to detect the content of Cd^{2+} in soil, and the Tessier five-step extraction method was employed to detect the bioavailability of Cd in soil. The adsorption and passivation effects of composite materials on Cd^{2+} in soil cultivation experiments were evaluated using atomic absorption (ThermoFisher ESCALAB 250Xi, USA).

4.6. Reagents

The reagents used in this study were shown in Table 3.

Table 3. Experimental reagents.

Reagent Name	Purity Level	Source
$C_6H_7NaO_6$ (SA)	Chemical pure	Sinopharm Chemical Reagent Co., Ltd., Shanghai, China
$CaCl_2$	Chemical pure	Bodi Chemical Industry Co., Ltd., Tianjin, China
NaOH	Analytial reagent	Sinopharm Chemical Reagent Co., Ltd., Shanghai, China
$Cd(NO_3)_2 \cdot 4H_2O$	Analytial reagent	Fuchen Chemical Reagent Co., Ltd., Tianjin, China
$NaNO_3$	Analytial reagent	Xinhao Chemical Industry Co., Ltd., Zibo, China
HNO_3	Analytial reagent	Luxi Chemical Industry Group Co., Ltd., Liaocheng, China
H_3PO_4	Analytial reagent	Taixi Chemical Industry Co., Ltd., Jinan, China
HCl	Analytial reagent	Beiyuan Chemical Industry Group Co., Ltd., Yulin, China

Author Contributions: Conceptualization, Y.W.; methodology, S.W.; software, S.W.; validation, Y.W. and D.L.; formal analysis, X.W. and S.S.; writing—original draft preparation, S.W.; writing—review and editing, Y.W., Y.Z. and W.J.; project administration, D.L. All authors have read and agreed to the published version of the manuscript.

Funding: This research was funded by the Basic research project of Chinese Academy of Agricultural Sciences, grant number Y2023LM12.

Institutional Review Board Statement: Not applicable.

Informed Consent Statement: Not applicable.

Data Availability Statement: The data presented in this study are openly available in article.

Conflicts of Interest: The authors declare no conflicts of interest.

References

1. Ghosh, A.; Manna, M.C.; Jha, S.; Singh, A.K.; Misra, S.; Srivastava, R.C.; Srivastava, P.P.; Laik, R.; Bhattacharyya, R.; Prasad, S.S.; et al. Impact of soil-water contaminants on tropical agriculture, animal and societal environment. *Adv. Agron.* **2022**, *176*, 209–274.
2. Sheydaei, M. Investigation of Heavy Metals Pollution and Their Removal Methods: A Review. *Geomicrobiol. J.* **2024**, *41*, 213–230. [CrossRef]
3. Cirovic, A.; Satarug, S. Toxicity Tolerance in the Carcinogenesis of Environmental Cadmium. *Int. J. Mol. Sci.* **2024**, *25*, 1851. [CrossRef] [PubMed]
4. Li, Y.; Liu, M.; Wang, H.; Li, C.; Zhang, Y.; Dong, Z.; Fu, C.; Ye, Y.; Wang, F.; Chen, X.; et al. Effects of different phosphorus fertilizers on cadmium absorption and accumulation in rice under low-phosphorus and rich-cadmium soil. *Environ. Sci. Pollut. Res.* **2024**, *31*, 11898–11911. [CrossRef] [PubMed]
5. Ding, R.; Wei, D.; Wu, Y.; Liao, Z.; Lu, Y.; Chen, Z.; Gao, H.; Xu, H.; Hu, H. Profound regional disparities shaping the ecological risk in surface waters: A case study on cadmium across China. *J. Hazard. Mater.* **2024**, *465*, 133450. [CrossRef] [PubMed]
6. Huang, W.; Sun, D.; Zhao, T.; Long, K.; Zhang, Z. Spatial-temporal distribution and source analysis of atmospheric particulate-bound cadmium from 1998 to 2021 in China. *Environ. Geochem. Health* **2024**, *46*, 44. [CrossRef] [PubMed]
7. Khan, M.S.; Zaidi, A.; Wani, P.A.; Oves, M. Role of Plant Growth Promoting Rhizobacteria in the Remediation of Metal Contaminated Soils: A Review. In *Organic Farming, Pest Control and Remediation of Soil Pollutants*; Lichtfouse, E., Ed.; Springer: Dordrecht, The Netherlands, 2009; Volume 1, pp. 319–350.

8. Kayiranga, A.; Li, Z.; Isabwe, A.; Ke, X.; Simbi, C.H.; Ifon, B.E.; Yao, H.-f.; Wang, B.; Sun, X. The effects of heavy metal pollution on Collembola in urban soils and associated recovery using biochar remediation: A review. *Int. J. Environ. Res. Public Health* **2023**, *20*, 3077. [CrossRef] [PubMed]
9. Mortensen, L.H.; Ronn, R.; Vestergard, M. Bioaccumulation of cadmium in soil organisms—With focus on wood ash application. *Ecotoxicol. Environ. Saf.* **2018**, *156*, 452–462. [CrossRef] [PubMed]
10. Cheng, Y.; Qiu, L.; Shen, P.; Wang, Y.; Li, J.; Dai, Z.; Qi, M.; Zhou, Y.; Zou, Z. Transcriptome studies on cadmium tolerance and biochar mitigating cadmium stress in muskmelon. *Plant Physiol. Biochem.* **2023**, *197*, 107661. [CrossRef]
11. Li, Y.; Xu, R.; Ma, C.; Yu, J.; Lei, S.; Han, Q.; Wang, H. Potential functions of engineered nanomaterials in cadmium remediation in soil-plant system: A review. *Environ. Pollut.* **2023**, *336*, 122340. [CrossRef]
12. Wang, A.; Wang, Y.; Zhao, P.; Huang, Z. Effects of composite environmental materials on the passivation and biochemical effectiveness of Pb and Cd in soil: Analyses at the ex-planta of the Pak-choi root and leave. *Environ. Pollut.* **2022**, *309*, 119812. [CrossRef] [PubMed]
13. Wang, Z.; Zhang, T.; Zhao, Y.; Miao, Y.; Zhang, L.; Sarocchi, D.; Song, S.; Zhang, Q. Immobilization of Cd in contaminated soil by mechanically activated calcite: Sustained release activity-depended performance and mechanisms. *Chem. Eng. J.* **2024**, *482*, 149024. [CrossRef]
14. Zong, Y.; Chen, H.; Malik, Z.; Xiao, Q.; Lu, S. Comparative study on the potential risk of contaminated-rice straw, its derived biochar and phosphorus modified biochar as an amendment and their implication for environment. *Environ. Pollut.* **2022**, *293*, 118515. [CrossRef] [PubMed]
15. Yang, Y.; Li, Y.; Wang, M.; Chen, W.; Dai, Y. Limestone dosage response of cadmium phytoavailability minimization in rice: A trade-off relationship between soil pH and amorphous manganese content. *J. Hazard. Mater.* **2021**, *403*, 123664. [CrossRef] [PubMed]
16. Zhan, J.; Wen, Y.; Wang, Y.; Zhu, H.; Li, S.; Chen, Y.; Chen, X.; Wang, Y.; Shang, Q. Synergistic regulatory effects of oyster shell powder on soil acidification and cadmium pollution in paddy fields. *Acta Agric. Univ. Jiangxiensis* **2023**, *45*, 787–794.
17. Lin, C.F.; Lo, S.S.; Lin, H.Y.; Lee, Y.C. Stabilization of cadmium contaminated soils using synthesized zeolite. *J. Hazard. Mater.* **1998**, *60*, 217–226. [CrossRef]
18. Li, Y.; Li, X.; Kang, X.; Zhang, J.; Sun, M.; Yu, J.; Wang, H.; Pan, H.; Yang, Q.; Lou, Y.; et al. Effects of a novel Cd passivation approach on soil Cd availability, plant uptake, and microbial activity in weakly alkaline soils. *Ecotoxicol. Environ. Saf.* **2023**, *253*, 114631. [CrossRef]
19. Ahmad, M.; Rajapaksha, A.U.; Lim, J.E.; Zhang, M.; Bolan, N.; Mohan, D.; Vithanage, M.; Lee, S.S.; Ok, Y.S. Biochar as a sorbent for contaminant management in soil and water: A review. *Chemosphere* **2014**, *99*, 19–33. [CrossRef]
20. Li, X.; Li, R.; Zhan, M.; Hou, Q.; Zhang, H.; Wu, G.; Ding, L.; Lv, X.; Xu, Y. Combined magnetic biochar and ryegrass enhanced the remediation effect of soils contaminated with multiple heavy metals. *Environ. Int.* **2024**, *185*, 108498. [CrossRef]
21. Peng, J.; Zhang, Z.; Wang, Z.; Zhou, F.; Yu, J.; Chi, R.; Xiao, C. Adsorption of Pb^{2+} in solution by phosphate-solubilizing microbially modified biochar loaded with Fe_3O_4. *J. Taiwan Inst. Chem. Eng.* **2024**, *156*, 105363. [CrossRef]
22. Wu, G.; Wang, B.; Xiao, C.; Huang, F.; Long, Q.; Tu, W.; Chen, S. Effect of montmorillonite modified straw biochar on transfer behavior of lead and copper in the historical mining areas of dry-hot valleys. *Chemosphere* **2024**, *352*, 141344. [CrossRef]
23. Tan, L.; Nie, Y.; Chang, X.; Zhu, L.; Guo, K.; Ran, X.; Zhong, N.; Zhong, D.; Xu, Y.; Ho, S.-H. Adsorption performance of Ni(II) by KOH-modified biochar derived from different microalgae species. *Bioresour. Technol.* **2024**, *394*, 130287. [CrossRef] [PubMed]
24. Li, X.; Shi, J. Simultaneous adsorption of tetracycline, ammonium and phosphate from wastewater by iron and nitrogen modified biochar: Kinetics, isotherm, thermodynamic and mechanism. *Chemosphere* **2022**, *293*, 133574. [CrossRef] [PubMed]
25. Wang, M.; Hu, S.; Wang, Q.; Liang, Y.; Liu, C.; Xu, J.; Ye, Q. Enhanced nitrogen and phosphorus adsorption performance and stabilization by novel panda manure biochar modified by CMC stabilized nZVZ composite in aqueous solution: Mechanisms and application potential. *J. Clean. Prod.* **2021**, *291*, 125221. [CrossRef]
26. Jin, H.; Capareda, S.; Chang, Z.; Gao, J.; Xu, Y.; Zhang, J. Biochar pyrolytically produced from municipal solid wastes for aqueous As(V) removal: Adsorption property and its improvement with KOH activation. *Bioresour. Technol.* **2014**, *169*, 622–629. [CrossRef] [PubMed]
27. Han, Y.; Cao, X.; Ouyang, X.; Sohi, S.P.; Chen, J. Adsorption kinetics of magnetic biochar derived from peanut hull on removal of Cr (VI) from aqueous solution: Effects of production conditions and particle size. *Chemosphere* **2016**, *145*, 336–341. [CrossRef] [PubMed]
28. Liang, J.; Li, X.; Yu, Z.; Zeng, G.; Luo, Y.; Jiang, L.; Yang, Z.; Qian, Y.; Wu, H. Amorphous MnO_2 Modified Biochar Derived from Aerobically Composted Swine Manure for Adsorption of Pb(II) and Cd(II). *ACS Sustain. Chem. Eng.* **2017**, *5*, 5049–5058. [CrossRef]
29. Liang, X.; Qin, X.; Huang, Q.; Huang, R.; Yin, X.; Wang, L.; Sun, Y.; Xu, Y. Mercapto functionalized sepiolite: A novel and efficient immobilization agent for cadmium polluted soil. *RSC Adv.* **2017**, *7*, 39955–39961. [CrossRef]
30. Jiang, C.; Wang, X.; Wang, G.; Hao, C.; Li, X.; Li, T. Adsorption performance of a polysaccharide composite hydrogel based on crosslinked glucan/chitosan for heavy metal ions. *Compos. Part B—Eng.* **2019**, *169*, 45–54. [CrossRef]
31. Jiang, H.; Yang, Y.; Lin, Z.; Zhao, B.; Wang, J.; Xie, J.; Zhang, A. Preparation of a novel bio-adsorbent of sodium alginate grafted polyacrylamide/graphene oxide hydrogel for the adsorption of heavy metal ion. *Sci. Total Environ.* **2020**, *744*, 140653. [CrossRef]

32. Vinh Van, T.; Park, D.; Lee, Y.-C. Hydrogel applications for adsorption of contaminants in water and wastewater treatment. *Environ. Sci. Pollut. Res.* **2018**, *25*, 24569–24599.
33. Tassanapukdee, Y.; Prayongpan, P.; Songsrirote, K. Removal of heavy metal ions from an aqueous solution by CS/PVA/PVP composite hydrogel synthesized using microwaved-assisted irradiation. *Environ. Technol. Innov.* **2021**, *24*, 101898. [CrossRef]
34. Sinha, V.; Chakma, S. Advances in the preparation of hydrogel for wastewater treatment: A concise review. *J. Environ. Chem. Eng.* **2019**, *7*, 103295. [CrossRef]
35. Khan, M.; Lo, I.M.C. A holistic review of hydrogel applications in the adsorptive removal of aqueous pollutants: Recent progress, challenges, and perspectives. *Water Res.* **2016**, *106*, 259–271. [CrossRef] [PubMed]
36. Wang, B.; Gao, B.; Wan, Y. Entrapment of ball-milled biochar in Ca-alginate beads for the removal of aqueous Cd(II). *J. Ind. Eng. Chem.* **2018**, *61*, 161–168. [CrossRef]
37. Xue, Y.; Gao, B.; Yao, Y.; Inyang, M.; Zhang, M.; Zimmerman, A.R.; Ro, K.S. Hydrogen peroxide modification enhances the ability of biochar (hydrochar) produced from hydrothermal carbonization of peanut hull to remove aqueous heavy metals: Batch and column tests. *Chem. Eng. J.* **2012**, *200*, 673–680. [CrossRef]
38. Chan, K.; Morikawa, K.; Shibata, N.; Zinchenko, A. Adsorptive Removal of Heavy Metal Ions, Organic Dyes, and Pharmaceuticals by DNA-Chitosan Hydrogels. *Gels* **2021**, *7*, 112. [CrossRef] [PubMed]
39. Yetimoglu, E.K.; Firlak, M.; Kahraman, M.V.; Deniz, S. Removal of Pb^{2+} and Cd^{2+} ions from aqueous solutions using guanidine modified hydrogels. *Polym. Adv. Technol.* **2011**, *22*, 612–619. [CrossRef]
40. Yang, J.; Mosby, D. Field assessment of treatment efficacy by three methods of phosphoric acid application in lead-contaminated urban soil. *Sci. Total Environ.* **2006**, *366*, 136–142. [CrossRef]
41. Xi, H.; Zhang, X.; Zhang, A.H.; Guo, F.; Yang, Y.; Lu, Z.; Ying, G.; Zhang, J. Concurrent removal of phosphate and ammonium from wastewater for utilization using Mg-doped biochar/bentonite composite beads. *Sep. Purif. Technol.* **2022**, *285*, 120399. [CrossRef]
42. Li, X.; Wang, X.; Han, T.; Hao, C.; Han, S.; Fan, X. Synthesis of sodium lignosulfonate-guar gum composite hydrogel for the removal of Cu^{2+} and Co^{2+}. *Int. J. Biol. Macromol.* **2021**, *175*, 459–472. [CrossRef] [PubMed]
43. Wong, S.M.; Zulkifli, M.Z.A.; Nordin, D.; Teow, Y.H. Synthesis of Cellulose/Nano-hydroxyapatite Composite Hydrogel Absorbent for Removal of Heavy Metal Ions from Palm Oil Mill Effluents. *J. Polym. Environ.* **2021**, *29*, 4106–4119. [CrossRef]
44. Yin, X.; Wu, J.; Zhao, Y.; Lin, X.; Pei, L.; Li, J.; Liu, X.; Jin, L. Preparation of polymer modified xanthan gum/hydroxyapatite composite hydrogel and its absorption for metal ions. *Acta Sci. Circumstantiae* **2017**, *37*, 633–641.
45. Zhang, L.; Lu, H.; Yu, J.; Fan, Y.; Yang, Y.; Ma, J.; Wang, Z. Synthesis of lignocellulose-based composite hydrogel as a novel biosorbent for Cu^{2+} removal. *Cellulose* **2018**, *25*, 7315–7328. [CrossRef]
46. Zhao, L.; Zheng, W.; Mašek, O.; Chen, X.; Gu, B.; Sharma, B.K.; Cao, X. Roles of phosphoric acid in biochar formation: Synchronously improving carbon retention and sorption capacity. *J. Environ. Qual.* **2017**, *46*, 393–401. [CrossRef] [PubMed]
47. Teng, D.; Zhang, B.; Xu, G.; Wang, B.; Mao, K.; Wang, J.; Sun, J.; Feng, X.; Yang, Z.; Zhang, H. Efficient removal of Cd (II) from aqueous solution by pinecone biochar: Sorption performance and governing mechanisms. *Environ. Pollut.* **2020**, *265*, 115001. [CrossRef] [PubMed]
48. Bandara, T.; Franks, A.; Xu, J.; Bolan, N.; Wang, H.; Tang, C. Chemical and biological immobilization mechanisms of potentially toxic elements in biochar-amended soils. *Crit. Rev. Environ. Sci. Technol.* **2020**, *50*, 903–978. [CrossRef]
49. Zhao, D.; Yang, X.; Zhang, H.; Chen, C.; Wang, X. Effect of environmental conditions on Pb (II) adsorption on β-MnO_2. *Chem. Eng. J.* **2010**, *164*, 49–55. [CrossRef]
50. Dada, A.; Olalekan, A.; Olatunya, A.; Dada, O. Langmuir, Freundlich, Temkin and Dubinin–Radushkevich isotherms studies of equilibrium sorption of Zn^{2+} unto phosphoric acid modified rice husk. *IOSR J. Appl. Chem.* **2012**, *3*, 38–45.
51. Ibrahim, A.G.; Saleh, A.S.; Elsharma, E.M.; Metwally, E.; Siyam, T. Chitosan-g-maleic acid for effective removal of copper and nickel ions from their solutions. *Int. J. Biol. Macromol.* **2019**, *121*, 1287–1294. [CrossRef]

Disclaimer/Publisher's Note: The statements, opinions and data contained in all publications are solely those of the individual author(s) and contributor(s) and not of MDPI and/or the editor(s). MDPI and/or the editor(s) disclaim responsibility for any injury to people or property resulting from any ideas, methods, instructions or products referred to in the content.

Efficacy of Chitosan-Carboxylic Acid Hydrogels in Reducing and Chelating Iron for the Removal of Rust from Stone Surface

Francesco Gabriele *, Cinzia Casieri and Nicoletta Spreti

Department of Physical and Chemical Sciences, University of L'Aquila, 67100 L'Aquila, Italy; cinzia.casieri@aquila.infn.it (C.C.); nicoletta.spreti@univaq.it (N.S.)
* Correspondence: francesco.gabriele@univaq.it

Abstract: In the field of stone conservation, the removal of iron stains is one of the most challenging issues due to the stability and low solubility of the ferrous species. In the present paper, three different chitosan-based hydrogels added with acetic, oxalic or citric acids are applied on different lithotypes, i.e., granite, travertine and marble, widely diffused in monumental heritages, and artificially stained by deposition of a rust dispersion. The reducing power of carboxylic acids is combined with the good chelating properties of chitosan to effectively remove rust from stone surfaces. As evidenced by colorimetry on three samples of each lithotype and confirmed by ^{1}H-NMR relaxometry and SEM/EDS analyses, the chitosan-oxalic acid hydrogel shows the best performance and a single application of 24 h is enough to get a good restoration of the stone original features. Lastly, the chitosan-oxalic acid hydrogel performs well when a rusted iron grid is placed directly on the lithic surfaces to simulate a more realistic pollution. Current work in progress is devoted to finding better formulations for marble, which is the most challenging to clean or, with a different approach, to developing protective agents to prevent rust deposition.

Keywords: iron rust; lithic surface; cleaning; chitosan hydrogel; reducing agent; chelating agent

1. Introduction

Removal of rust stains from stone materials, especially in the field of monumental heritage, is one of the most challenging cleaning actions due to the high thermodynamic stability of iron oxyhydroxides and hydrated oxides. Iron is often in contact with stone materials in cultural heritages; in fact, iron bars, nails, and other decorative or support objects are diffused elements in lithic works of art [1–3]. Moreover, iron can be found in minerals of lithic substrates, such as in the form of pyrite (FeS_2), siderite ($FeCO_3$) or biotite ($K(Mg,Fe^{2+})_3AlSi_3O_{10}(OH,F)_2$) and others [4]. Their continuous exposure to pollutants, particulate, acid rain, and humidity induces the formation of numerous iron oxidation products and the appearance of yellow-brownish and red-blackish stains on the stone substrate [5]. The presence of rust products not only affects the aesthetics of the artwork but also leads to degrading actions of the lithic supports and even to the formation of cracks. This phenomenon is caused by the "rust expansion" effect for which the oxidized metal undergoes an increase in its volume over time [6,7].

Water plays an important role for iron oxyhydroxide nanoparticles, both as a dispersing medium and as a carrier through the stone. Then, the nanoparticles can interact with the substrate and undergo a series of transformations before being converted into more stable iron compounds [8,9]. There is a plethora of iron-based compounds that form rust, which can interconvert each other according to the environmental conditions, i.e., pH, relative humidity (RH), and the counterions involved in the oxidation processes. In a dried oxygen-poor environment, a mixed ferrous/ferric oxide is mainly formed, and the so-called magnetite (Fe_3O_4) appears as a dark stain. In contrast, under high RH conditions, the iron-hydrated oxides and hydroxides are predominant.

The thermodynamic constants and the solubility of some of the main rust constituents, such as ferrihydrite, α-, β- and γ-FeOOH, were investigated and goethite (α-FeOOH) was found to be the most stable and insoluble (solubility product constant, $K_{sp} = 10^{-41}$) [5,10,11]. Therefore, being the most thermodynamically stable, an effective cleaning of rust stains must ensure the complete removal of goethite from the lithic substrate [5].

Nowadays, physical or chemical approaches are used to remove iron stains from stone surfaces, also in the field of cultural heritage. The first ones are generally mechanical methods, and recently, good results have been reached on granite by using laser cleaning. This technique reduces the typical loss of substrate layers due to the abrasive blasting and ensures a safer removal of rust products [12].

Chemical approaches, the most diffused in conservation practice, involve the use of chelating or reducing agents. Aqueous solutions of ammonium citrate or ethylenediaminetetraacetic acid (EDTA) are applied to bind iron and remove it from artworks [13]; differently, reducing agents, such as oxalic acid and sodium dithionite or thioglycolic acid are employed to increase the solubility of iron and allow its removal from the stone [14,15] or from old iron objects [16], respectively. However, chemical treatments often lead to the deposition of unwanted residues even if the substrate surface is rinsed with water, as well as to the removal of calcium ions from carbonate substrates. Poultices of cellulose, carboxymethyl cellulose (CMC), or other materials, commonly used to clean stone surfaces [17,18], can incorporate reducing and chelating agents to effectively remove iron oxyhydroxides and by-products from the artwork surface [3,5]. Nevertheless, by adopting this procedure on carbonate stones, the calcium ions of the substrate could also be sequestered, thus damaging the surface of the stone [3]. By comparing the side effects of the two chelating agents on marble under the same experimental condition, ammonium citrate turns out less aggressive than EDTA, albeit only partially removing the calcium ions [19]. A recent study has shown a new green methodology to effectively chelate and remove Fe (III) on artificially stained marble samples by using proteins, such as ovotransferrin and lactotransferrin [20]. Due to the high specificity of these proteins for ferric cations, they should not interact with the substrate, including carbonates, to ensure the integrity of the material. Oxalic acid used as a reducing agent proved to be a good strategy to effectively remove rust stains from sandstone when applied together with doped cellulose poultice [14]. In this way, Fe(III) species reduced to Fe(II) compounds are generally more soluble, i.e., Fe(OH)$_2$ shows a K_{sp} of about 10^{-14} and therefore could easily diffuse into the poultice support. More recently, a viscous DES composed of oxalic acid and choline chloride has shown high efficacy in removing iron oxides from both cellulosic and lithic-stained substrates [21].

To achieve a more effective rust removal, the treatment often involves the combined use of reducing and complexing agents. In fact, in the last two decades, several works have been aimed at finding the optimal pH and the more effective combination of reducing and chelating agents to successfully remove iron stains without compromising the integrity of the stone substrates. Sodium dithionite seems to be a good reducing agent that can easily reduce Fe(III) to Fe(II) and then can be coupled with different chelating agents to remove iron from stone substrates. Among the common ligands, cysteine forms strong complexes with Fe(II) that could be easily transferred into suitable cellulose poultice [5,22]. However, the optimum working pH of this last system is less than 10, which represents the threshold value for the partial dissolution of the calcium carbonate. To avoid this problem, cysteine can be replaced with hexadentate N,N,N',N'-tetrakis(2-pyridylmethyl)ethylenediamine (TPEN) that works efficiently even at a pH of about 10. Unlike EDTA, this chelating agent is highly selective towards heavy metal ions over calcium ones, making it less aggressive to carbonate stone substrates [23]. Another alternative is the biodegradable tetrasodium 3-hydroxy-2,2'-iminodisuccinate (HIDS), which has been shown to be effective in removing rust stains from cotton fabrics when combined with sodium dithionite as a reducing agent [24].

Although most reported investigations describe the use of pure or doped cellulose poultice as a medium to convey cleaning agents for the removal of pollutant species, the application of nanostructured fluids or hydrogels has increased significantly in recent years [25–27]. For example, polysaccharide-based hydrogels encapsulating biocides of different natures have been effectively employed to remove microbial colonizers from biodeteriorated stone surfaces, both in the laboratory and in situ [28–30]. The use of agar gels in the cleaning procedures for various conservation interventions has been summarized by Sansonetti and coworkers [31]. Inter alia, the review counts on the efficacy of agar-chelating agent gels in removing copper stains from marble surfaces [32]. Moreover, agar hydrogels, incorporating ionic liquids, have been applied to sequester iron and copper ions from stone materials, and the effectiveness of the formulations, selected by laboratory tests, have been validated in situ on naturally stained substrates [33]. Recently, an EDTA-loaded bacterial nanocellulose hydrogel has been able to completely remove copper stains from marble, proving to be an effective alternative to traditional hydrogels for the cleaning of cultural heritage materials [34].

In this paper, chitosan-based hydrogels were prepared for the removal of stable rust products from the surface of different stone lithotypes. Chitosan is a linear polysaccharide that derives from the partial deacetylation of chitin; it is composed of β (1-4)-linked D-glucosamine and N-acetyl-D-glucosamine disposed randomly along the chains [35]. It is soluble only in weakly acid-aqueous solutions ($pK_a \approx 6.5$) and, at high polysaccharide concentrations, its chains can become entangled, leading to the formation of a physical hydrogel [36,37]. Its high biocompatibility, biodegradability, and intrinsic bacteriostatic efficacy also show good chelating properties. In fact, thanks to the large number of hydroxyl groups, the active ammino groups, and the flexible polymeric structure, chitosan is generally suitable for the adsorption of heavy metal ions. To enhance the complexing ability of the hydrogel in the removal of silver from aqueous solutions, promising results have been obtained using chemically crosslinked chitosan encapsulating a siderophore [38]. In addition, physical chitosan hydrogel dissolved in an acid solution and added with thiourea dioxide as a reducing agent has been effective in removing manganese stains from both glass and marble [39].

Despite the good chelating properties of both pure, modified and crosslinked chitosan towards Fe(II) and Fe(III) [40–42], to the best of our knowledge, there are no papers dealing with the use of chitosan-based hydrogels for the removal of iron oxides from stone surfaces. For this purpose, acetic, oxalic, and citric acids have been selected for reducing insoluble Fe(III) species to more soluble Fe(II) ones and, at the same time, to dissolve chitosan, which was selected as a chelating agent capable of forming three different hydrogels: chitosan-acetic acid (CS-Ac), chitosan-oxalic acid (CS-Ox) and chitosan-citric acid (CS-Cit).

Three lithotypes, i.e., granite, travertine, and marble, widely diffused in-built heritages and characterized by different compositions and porosity, have been artificially stained and treated with each of the three chitosan-based hydrogels. Firstly, colorimetry has been used to select the best formulation on a macroscopic scale. Then, ^{1}H-NMR relaxometry, SEM-EDS analysis, or stereomicroscope observation was performed to evaluate the effectiveness of the selected hydrogel system.

2. Results and Discussion

The viscosity of the reducing-chelating hydrogels, composed of chitosan and acetic, oxalic, or citric acid, was measured to assess the possibility of applying them also on vertical surfaces. Despite being constituted by the same concentration of both polysaccharide and acid, the viscosity decreases significantly from CS-Ac (280 P) to CS-Ox (80 P) and CS-Cit (40 P), indicating that the aggregation between the chitosan chains strongly depends on the number of carboxyl groups, as also evident from their photo in Figure S1.

The FTIR analysis was carried out to investigate the nature of these hydrogels. Therefore, to avoid the superimposition of signals of both solvent and plasticizer, the gels were prepared without glycerol and then dried at room temperature. The spectra, shown in Figures S2–S4, revealed that only acid–base reaction occurs without the formation of covalent bonds between the organic acids and chitosan, indicating that physical hydrogels are formed [43].

2.1. Cleaning of Lithotypes Stained with Rust Dispersion

2.1.1. Photos and Colorimetry

Three samples of granite, travertine, and marble stained with a rust dispersion were treated with the hydrogels, and their effectiveness in removing the iron stains was evaluated by photographing the surface of the samples during the cleaning procedure. Figure 1 reports, for each CS-acid and lithotype, the images of one of the three specimens before the staining (reference), soon after the rust deposition (stained), and after the treatment (treated).

Figure 1. Photographs of the reference, stained (with a rust dispersion) and treated with Cs-Ac, CS-Ox, and CS-Cit specimens; (**A**) travertine, (**B**) marble, and (**C**) granite.

It is evident from the figure that the degree of staining depends on the nature of the lithotype. Marble is less susceptible to stains than travertine and granite due to its polished surface and low water absorption.

After the treatment, CS-Ox effectively removes rust products from all the samples, whereas with CS-Cit achieves only a partial, unsatisfying, cleaning. Differently, Cs-Ac appears effective on marble specimen, but leaves visible residues on travertine and granite surfaces.

After a cleaning intervention, colorimetric analysis allows a deeper evaluation of the restoration of the original chromaticity and the gloss of a substrate. It is reported that color variations $\Delta E^* < 3$ result imperceptible to the human eye, while any treatment showing $\Delta E^* < 5$ is considered acceptable for conservative purposes [44]. Table 1 lists the mean colorimetric parameters of the reference sample, L^*, a^*, and b^*, their variations after staining and treatment, as well as their corresponding ΔE^* values, calculated according to Equation (1) (Section 4.2.3).

Table 1. For each lithotype subjected to the treatment with CS-Ac, CS-Ox, or CS-Cit, the mean chromatic coordinates of the reference stone, mean chromatic coordinates differences of both stained (with a rust dispersion) and treated specimens, and the corresponding mean color differences.

	Lithotype		L^*	a^*	b^*	
CS-Ac	Travertine	Ref	77 ± 2	3.9 ± 0.4	10 ± 1	
			ΔL^*	Δa^*	Δb^*	ΔE^*
		Stained	-33 ± 3	5 ± 1	16 ± 2	37
		Treated	-11 ± 2	1.1 ± 0.5	2 ± 1	11
			L^*	a^*	b^*	
	Marble	Ref	76 ± 2	-1.4 ± 0.1	-2.1 ± 0.3	
			ΔL^*	Δa^*	Δb^*	ΔE^*
		Stained	-9 ± 3	2.4 ± 0.9	11 ± 2	15
		Treated	-1 ± 2	0.1 ± 0.1	0.4 ± 0.4	1
			L^*	a^*	b^*	
	Granite	Ref	65 ± 3	0.6 ± 0.6	2.4 ± 0.9	
			ΔL^*	Δa^*	Δb^*	ΔE^*
		Stained	-23 ± 3	8.2 ± 0.8	23 ± 1	34
		Treated	-19 ± 2	5.3 ± 0.7	14 ± 2	24
CS-Ox	Travertine		L^*	a^*	b^*	
		Ref	77 ± 1	3.3 ± 0.2	7.5 ± 0.5	
			ΔL^*	Δa^*	Δb^*	ΔE^*
		Stained	-27 ± 3	7.3 ± 0.6	21 ± 1	35
		Treated	-1 ± 1	-0.2 ± 0.3	0.6 ± 0.7	1
			L^*	a^*	b^*	
	Marble	Ref	75 ± 2	-1.3 ± 0.2	-1.7 ± 0.3	
			ΔL^*	Δa^*	Δb^*	ΔE^*
		Stained	-11 ± 4	3 ± 1	13 ± 2	17
		Treated	-1 ± 2	0.0 ± 0.2	0.5 ± 0.4	1
			L^*	a^*	b^*	
	Granite	Ref	69 ± 3	0.2 ± 0.8	1 ± 1	
			ΔL^*	Δa^*	Δb^*	ΔE^*
		Stained	-25 ± 3	8 ± 1	22 ± 2	34
		Treated	-1 ± 2	0.1 ± 0.6	0.1 ± 0.7	1

Table 1. *Cont.*

			L*	a*	b*	
CS-Cit	Travertine	Ref	81 ± 3	3.1 ± 0.3	8 ± 1	
			ΔL*	Δa*	Δb*	ΔE*
		Stained	−34 ± 3	6.7 ± 0.6	19 ± 1	40
		Treated	−15 ± 6	2 ± 1	4 ± 2	16
	Marble		L*	a*	b*	
		Ref	74 ± 3	−1.4 ± 0.2	−1.9 ± 0.3	
			ΔL*	Δa*	Δb*	ΔE*
		Stained	−11 ± 3	3 ± 1	13 ± 2	17
		Treated	−4 ± 4	2 ± 1	9 ± 3	10
	Granite		L*	a*	b*	
		Ref	67 ± 4	0.3 ± 0.8	2 ± 1	
			ΔL*	Δa*	Δb*	ΔE*
		Stained	−17 ± 2	6.2 ± 0.7	18 ± 2	26
		Treated	−11 ± 3	2.5 ± 0.9	7 ± 2	13

In stained granite and travertine samples, the global chromatic differences show values $\Delta E^* \approx 35$, while marble specimens exhibit $\Delta E^* \approx 17$, a value halved compared to the other two stones. Whatever the extent of the chromatic variation on each specimen, it is mainly due to the darkening, reddening, and yellowing of the surfaces as highlighted by the signs of the values of ΔL^*, Δa^*, and Δb^*.

After the treatment with CS-Cit, the colorimetric coordinate values for all materials decrease, but the resulting chromatic variation turns out to always be higher than 10, a value not acceptable for conservative purposes. Similar results are observed following the treatment with the hydrogel containing acetic acid, with the sole exception of the marble specimens. By applying CS-Ac, the granite remains almost unaffected, and the travertine results only partially cleaned. On the contrary, the chromaticity of marble results is completely restored after only one treatment, with the recovery of all the colorimetric coordinates that provide accordingly $\Delta E^* = 1$, a value well below the perception limit of the human eye. Excellent results are obtained using the formulation containing the oxalic acid; indeed, a single treatment results enough to effectively restore the original chroma of the substrates regardless of the nature of the lithotype. As highlighted in Table 1, all the colorimetric coordinates return to the original reference values with a consequent negligible ΔE^*.

Recently, it has been reported that 1.6 M aqueous solution of oxalic acid, applied for 24 h in a cellulose poultice, is capable of effectively removing iron crusts from sandstone [14]. In this work we report similar results, but with an acid concentration six times lower, i.e., 0.25 M. Scheme 1 reports the hypothesized mechanism, in which the reducing ability of oxalic acid is combined with the chelating properties of chitosan.

As evident from the scheme, the complexation of Fe(II) involves both the amino and the C6-hydroxyl groups of chitosan, according to a previous study [45]. Therefore, the greater effectiveness of the CS-Ox hydrogel, compared to the poultice, could be attributed to the better ability of chitosan to chelate iron ions.

Based on these results, the CS-Ox hydrogel was selected to investigate its efficacy also at a microscopic level.

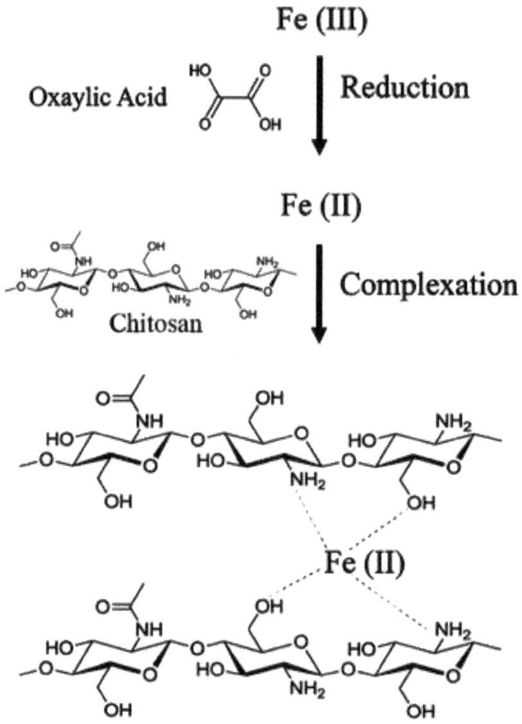

Scheme 1. Proposed mechanism of action of the chitosan-oxalic acid hydrogel in reducing Fe(III) and subsequent chelation of Fe(II) by chitosan.

2.1.2. ^{1}H-NMR Relaxometry

Relaxometry is a powerful technique for studying water confined in porous materials [46–49]. In fact, according to NMR theory, when the sample is fully water-saturated, there exists a correlation between the inverse Laplace transform of the proton T_2-decay of water and the pore-size distribution of the material [50].

In our experiment, the transverse relaxation decays of water-saturated reference, stained, and treated stone samples were acquired and directly compared, avoiding the inverse Laplace transform. Moreover, as the T_2-decays were recorded under stationary conditions of water absorption, there are no effects due to the time of water absorption are possible. Unfortunately, due to the presence of a high amount of intrinsic paramagnetic species in granite and the very low porosity of marble, it was only possible to apply this technique to travertine.

The travertine, which is mainly composed of calcium carbonate and characterized by large cavities, was successfully analyzed by ^{1}H-NMR T_2 relaxometry and the normalyzed signal decays of reference, stained and cleaned with CS-Ox samples are compared in Figure 2.

The red curve, corresponding to the stained samples, shows a faster decay than that of the reference (blue curve), suggesting the presence of rust products on the stone surface, which affects the relaxation process of the water molecules. In fact, being constant the surface-to-volume ratio of the sample as well as the echo time, the faster relaxation time of water can be imputed to the paramagnetic nuclei deposited within the porous structure. In fact, the presence of a few ppm of paramagnetic impurities, such as Fe(III) ions, can significantly modify the transverse relaxation times of confined water molecules. These impurities act as relaxation centers for protons, causing a reduction in the magnitude of the relaxation processes. In fact, their unpaired electrons generate local magnetic fields that interact with the nuclear spins of nearby water molecules, leading to a faster T_2-decay [46,48,51].

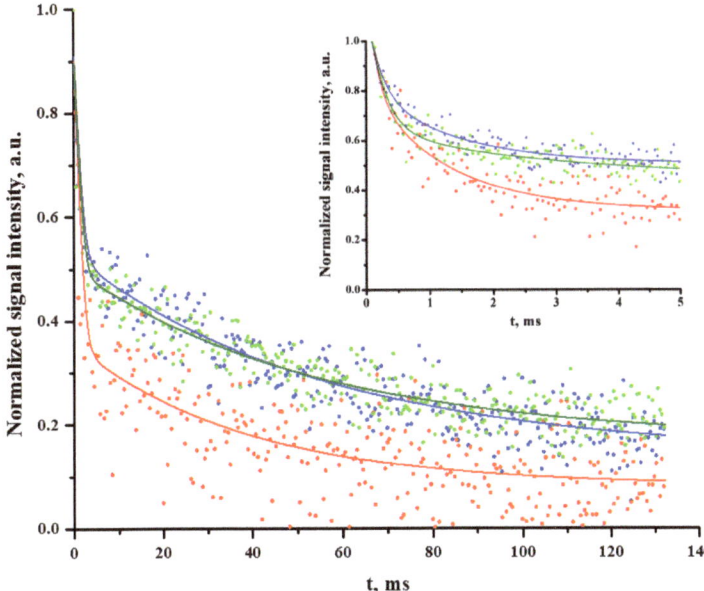

Figure 2. Water T_2 decays by NMR relaxometry on travertine for reference (blue), stained with rust dispersion (red), and treated with CS-Ox (green) samples. In the inset, the curves are plotted in the first 5 ms of decay.

After the cleaning procedure with CS-Ox, the decay signal (green curve) appears restored and overlaps almost completely with the reference signal. The differences highlighted here could be better appreciated in the inset of Figure 2, in which the decay signals in the first 5 ms are shown.

The experimental evidence of the T_2 decays suggest that iron products are completely removed from the stone surface, confirming the above discussed macroscopic observations.

2.1.3. SEM-EDS Analysis

All the lithotypes were observed through an electronic microscope to identify four areas of 1 mm² on the stone samples. EDS microanalyses were performed on each of them before the deposition of rust suspension (reference), after the staining, and after the treatment with CS-Ox. For each lithotype, Figure 3 shows the SEM images, with iron highlighted in orange, of the reference, stained, and treated areas and the 3D comparison between the corresponding EDS spectra.

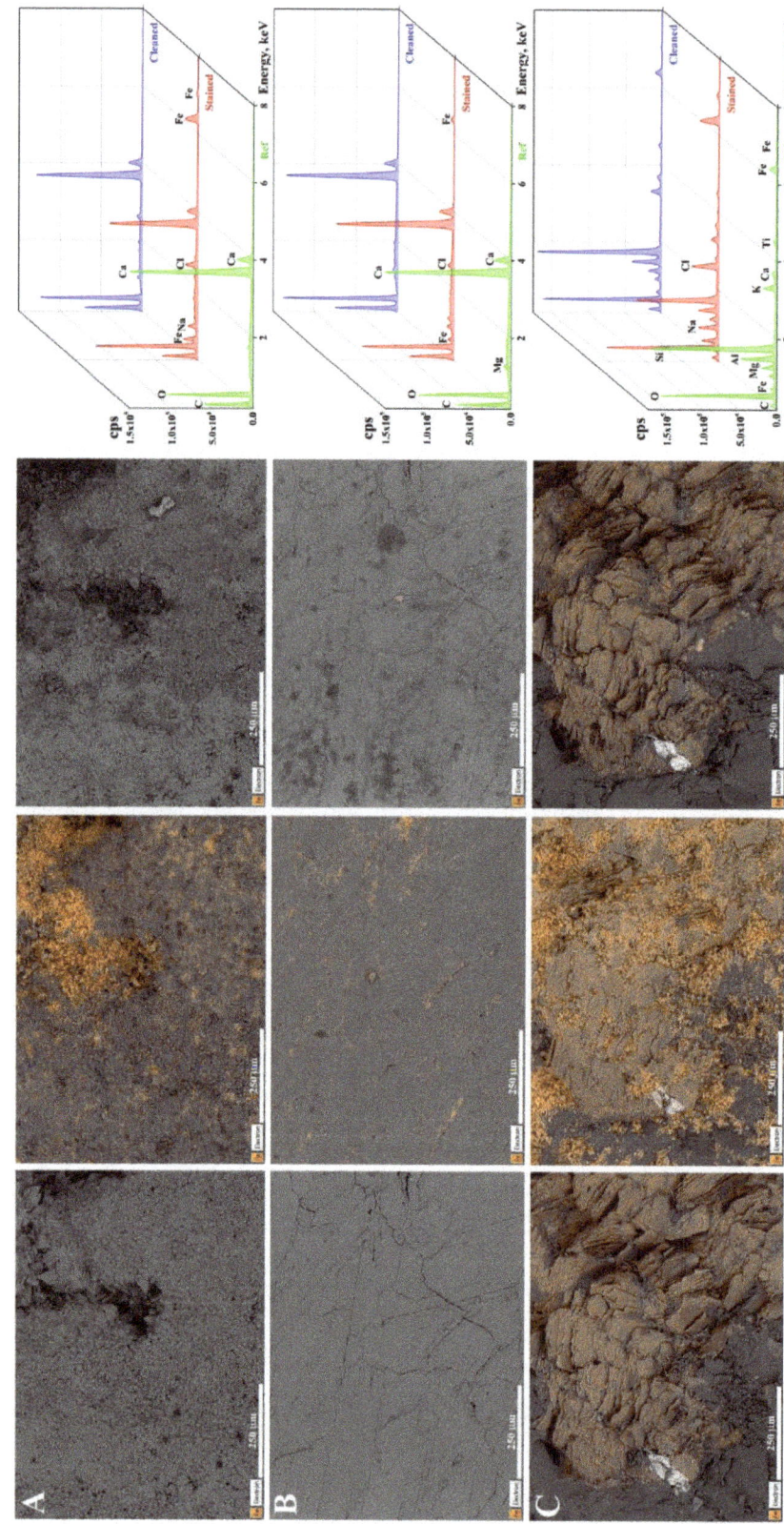

Figure 3. SEM-EDS analyses of 1 mm² area of travertine (**A**), marble (**B**) and granite (**C**), in which the presence of Fe is highlighted in orange, before rust deposition (on the **left**), after the staining (in the **middle**) and after the treatment with CS-Ox (on the **right**). In the last column, for each lithotype, the graphics 3D of EDS spectra in the three phases of the procedure.

As expected, before the deposition of rust suspension, the two carbonate-based substrates (Figure 3A,B) do not exhibit appreciable iron content; on the contrary, several Fe-domains are almost ubiquitous in the granite sample (Figure 3C).

After the staining of the stone surfaces, a significant presence of iron is observed also on travertine and marble, as evidenced by the presence of orange spots covering the investigated areas and semi-quantified by their corresponding EDS spectra. Marble is characterized by the lowest content of deposited ferrous species, which is in full agreement with its morphology and macroscopic analysis.

On all stained samples, the presence of Na and Cl are also detected, which is due to the sodium chloride, added for accelerating the iron oxidation process.

After the treatment with the CS-Ox, the deposited iron appears completely removed from the surface of the three specimens, as evidenced by the lack of the orange spots on SEM images and the near overlap of the EDS spectra of the treated and reference samples. In addition, the absence of nitrogen, detectable at 0.39 KeV, sodium and chlorine in the EDS spectra of the treated samples, indicates the complete removal of the hydrogel that, evidently, is able to adsorb also the sodium chloride deposited during the staining process.

To better interpret the results, the weight percentages of the most characterizing elements for each lithotype, as well as of iron, were evaluated on four selected areas of all the specimens, in the three phases of the procedure (Table 2).

Reference samples of travertine and marble show high reproducibility in their main constituting element, i.e., carbon and calcium, according to their carbonate nature. Differently, due to its coarse and multicomponent nature, granite shows different distributions of silicon and aluminum depending on the investigated area. Moreover, it is evident that the reference sample of travertine presents only traces of iron (0.1% wt), while for all the investigated areas of marble, the metal is not present at all. In contrast, the iron content of the different observed areas of granite ranges between 0.6 and 12%, showing an inhomogeneous distribution of ferrous compounds in the material.

Table 2. Weight percentages of most abundant constituents, including Fe, of travertine, marble, and granite, determined in four areas of each reference, stained with rust and cleaned sample and the corresponding mean values.

	Travertine					
	Element	Area 1	Area 2	Area 3	Area 4	Mean
Reference	Ca	40.2	40.1	39.0	40.1	39.9 ± 0.6
	C	13.7	13.7	14.3	13.9	13.9 ± 0.3
	Fe	0.1	0.1	0.1	0.1	0.1 ± 0.0
Stained	Ca	32.3	30.5	26.3	30.8	30 ± 3
	C	13.2	13.5	14.0	16.8	14 ± 2
	Fe	9.0	11.1	15.9	6.2	11 ± 4
Treated	Ca	34.8	36.9	33.1	35.5	35 ± 2
	C	16.7	15.3	17.2	15.8	16.3 ± 0.9
	Fe	0.1	0.2	0.2	0.2	0.2 ± 0.1
	Marble					
	Element	Area 1	Area 2	Area 3	Area 4	Mean
Reference	Ca	37.2	38.3	37.6	38.0	37.8 ± 0.5
	C	16.2	15.0	15.8	15.3	15.6 ± 0.5
	Fe	0.0	0.0	0.0	0.0	0.0
Stained	Ca	36.9	35.9	36.1	37.0	36.5 ± 0.6
	C	14.4	13.9	13.9	13.5	13.9 ± 0.4
	Fe	2.0	3.6	2.9	2.4	2.7 ± 0.7
Treated	Ca	31.5	32.6	32.3	32.8	32.3 ± 0.6
	C	17.5	16.7	16.5	16.8	16.9 ± 0.4
	Fe	0.1	0.2	0.2	0.4	0.2 ± 0.1

Table 2. *Cont.*

		Granite				
	Element	Area 1	Area 2	Area 3	Area 4	Mean
Reference	Si	32.4	23.5	26.6	30.8	28 ± 4
	Al	5.3	6.0	9.7	3.0	6 ± 3
	Fe	0.9	12.0	0.6	6.7	5 ± 5
Stained	Si	24.8	14.3	17.5	24.7	20 ± 5
	Al	4.3	3.2	6.4	2.3	4 ± 2
	Fe	14.0	27.3	16.0	15.3	18 ± 6
Treated	Si	31.5	21.6	24.8	28.5	27 ± 4
	Al	4.9	4.6	9.2	2.2	5 ± 3
	Fe	0.8	10.2	0.5	7.1	5 ± 5

Although the Fe content of all samples increases significantly after the deposition of rust, marble appears to be the less contaminated specimen with an overall iron abundance of only 3%, compared to travertine and granite, whose amounts are over 10%. Following the treatment, the 'natural' amount of iron of the pristine stone substrates is almost completely restored.

SEM-EDS analysis confirms not only the good ability of CS-Ox in removing iron stains of all stones, but also that the hydrogel is capable of removing other hygroscopic contaminants from the surface of the substrates without significantly affecting their morphology.

2.2. Cleaning of Lithotypes Stained with Iron Grid

2.2.1. Photos and Colorimetry

To more faithfully reproduce the staining process that stones undergo when in contact with iron elements, three specimens for each lithotype were artificially polluted by placing a rusty iron grid in contact with one of their larger surfaces. For each lithotype, Figure 4 shows the photos of one of the three samples before the staining (reference), after the staining and following the treatment with CS-Ox.

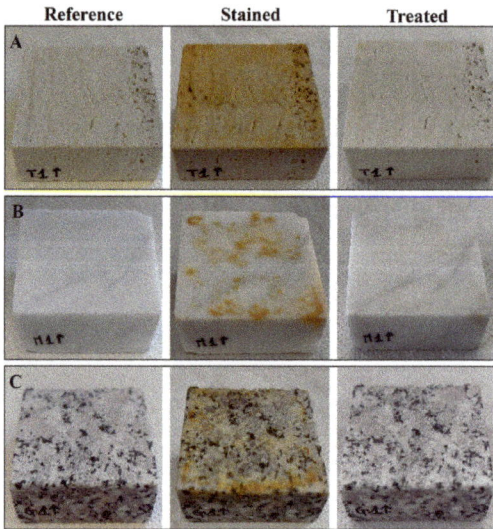

Figure 4. Photographs of the reference, stained with rusty iron grid and treated specimens subjected to the treatment with CS-Ox; (**A**) travertine, (**B**) marble, and (**C**) granite.

The marble sample exhibits a spotted deposition of rust on its surface, in contrast to travertine and granite, whose surfaces are more homogeneously stained. Nevertheless, one treatment is sufficient to remove the rust formed on the travertine and granite samples, while a residual light orange color is still perceptible on the marble surface.

The colorimetric analysis was then performed to evaluate the changes in the substrate color and Table 3 reports the mean values of chromatic coordinates of each reference stone, their difference in the stained and treated surfaces with their respective ΔE^* values.

Table 3. For each lithotype subjected to the treatment with CS-Ox, the mean chromatic coordinates of the reference stone, mean chromatic coordinate differences of both stained (with rusty iron grid) and cleaned specimens, and the corresponding mean color differences.

		L*	a*	b*	
Travertine	Ref	78 ± 4	3.6 ± 0.3	9 ± 1	
		ΔL^*	Δa^*	Δb^*	ΔE^*
	Stained	−7 ± 3	6 ± 3	16 ± 5	19
	Treated	0 ± 2	0.8 ± 0.4	2 ± 1	3
		L*	a*	b*	
Marble	Ref	79 ± 4	−0.8 ± 0.2	−2.1 ± 0.7	
		ΔL^*	Δa^*	Δb^*	ΔE^*
	Stained	−5 ± 4	3 ± 2	11 ± 5	13
	Treated	0 ± 1	0.4 ± 0.4	3 ± 2	3
		L*	a*	b*	
Granite	Ref	70 ± 5	−0.2 ± 0.8	1 ± 2	
		ΔL^*	Δa^*	Δb^*	ΔE^*
	Stained	−11 ± 9	6 ± 3	16 ± 6	21
	Treated	−1 ± 2	−0.1 ± 0.6	−0.3 ± 0.6	2

Although the color alteration of the samples stained with the iron grid results is slightly lower than that obtained with the deposition of the rust dispersion (Table 1), all stones result darker and yellower than the reference. The colorimetric coordinates of granite, after a single treatment with CS-Ox, appear to be similar to the reference samples, and the ΔE^* values turn out to be below the perception limit of the human eye. Although the stains on travertine samples appear completely removed from the treated surface (see Figure 4A), colorimetric analysis reveals a slightly perceptible color alteration, comparable to that of marble specimens. In fact, for both carbonate stones, the ΔE^* is equal to 3, and it is mainly attributable to the positive shift of the b* coordinate, indicating a partial yellowing of their surfaces compared to the pristine material.

Colorimetric and photographic analyses have shown that a single treatment with CS-Ox is sufficient to completely restore the original color of the granite surface and to reduce the color change caused by rust stains on travertine and marble to values acceptable for restoration purposes, less than 5.

2.2.2. Stereomicroscopy

Microscopic observations were made on the treated surface of all samples in order to study the residual staining of carbonate stones and to confirm the high effectiveness of the treatment on granite. For each lithotype, Figure 5 reports micrographs of the reference, stained and treated surfaces at 56× of magnification.

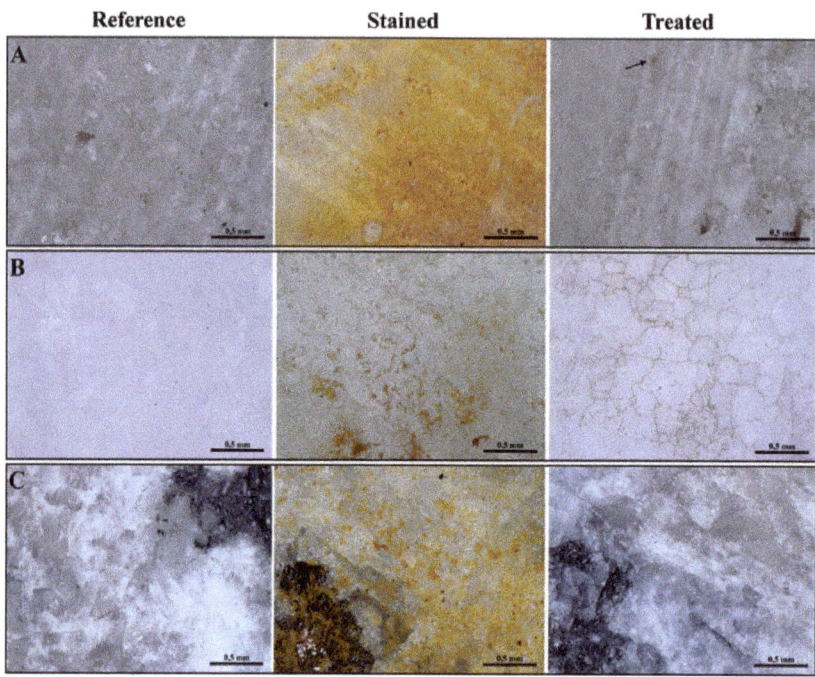

Figure 5. Stereomicroscope images at 56× of magnification of travertine (**A**), marble (**B**), and granite (**C**), reference, stained and treated with CS-Ox surfaces. In panel A (right), an evident spot of rust on the travertine-treated surface is indicated by a black arrow. Bar = 0.5 mm.

The micrographs of the stained samples reflect the evidence highlighted by photography and colorimetry. In fact, on travertine and granite, a homogeneous deposition of rust can be observed, while on marble, the less stained stone, a thinner and more spotted layer can be seen.

The travertine surface gives results almost entirely comparable to the reference, except for a few small rust spots, approximately 50 µm in size, one of which is indicated by the arrow in Figure 5A. Differently, on marble surface (Figure 5B), the rust residues are confined between the boundary areas of the interlocking grains of calcite. This evidence can explain the slight coloration revealed by the colorimetric analysis, which is also partially visible to the naked eye. Anyway, these residues are strongly bound to the marble surface and are the most challenging to remove, as also already reported in the literature [5,20]. Therefore, further studies will be necessary to find new formulations capable of completely cleaning marble stone or, alternately, to develop protective agents capable of avoiding the deposition of such residues.

Regarding granite, the microscope analysis (Figure 5C) confirms the absence of rust residues on the treated surface and the complete restoration of its original features, as previously obtained when the lithotype was stained with the dispersion of corrosion products.

Comparing the results of carbonate stones, CS-Ox hydrogel is able to clean both stones stained with the rust dispersion better than those stained with the iron grids, despite they resulted more contaminated (see ΔE^* values in Tables 1 and 3). This phenomenon could be attributed to the different sizes of the rust particles involved in the staining processes. In fact, when the iron grid is immersed in an aqueous salt solution, nanoparticles of corrosion products are formed, which then interact to form large aggregates. Thus, the rust dispersion used for the first experiments should consist of large aggregates that can effectively adhere to the sample surface but are unable to penetrate into the material. Differently, the rusted iron grid could transfer nanometer-sized particles, which spread more easily and deeper into the stone [9].

3. Conclusions

Both the stability and low solubility of ferrous species make rust removal from monumental stone surfaces a crucial issue. Chitosan-carboxylic acid hydrogels, optimized by combining the reducing ability of carboxylic acids with the intrinsic chelating properties of the polysaccharide, have been applied on granite, travertine and marble samples, artificially stained with rust dispersion. Photos and colorimetric parameters have shown that the formulation containing oxalic acid is the most effective in removing rust stains from all the specimens with just one treatment. The good cleaning performance of CS-Ox has also been confirmed by NMR and SEM/EDS analysis, which highlighted the ability of the hydrogel to remove iron and undesired hygroscopic pollutants from the stone surface. Moreover, similar results have been achieved on samples stained by placing a rusty iron grid on their surfaces to more faithfully reproduce a natural staining process. Colorimetry has highlighted that, despite the color difference between the treated and reference samples being slightly higher than those artificially stained with rust dispersion, they are well below the acceptable limit for a restoration intervention. In particular, small rust residues on the surfaces of travertine and between the grains of marble have been evidenced by the stereomicroscope observations.

The here optimized cleaning protocol will be tested "in situ" on naturally stained artworks. Nevertheless, further studies will be aimed at finding other formulations that can be applied to marble and travertine to completely remove iron residues. In addition, strategies aimed at protecting stone surfaces in contact with iron elements will be investigated to prevent the deposition of rust and to facilitate its removal.

4. Materials and Methods

4.1. Materials

Three lithotypes were selected for this study among the most used stone materials in built heritage i.e., granite, travertine and marble. Cut specimens of approximately $5 \times 5 \times 2$ cm^3 were purchased from Elia Marmi s.n.c. L'Aquila (AQ), Italy.

Sardinian pink granite is a medium-grained (0.5–2 cm) material with an open porosity that ranges from 0.5 to 1%. It is mainly composed of silicates and aluminosilicates, such as plagioclase (35%), quartz (31%), which confers the typical shine of the material, K-feldspar (24%), responsible of the pinkish inclusions, and biotite (10%) [52,53].

Roman travertine is a calcareous sedimentary rock mainly composed of CaCO$_3$ (97–99%) and appears as a white-beige stone characterized by 5 to 15% open porosity, mostly due to the presence of cavities and macropores [54,55].

Carrara marble is a metamorphic stone, with a grain size ranging from 0.12 to 0.35 mm and a very low open porosity of about 0.5% [56]. It is primarily composed of calcium carbonate with low content of magnesium carbonate and other trace elements. White Carrara marble, also called "Ordinary white", the widely diffused and commercially available, appears as a pearl-white stone with typical grey veins and spots [57].

Low viscosity chitosan with 77% deacetylation degree and 590 kDa average molecular weight [58], acetic acid, oxalic acid dihydrate, citric acid monohydrate and glycerol were supplied by Sigma-Aldrich (St. Louise, MO, USA) and were used as received.

4.2. Methods

4.2.1. Staining of Stone Surfaces

Iron grids were immersed in a solution containing sodium chloride ($\approx$1 M) to accelerate the oxidation process and obtain a dispersion of corrosion products. To reach a reproducible staining of the lithic substrates, regardless of their nature, 1 mL of the rust dispersion, containing about 28 mg of the oxidized compounds, was deposited on one of the 5 × 5 cm² surfaces of three specimens for each lithotype.

The other three samples for lithotype were stained by placing the rusted grids directly on one of their surfaces to simulate a more realistic pollution. All stone specimens were kept submerged in distilled water for about a week to favor the deposition of corrosion products on their surfaces.

4.2.2. Hydrogel Preparation

Three chitosan-based hydrogels containing acetic (CS-Ac), oxalic (CS-Ox), and citric (CS-Cit) acids were prepared.

Chitosan (5% w/w) was dissolved in an aqueous solution acidified with the minimum amount of carboxylic acid capable of completely dispersing the polysaccharide, i.e., 1.5%, 3% and 4% w/w of acetic, oxalic, or citric acid, respectively, all corresponding to about 0.25 M. The viscosity of the hydrogels was measured at room temperature using a Fungilab Viscolead mod. ADV L rotational viscometer (Fungilab, Barcelona, Spain) (spindle L4 at 20 rpm).

To peel off the dried gels more easily and avoid several rinses, 3% w/w of glycerol, which is commonly used as a plasticizer, was added to all the formulations. The hydrogels were then applied to the stained stone surfaces with the aid of cotton gauze to facilitate their removal after approximately 24 h, the time required for the gel to dry at room temperature.

4.2.3. Colorimetry

Chromatic changes of both stained and treated surfaces were assessed by colorimetric analysis and compared to data before staining. The measurements were performed by means of a Sama Tools SA230 (Sama Tools, Viareggio, LU, Italy) portable colorimeter working in SCE mode with an 8° standard angle observer, light D65, and temperature of 6504 K (average daylight, including the UV region). For all specimens, 25 points were acquired by using a 5 × 5 grid to cover approximately 60% of the total area. The colorimetric coordinates, L*, a*, and b*, were acquired in the CIELAB color space proposed in 1976 by the International Lighting Commission (CIE) [59]. Color differences, expressed in terms of ΔL*, Δa*, and Δb*, were calculated between both the stained and treated surfaces and their reference (surface before staining). By means of a vector sum, the color alteration values, expressed as ΔE*, were calculated using Equation (1):

$$\Delta E^* = \sqrt{\Delta L^{*2} + \Delta a^{*2} + \Delta b^{*2}} \qquad (1)$$

4.2.4. ¹H-NMR Relaxometry

Non-destructive relaxometry was performed using the NMR equipment mq-Profiler (Bruker, Milan, Italy), which consisted of a surface probe with a portable electronic apparatus. The coil in use works at a Larmor frequency of 17.8 MHz, can be put in contact with the sample surface of whatever dimension and can excite water protons up to 2 mm deep from its surface with a sensitive volume of 2 × 0.2 × 0.8 cm³ (x, y, z). Fully water-saturated conditions for the stones were obtained by placing the air-dried samples under a vacuum for 30 min and then keeping them submerged in distilled water for another 30 min, after which they were wrapped in polyethylene foils to avoid water evaporation during the NMR measurement. The T_2 signal decays were acquired by means of a CPMG pulse sequence of 3000 echoes with the shortest possible echo time of 44 µs to reduce the diffusion effect. In addition, 512 scans were performed by repeating the sequence every 2 ss to improve the signal/noise ratio. Signal decays of water absorbed in stained and treated samples were compared with that of the reference.

4.2.5. Microscopy Analysis

SEM-EDS analyses or stereomicroscopy were performed to evaluate, at a microscopic level, the efficacy of hydrogels in removing iron oxides from the surface of lithotypes. For

SEM/EDS measurements, a Zeiss GeminiSEM 500 (Zeiss, Jena, Germany) equipped with EDS OXFORD Aztec Energy with INCA X-ACT detector was used; all measures were carried out in variable pressure mode (VP) with an accelerating voltage of 15 keV and a working distance of 8.5 mm. Four areas of about 1 mm^2 of each stone were selected, and their position was saved to correlate the elemental composition changes during the experimental procedure. Therefore, EDS microanalysis was performed on all the investigated areas before, soon after the staining process, and after the cleaning treatment.

Micrographs were acquired by using an AxioZoom V16 (Zeiss, Jena, Germany) stereomicroscope at a 56× magnification. The images were then elaborated by means of the Zen Blue 3.3 software.

Supplementary Materials: The following supporting information can be downloaded at https://www.mdpi.com/article/10.3390/gels10060359/s1, Figure S1: Photograph of the physical hydrogels Cs-Ac, CS-Ox and CS-Cit; Figure S2: FTIR spectra of acetic acid (red), pure chitosan (black) and CS-Ac dried hydrogel (blue); Figure S3: FTIR spectra of oxalic acid (red), pure chitosan (black) and CS-Ox dried hydrogel (blue); Figure S4: FTIR spectra of citric acid (red), pure chitosan (black) and CS-Cit dried hydrogel (blue). References [43,58].

Author Contributions: Conceptualization, F.G., C.C. and N.S.; Investigation, F.G.; Data Curation, F.G.; Writing—Original Draft Preparation, F.G.; Writing—Review and Editing, F.G., C.C. and N.S.; Visualization, F.G.; Supervision, C.C. and N.S.; Funding Acquisition, N.S. All authors have read and agreed to the published version of the manuscript.

Funding: This work was supported by the Italian Ministry of Education, Universities and Research (MIUR): project Smart Cities and Communities and Social Innovation on Cultural Heritage (SCN_00520).

Institutional Review Board Statement: Not applicable.

Informed Consent Statement: Not applicable.

Data Availability Statement: The data presented in this study are openly available in the article.

Acknowledgments: The authors acknowledge Maria Giammatteo and Lorenzo Arrizza (Centre of Microscopy, University of L'Aquila) for stereomicroscopy and SEM-EDS analyses.

Conflicts of Interest: The authors declare no conflicts of interest.

References

1. Pinna, D.; Galeotti, M.; Rizzo, A. Brownish alterations on the marble statues in the church of Orsanmichele in Florence: What is their origin? *Herit. Sci.* **2015**, *3*, 7. [CrossRef]
2. Bams, V.; Dewaele, S. Staining of white marble. *Mater. Charact.* **2007**, *58*, 1052–1062. [CrossRef]
3. Henry, A. *Stone Conservation: Principles and Practice*, 1st ed.; Donhead: Shaftesbury, UK, 2006.
4. Winkler, E.M. Iron in minerals and the formation of rust in stone. In *Stone in Architecture*; Winkler, E.M., Ed.; Springer: Berlin/Heidelberg, Germany, 1997; pp. 233–240. [CrossRef]
5. Spile, S.; Suzuki, T.; Bendix, J.; Simonsen, K.P. Effective cleaning of rust stained marble. *Herit. Sci.* **2016**, *4*, 12. [CrossRef]
6. Andrade, C.; Alonso, C.; Molina, F.J. Cover cracking as a function of bar corrosion: Part I-Experimental test. *Mater. Struct.* **1993**, *26*, 453–464. [CrossRef]
7. Caré, S.; Nguyen, Q.T.; L'Hostis, V.; Berthaud, Y. Mechanical properties of the rust layer induced by impressed current method in reinforced mortar. *Cem. Concr. Res.* **2008**, *38*, 1079–1091. [CrossRef]
8. Beltran, M.; Playà, E.; Artigau, M.; Arroyo, P.; Guinea, A. Iron patinas on alabaster surfaces (Santa Maria de Poblet Monastery, Tarragona, NE Spain). *J. Cult. Herit.* **2016**, *18*, 370–374. [CrossRef]
9. Reale, R.; Andreozzi, G.B.; Sammartino, M.P.; Salvi, A.M. Analytical investigation of iron-based stains on carbonate stones: Rust formation, diffusion mechanisms, and speciation. *Molecules* **2023**, *28*, 1582. [CrossRef]
10. Dillmann, P.; Mazaudier, F.; Hœrlé, S. Advances in understanding atmospheric corrosion of iron. I. Rust characterisation of ancient ferrous artefacts exposed to indoor atmospheric corrosion. *Corros. Sci.* **2004**, *46*, 1401–1429. [CrossRef]
11. Hœrlé, S.; Mazaudier, F.; Dillmann, P.; Santarini, G. Advances in understanding atmospheric corrosion of iron. II. Mechanistic modelling of wet–dry cycles. *Corros. Sci.* **2004**, *46*, 1431–1465. [CrossRef]
12. Brand, J.; Wain, A.; Rode, A.V.; Madden, S.; Rapp, L. Towards safe and effective femtosecond laser cleaning for the preservation of historic monuments. *Appl. Phys. A* **2023**, *129*, 246. [CrossRef]

13. Matero, F.G.; Tagle, A.A. Cleaning, iron stain removal, and surface repair of architectural marble and crystalline limestone: The Metropolitan Club. *J. Am. Inst. Conserv.* **1995**, *34*, 49–68. [CrossRef]
14. Franzen, C.; Fischer, T. Removal of iron crusts from sandstone sculptures in a fountain. *Environ. Earth Sci.* **2022**, *81*, 216. [CrossRef]
15. Selwyn, L.; Tse, S. The chemistry of sodium dithionite and its use in conservation. *Stud. Conserv.* **2014**, *53*, 61–73. [CrossRef]
16. Stambolov, T.; Van Rheeden, B. Note on the removal of rust from old iron with thioglycolic acid. *Stud. Conserv.* **1968**, *13*, 142–144. [CrossRef]
17. Vergès-Belmin, V.; Heritage, A.; Bourgès, A. Powdered cellulose poultices in stone and wall painting conservation—Myths and realities. *Stud. Conserv.* **2011**, *56*, 281–297. [CrossRef]
18. Lauffenburger, J.A.; Grissom, C.A.; Charola, A.E. Changes in gloss of marble surfaces as a result of methylcellulose poulticing. *Stud. Conserv.* **1992**, *37*, 155–164. [CrossRef]
19. Gervais, C.; Grissom, C.A.; Little, N.; Wachowiak, M.J. Cleaning marble with ammonium citrate. *Stud. Conserv.* **2010**, *55*, 164–176. [CrossRef]
20. Campanella, L.; Cardellicchio, F.; Dell'Aglio, E.; Reale, R.; Salvi, A.M. A green approach to clean iron stains from marble surfaces. *Herit. Sci.* **2022**, *10*, 79. [CrossRef]
21. Gabriele, F.; Casieri, C.; Spreti, N. Natural deep eutectic solvents as rust removal agents from lithic and cellulosic substrates. *Molecules* **2024**, *29*, 624. [CrossRef]
22. Macchia, A.; Ruffolo, S.A.; Rivaroli, L.; La Russa, M.F. The treatment of iron-stained marble: Toward a "green" solution. *Int. J. Conserv. Sci.* **2016**, *7*, 323–332.
23. Cushman, M.; Wolbers, R. A new approach to cleaning iron-stained marble surfaces. *WAAC Newslett.* **2007**, *29*, 23–28.
24. Nakamura, T.; Tsukizawa, T.; Oya, M. Combined use of reducing agents and biodegradable chelating agent for iron rust removal. *J. Oleo Sci.* **2022**, *71*, 493–504. [CrossRef]
25. Chelazzi, D.; Baglioni, P. From nanoparticles to gels: A breakthrough in art conservation science. *Langmuir* **2023**, *39*, 10744–10755. [CrossRef] [PubMed]
26. Chelazzi, D.; Bordes, R.; Giorgi, R.; Holmberg, K.; Baglioni, P. The use of surfactants in the cleaning of works of art. *Curr. Opin. Colloid Interface Sci.* **2020**, *45*, 108–123. [CrossRef]
27. Chelazzi, D.; Giorgi, R.; Baglioni, P. Microemulsions, micelles, and functional gels: How colloids and soft matter preserve works of art. *Angew. Chem. Int. Ed.* **2018**, *57*, 7296–7303. [CrossRef]
28. Gabriele, F.; Vetrano, A.; Bruno, L.; Casieri, C.; Germani, R.; Rugnini, L.; Spreti, N. New oxidative alginate-biocide hydrogels against stone biodeterioration. *Int. Biodeter. Biodegr.* **2021**, *163*, 105281. [CrossRef]
29. Ranaldi, R.; Rugnini, L.; Gabriele, F.; Spreti, N.; Casieri, C.; Di Marco, G.; Gismondi, A.; Bruno, L. Plant essential oils suspended into hydrogel: Development of an easy-to-use protocol for the restoration of stone cultural heritage. *Int. Biodeter. Biodegr.* **2022**, *172*, 105436. [CrossRef]
30. Gabriele, F.; Bruno, L.; Casieri, C.; Ranaldi, R.; Rugnini, L.; Spreti, N. Application and monitoring of oxidative alginate-biocide hydrogels for two case studies in "the Sassi and the Park of the Rupestrian Churches of Matera". *Coatings* **2022**, *12*, 462. [CrossRef]
31. Sansonetti, A.; Bertasa, M.; Canevali, C.; Rabbolini, A.; Anzani, M.; Scalarone, D. A review in using agar gels for cleaning art surfaces. *J. Cult. Herit.* **2020**, *44*, 285–296. [CrossRef]
32. Canevali, C.; Fasoli, M.; Bertasa, M.; Botteon, A.; Colombo, A.; Di Tullio, V.; Capitani, D.; Proietti, N.; Scalarone, D.; Sansonetti, A. A multi-analytical approach for the study of copper stain removal by agar gels. *Microchem. J.* **2016**, *129*, 249–258. [CrossRef]
33. Irizar, P.; Gomez-Laserna, O.; Arana, G.; Madariaga, J.M.; Martínez-Arkarazo, I. Ionic liquids (ILs)-loaded hydrogels as a potential cleaning method of metallic stains for stone conservation. *J. Cult. Herit.* **2023**, *64*, 12–22. [CrossRef]
34. Sonaglia, E.; Schifano, E.; Sharbaf, M.; Uccelletti, D.; Felici, A.C.; Santarelli, M.L. Bacterial nanocellulose hydrogel for the green cleaning of copper stains from marble. *Gels* **2024**, *10*, 150. [CrossRef] [PubMed]
35. Younes, I.; Rinaudo, M. Chitin and chitosan preparation from marine sources. Structure, properties and applications. *Mar. Drugs* **2015**, *13*, 1133–1174. [CrossRef] [PubMed]
36. Sacco, P.; Furlani, F.; De Marzo, G.; Marsich, E.; Paoletti, S.; Donati, I. Concepts for developing physical gels of chitosan and of chitosan derivatives. *Gels* **2018**, *4*, 67. [CrossRef] [PubMed]
37. Yang, Y.; Wu, D. Energy dissipative and soften resistant hydrogels based on chitosan physical network: From construction to application. *Chin. J. Chem.* **2022**, *40*, 2118–2134. [CrossRef]
38. Cuvillier, L.; Passaretti, A.; Guilminot, E.; Joseph, E. Agar and chitosan hydrogels' design for metal-uptaking treatments. *Gels* **2024**, *10*, 55. [CrossRef] [PubMed]
39. Campos, B.; Marco, A.; Cadeco, G.; Freire-Lista, D.M.; Silvestre-Albero, J.; Algarra, M.; Vieira, E.; Pintado, M.; Moreira, P. Green chitosan: Thiourea dioxide cleaning gel for manganese stains on granite and glass substrates. *Herit. Sci.* **2021**, *9*, 160. [CrossRef]
40. Hernández, R.B.; Franco, A.P.; Yola, O.R.; López-Delgado, A.; Felcman, J.; Recio, M.A.L.; Mercê, A.L.R. Coordination study of chitosan and Fe^{3+}. *J. Mol. Struct.* **2008**, *877*, 89–99. [CrossRef]
41. Burke, A.; Yilmaz, E.; Hasirci, N.; Yilmaz, O. Iron(III) ion removal from solution through adsorption on chitosan. *J. Appl. Polym. Sci.* **2002**, *84*, 1185–1192. [CrossRef]
42. Radnia, H.; Ghoreyshi, A.A.; Younesi, H.; Najafpour, G.D. Adsorption of Fe(II) ions from aqueous phase by chitosan adsorbent: Equilibrium, kinetic, and thermodynamic studies. *Desalin. Water Treat.* **2012**, *50*, 348–359. [CrossRef]
43. Bellamy, L.J.; Pace, R.J. Hydrogen bonding in carboxylic acids—I. Oxalic acids. *Spectrochim. Acta* **1963**, *19*, 435–442. [CrossRef]

44. Chatzigrigoriou, A.; Karapanagiotis, I.; Poulios, I. Superhydrophobic coatings based on siloxane resin and calcium hydroxide nanoparticles for marble protection. *Coatings* **2020**, *10*, 334. [CrossRef]
45. Zia, J.; Pandey, J. Preparation and characterization of iron-metal nanocomposite of chitosan: Towards heterogeneous biocatalysts for organic reactions. *JBAER* **2018**, *5*, 101–105.
46. Brownstein, K.R.; Tarr, C.E. Importance of classical diffusion in NMR studies of water in biological cells. *Phys. Rev. A* **1979**, *19*, 2446–2453. [CrossRef]
47. Meyer, M.; Buchmann, C.; Schaumann, G.E. Determination of quantitative pore-size distribution of soils with ^{1}H NMR relaxometry. *Eur. J. Soil Sci.* **2018**, *69*, 393–406. [CrossRef]
48. Camaiti, M.; Bortolotti, V.; Fantazzini, P. Stone porosity, wettability changes and other features detected by MRI and NMR relaxometry: A more than 15-year study. *Magn. Reson. Chem.* **2015**, *53*, 34–47. [CrossRef] [PubMed]
49. Tortora, M.; Chiarini, M.; Spreti, N.; Casieri, C. ^{1}H-NMR-relaxation and colorimetry for evaluating nanopolymeric dispersions as stone protective coatings. *J. Cult. Herit.* **2020**, *44*, 204–210. [CrossRef]
50. Bortolotti, V.; Camaiti, M.; Casieri, C.; De Luca, F.; Fantazzini, P.; Terenzi, C. Water absorption kinetics in different wettability conditions studied at pore and sample scale in porous media by portable single-sided and laboratory imaging devices. *J. Magn. Reson.* **2006**, *181*, 287–295. [CrossRef]
51. Blümich, B.; Perlo, J.; Casanova, F. Mobile single-sided NMR. *Prog. Nucl. Magn. Reason. Spectrosc.* **2008**, *52*, 197–269. [CrossRef]
52. Careddu, N.; Grillo, S. Rosa Beta granite (Sardinian Pink Granite): A heritage stone of international significance from Italy. *Geol. Soc. Spec. Publ.* **2015**, *407*, 155–172. [CrossRef]
53. Cuccuru, S.; Puccini, A. Petrographic, physical–mechanical and radiological characterisation of the rosa beta granite (Corsica-sardinia batholith). In *Engineering Geology for Society and Territory*; Lollino, G., Manconi, A., Guzzetti, F., Culshaw, M., Bobrowsky, P., Luino, F., Eds.; Springer: Cham, Switzerland, 2015; Volume 5, pp. 233–236. [CrossRef]
54. Mancini, A.; Frondini, F.; Capezzuoli, E.; Galvez Mejia, E.; Lezzi, G.; Matarazzi, D.; Brogi, A.; Swennen, R. Porosity, bulk density and $CaCO_3$ content of travertines. A new dataset from Rapolano, Canino and Tivoli travertines (Italy). *Data Br.* **2019**, *25*, 104158. [CrossRef]
55. Mancini, A.; Frondini, F.; Capezzuoli, E.; Galvez Mejia, E.; Lezzi, G.; Matarazzi, D.; Brogi, A.; Swennen, R. Evaluating the geogenic CO_2 flux from geothermal areas by analysing quaternary travertine masses. New data from western central Italy and review of previous CO_2 flux data. *Quat. Sci. Rev.* **2019**, *215*, 132–143. [CrossRef]
56. Alber, M.; Hauptfleisch, U. Generation and visualization of microfractures in Carrara marble for estimating fracture toughness, fracture shear and fracture normal stiffness. *Int. J. Rock Mech. Min. Sci.* **1999**, *8*, 1065–1071. [CrossRef]
57. Meccheri, M.; Molli, G.; Conti, P.; Blasi, P.; Vaselli, L. The Carrara Marbles (Alpi Apuane, Italy): A geological and economical updated review. *Z. Ges. Geowiss.* **2007**, *158*, 719–736. [CrossRef]
58. Gabriele, F.; Donnadio, A.; Casciola, M.; Germani, R.; Spreti, N. Ionic and covalent crosslinking in chitosan-succinic acid membranes: Effect on physicochemical properties. *Carbohydr. Polym.* **2021**, *251*, 117106. [CrossRef]
59. UNI EN 15886:2010; Conservation of Cultural Property—Test Methods—Color Measurement of Surfaces. Ente Nazionale Italiano di Unificazione (UNI): Roma, Italy, 2010.

Disclaimer/Publisher's Note: The statements, opinions and data contained in all publications are solely those of the individual author(s) and contributor(s) and not of MDPI and/or the editor(s). MDPI and/or the editor(s) disclaim responsibility for any injury to people or property resulting from any ideas, methods, instructions or products referred to in the content.

Article

Preparation of Peptide-Based Magnetogels for Removing Organic Dyes from Water

Farid Hajareh Haghighi [1,†], Roya Binaymotlagh [1,†], Paula Stefana Pintilei [1], Laura Chronopoulou [1,2,*] and Cleofe Palocci [1,2,*]

1 Department of Chemistry, Sapienza University of Rome, Piazzale Aldo Moro 5, 00185 Rome, Italy
2 Research Center for Applied Sciences to the Safeguard of Environment and Cultural Heritage (CIABC), Sapienza University of Rome, Piazzale Aldo Moro 5, 00185 Rome, Italy
* Correspondence: laura.chronopoulou@uniroma1.it (L.C.); cleofe.palocci@uniroma1.it (C.P.); Tel.: +39-06-4991-3317 (C.P.)
† These authors have contributed equally to this work.

Citation: Hajareh Haghighi, F.; Binaymotlagh, R.; Pintilei, P.S.; Chronopoulou, L.; Palocci, C. Preparation of Peptide-Based Magnetogels for Removing Organic Dyes from Water. *Gels* **2024**, *10*, 287. https://doi.org/10.3390/gels10050287

Academic Editors: Shiyang Li, Zhenxing Fang and Kaiming Peng

Received: 28 March 2024
Revised: 18 April 2024
Accepted: 22 April 2024
Published: 24 April 2024

Copyright: © 2024 by the authors. Licensee MDPI, Basel, Switzerland. This article is an open access article distributed under the terms and conditions of the Creative Commons Attribution (CC BY) license (https://creativecommons.org/licenses/by/4.0/).

Abstract: Water pollution by organic dyes represents a major health and environmental issue. Despite the fact that peptide-based hydrogels are considered to be optimal absorbents for removing such contaminants, hydrogel systems often suffer from a lack of mechanical stability and complex recovery. Recently, we developed an enzymatic approach for the preparation of a new peptide-based magnetogel containing polyacrylic acid-modified γ-Fe$_2$O$_3$ nanoparticles (γ-Fe$_2$O$_3$NPs) that showed the promising ability to remove cationic metal ions from aqueous phases. In the present work, we tested the ability of the magnetogel formulation to remove three model organic dyes: methyl orange, methylene blue, and rhodamine 6G. Three different hydrogel-based systems were studied, including: (1) Fmoc-Phe$_3$ hydrogel; (2) γ-Fe$_2$O$_3$NPs dispersed in the peptide-based gel (Fe$_2$O$_3$NPs@gel); and (3) Fe$_2$O$_3$NPs@gel with the application of a magnetic field. The removal efficiencies of such adsorbents were evaluated using two different experimental set-ups, by placing the hydrogel sample inside cuvettes or, alternatively, by placing them inside syringes. The obtained peptide magnetogel formulation could represent a valuable and environmentally friendly alternative to currently employed adsorbents.

Keywords: magnetogels; magnetic γ-Fe$_2$O$_3$ nanoparticles; peptide-based hydrogels; water purification; methyl orange; methylene blue; rhodamine 6G

1. Introduction

Safe water is a global need for all aspects of life (e.g., household, industrial, and agricultural purposes) and, considering the current limitation of water resources, the importance of effective water recycling and purification is constantly growing [1]. Commonly used technologies for wastewater treatment are helpful but should be improved in terms of their economic feasibility, efficiency, and environmental footprint [2]. Due to the absence of adequate wastewater treatment methods, different dangerous materials constantly enter natural water resources [3]. Every year, tons of chemically stable synthetic dyes are wasted from the pharmaceutical [4,5], garment [6], textile [7], leather, ink, paper [8], and plastic industries [9], negatively affecting aquatic life and also threatening human health, as many of these chemicals are highly hazardous and have been classified as mutagenic, carcinogenic, or genotoxic [10]. Furthermore, these water contaminants decrease sunlight penetration into bodies of water, which influences photosynthetic processes and affects aquatic flora and fauna [11]. In particular, sunlight plays a crucial role in the photosynthesis of aquatic plants. It provides the energy needed for the conversion of inorganic carbon to organic carbon compounds, which is the fundamental biological process on Earth. Synthetic dyes such as methylene blue (MB), rhodamine 6G (Rh6G) (model cationic dyes), and methyl orange

(MO) (a model anionic dye) cause several health disorders like respiratory tract infections, skin disease, and eye irritation, so their removal from water resources is necessary [12].

There are standard methods for wastewater treatment, including chemical precipitation, adsorption, ion exchange, electrochemical separation, and coagulation-flocculation methods. Nonetheless, all of these methods have some drawbacks, e.g., high energy consumption, partial contaminant elimination, high operation costs, and the production of toxic sludges [13]. Today, the removal of contaminants from wastewater is considered a crucial step towards achieving sustainable development. Among the approaches listed [14], adsorption is still the most commonly used method, and an ideal adsorbent removes most pollutants and can be recovered easily using cost-effective methods like electrochemical or solvent treatments [15,16]. However, common commercial adsorbents such as biochar, zeolites, and activated carbon pose a high risk of water contamination by themselves.

Recently, hydrogel nanohybrids have become increasingly used in numerous adsorption-based water treatment applications [17]: generally, a polymer forms a hydrophilic porous network and nanomaterials are used as modifiers to enhance the adsorption properties [18]. For water remediation, the possibility of using biocompatible materials is very important. Peptide hydrogels can be considered as promising candidates to decrease the self-contamination risk of adsorbent systems [19]. Compared with commercially available adsorbents, these hydrogels benefit from being porous and highly hydrophilic, as well as from possessing large surface areas and numerous functional groups. In particular, short self-assembling peptides are an interesting class of biocompatible and biodegradable hydrogel-forming materials. Their synthesis is generally simple and scalable. Moreover, both the properties and self-assembly of peptides can be tuned by altering the peptide sequence [20].

Magnetic nanoparticles (NPs) can be incorporated into hydrogels to provide a magnetic property to the resulting hybrid system. Among magnetic nanostructures, magnetite (Fe_3O_4) and maghemite (γ-Fe_2O_3) NPs have been extensively used for environmental applications because of their biocompatibility, well-assessed synthesis methods, and on/off superparamagnetic properties [21]. Although γ-Fe_2O_3 NPs show a slightly smaller magnetic moment than Fe_3O_4 NPs, they are more stable in air [22], and more reliable for practical applications. To date, various nanohybrid-based γ-Fe_2O_3 NPs have been used as adsorbents to remove several organic dyes (e.g., MO, rose bengal, MB, brilliant cresyl blue, Congo red, thionine, and Janus green B) from aqueous solutions with significant adsorption efficiencies. Aside from the numerous advantages of peptide-based hydrogel nanohybrids, their poor mechanical properties and low elasticity are their two main limitations in water remediation applications; however, these drawbacks may be overcome by using appropriate cross-linkers.

In a recent work [23], we reported the preparation of a new peptide composite magnetogel made of a Fmoc-Phe$_3$ hydrogel matrix containing γ-Fe_2O_3-polyacrylic acid (PAA) NPs (γ-Fe_2O_3NPs) and its application for removing Cr(III), Ni(II), and Co(II) from water, demonstrating that both the native hydrogel as well as the magnetogel can effectively eliminate all the examined metal ions from water. Moreover, thanks to the presence of γ-Fe_2O_3NPs, the efficiency of this system can be promoted by the application of a magnetic field. In the present study, we tested the ability of the magnetogel formulation to remove three model dyes: MO, MB, and Rh6G. On this basis, three different hydrogel-based systems, including (1) Fmoc-Phe$_3$ hydrogel (gel); (2) γ-Fe_2O_3NPs loaded in the hydrogel (γ-Fe_2O_3NPs@gel); and (3) γ-Fe_2O_3NPs@gel in the presence of an external magnetic field (γ-Fe_2O_3NPs@gel + mf), were studied. The removal efficiency of these gels was evaluated using two experimental set-ups, by placing hydrogel samples: (1) inside cuvettes or (2) inside syringes.

2. Results and Discussion

2.1. Preparation of γ-Fe$_2$O$_3$NPs and Magnetogels

The magnetogel nanohybrids were synthesized in two steps, as described in our previous publication [23]. Firstly, PAA-stabilized γ-Fe$_2$O$_3$NPs were prepared by a co-precipitation method to obtain small and colloidally stable NPs. Co-precipitation is an uncomplicated and cost-effective way to prepare γ-Fe$_2$O$_3$NPs, and involves starting from aqueous solutions of Fe(II) and Fe(III) and adding a base as a precipitating agent. This method requires mild temperatures and no organic solvents or toxic precursors, and allows for the preparation of large amounts of NPs. Also, thanks to its simpleness, reliability, and environmentally friendly conditions, it can be adjusted and scaled up to an industrial scale.

For the magnetogel synthesis, γ-Fe2O3NPs were suspended in an aqueous phase containing the hydrogel precursors and the gelation process was carried out. The main advantage of such a blending method is its simplicity; however γ-Fe2O3NPs must be stabilized (i.e., with PAA) to prevent their heterogeneous distribution within the hydrogel network. The hydrogel formation was conducted using a biotechnological approach that employs a microbial lipase to catalyze the formation of the Fmoc-Phe$_3$ hydrogelator, as previously reported [24]. The resulting magnetogel was characterized in a previous work through FT-IR, Raman, and XPS experiments [23].

2.2. Rheology and Swelling Ability Studies

Rheological studies were conducted to unravel the influence of the modifiers used on the tensile strength of peptide hydrogels. The presence of γ-Fe$_2$O$_3$NPs was able to enhance the rheological properties of the hydrogel, as previously reported (Figure 1a).

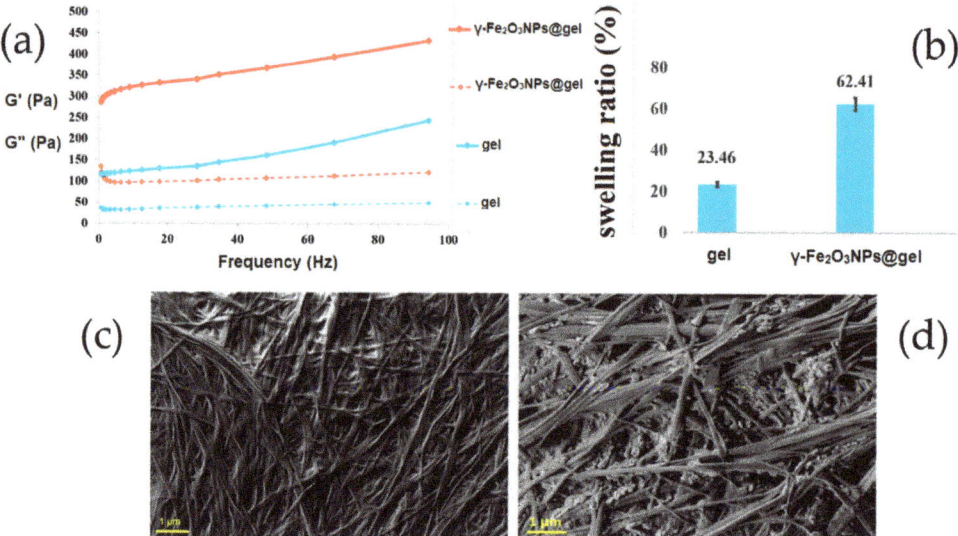

Figure 1. (**a**) Frequency sweep of the gel and γ-Fe$_2$O$_3$NPs@gel; (**b**) swelling abilities of the gel and γ-Fe$_2$O$_3$NPs@gel; SEM images of: (**c**) the gel and (**d**) γ-Fe$_2$O$_3$NPs@gel.

After the addition of γ-Fe$_2$O$_3$NPs to the pristine hydrogel, its swelling ability increased up to ≈62%, which may be related to the presence of NPs that provide porosity inside the hydrogel, increasing its swelling properties (Figure 1b).

2.3. SEM Experiments

SEM was employed to investigate the morphology of the pristine gel as well as that of the magnetogel. Figure 1c shows the typical fibrillary structure of the pristine hydrogel.

The fibrils are several micrometers long while their width is between 100 and 150 nm. The addition of magnetic NPs to the gel did not seem to have any significant effect on its fibrillary structure (Figure 1d). On the basis of SEM experiments, the average size of γ-Fe$_2$O$_3$NPs was determined to be in a range between 15 and 60 nm.

2.4. Removal Studies of MB, Rh6G and MO from Water

Two different experimental set-ups were developed for the removal of MB, Rh6G, and MO by using different hydrogel samples (Figure 2). In the first set of experiments, the gel samples were prepared inside cuvettes, then the aqueous dye solution was placed on top of them. With this configuration, the dyes diffuse into the gel matrix by gravity (without the need to apply any external forces). In the other set of experiments, the gel samples were prepared inside syringes. In the removal experiments, the dye solutions were allowed to flow inside the syringes, placed vertically, either by gravity or by applying an external force. Therefore, in this set-up, the dyes pass through the gel matrix and exit from the lower inlet of the syringe.

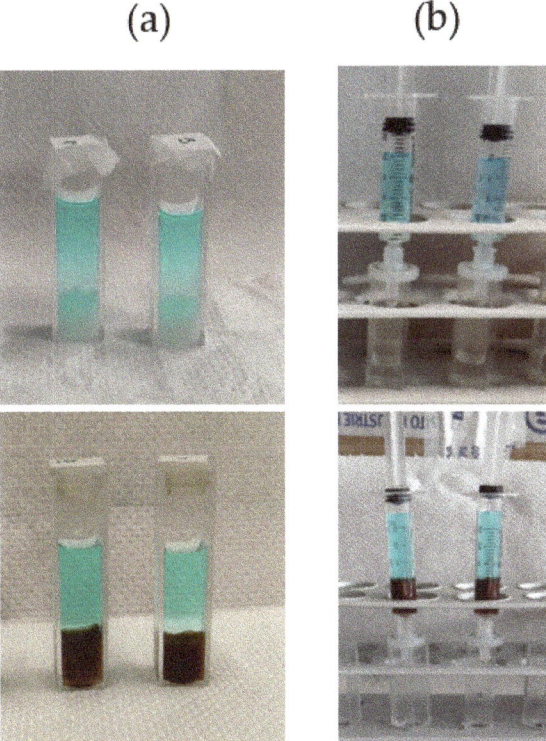

Figure 2. The two different set-ups used for the removal studies of organic dyes from water: (**a**) employing cuvettes and (**b**) employing syringes.

2.4.1. Experimental Set-Up Employing Cuvettes

MB has a heterocyclic aromatic structure (Figure 3) with characteristic absorbance peaks at 293 and 664 nm, related to π—π^* and n—π^* transitions, respectively [25]. The visible absorption (664 nm) was monitored during the MB removal studies. Regarding Rh6G, it contains a xanthene moiety that is responsible for its π—π^* absorption at visible wavelengths [26]. The visible absorption at 527 nm was monitored to evaluate the Rh6G removal by the adsorbents. The anionic dye MO has a strong absorption at around 500 nm, which was selected for the adsorption experiments.

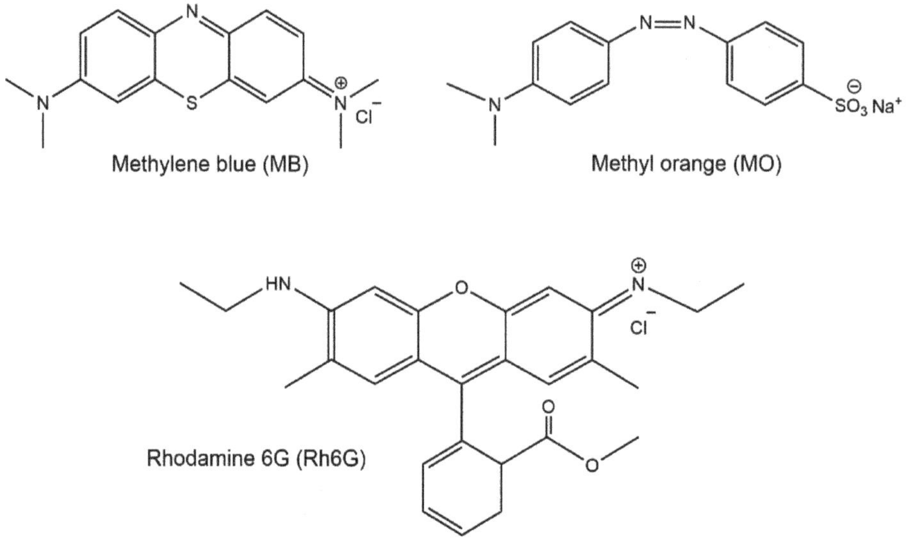

Figure 3. Chemical structures of methylene blue (MB), rhodamine 6G (Rh6G) (model cationic dyes), and methyl orange (MO) (model anionic dye).

Three different adsorbents were used to remove the three organic dyes described above from aqueous solutions: the native gel, γ-Fe$_2$O$_3$NPs@gel, and γ-Fe$_2$O$_3$NPs@gel + mf (mf: external magnetic field).

In particular, the adsorption abilities of these materials were evaluated at different contact times with the solution of the selected dye and expressed as q_t (mg of dye absorbed by 1 g of dry hydrogel). The evolution of the q_t values as a function of time for each dye + adsorbent system is reported in the plots in Figure 4a–c, for contact times ranging from 0 to 480 min. Also, Tables 1 and 2 show the removal efficiencies/capacities and calculated kinetic parameters of the three adsorbent systems for the three dyes, respectively.

Table 1. Removal efficiencies and capacities (q_e) of different hydrogel systems for MB, Rh6G, and MO.

	Adsorbent	Removal Efficiency% (RE%)	q_e (mg g^{-1})
	gel	30.8 ± 1	0.11 ± 0.0
MB	γ-Fe$_2$O$_3$NPs@gel	22.7 ± 2	0.08 ± 0.0
	γ-Fe$_2$O$_3$NPs@gel + mf	27.7 ± 1	0.09 ± 0.0
	gel	45.7 ± 1	0.16 ± 0.0
Rh6G	γ-Fe$_2$O$_3$NPs@gel	47.5 ± 1	0.17 ± 0.0
	γ-Fe$_2$O$_3$NPs@gel + mf	32.4 ± 2	0.11 ± 0.0
	gel	13.3 ± 1	0.09 ± 0.0
MO	γ-Fe$_2$O$_3$NPs@gel	14.1 ± 0	0.10 ± 0.0
	γ-Fe$_2$O$_3$NPs@gel + mf	13.4 ± 0	0.09 ± 0.0

Table 2. Kinetic data obtained from pseudo-first- and pseudo-second-order models for different hydrogel systems and MB, Rh6G, or MO at RT.

	Adsorbent	Pseudo-First Order			Pseudo-Second Order		
		k_1 (min^{-1})	q_e (mg g^{-1})	R^2	k_2 (g mg^{-1} min^{-1})	q_e (mg g^{-1})	R^2
MB	gel	0.011	0.14	0.9715	0.037	0.16	0.9713
	γ-Fe$_2$O$_3$NPs@gel	0.012	0.12	0.9493	0.0028	0.35	0.1329
	γ-Fe$_2$O$_3$NPs@gel + mf	0.011	0.15	0.9323	0.017	0.18	0.8257
Rh6G	gel	0.0074	0.17	0.992	0.023	0.23	0.9822
	γ-Fe$_2$O$_3$NPs@gel	0.0080	0.18	0.9881	0.026	0.23	0.9921
	γ-Fe$_2$O$_3$NPs@gel + mf	0.0089	0.14	0.9696	0.018	0.20	0.8597
MO	gel	0.0069	0.10	0.9712	0.048	0.13	0.954
	γ-Fe$_2$O$_3$NPs@gel	0.010	0.13	0.9328	0.037	0.15	0.9832
	γ-Fe$_2$O$_3$NPs@gel + mf	0.0080	0.11	0.9638	0.024	0.15	0.8149

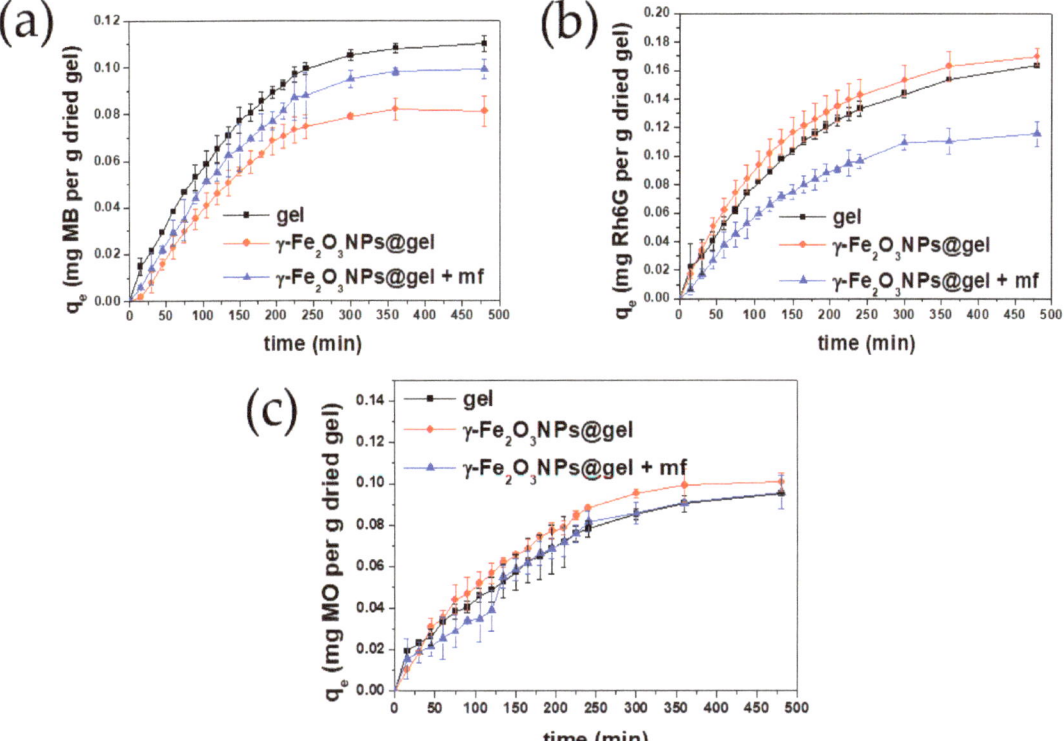

Figure 4. Adsorption capacities (q_t, mg/g) of the native gel, γ-Fe$_2$O$_3$NPs@gel, and γ-Fe$_2$O$_3$NPs@gel + mf, versus time for MB (**a**), Rh6G (**b**), and MO (**c**).

The results obtained with this experimental set-up show that the magnetic NPs change the adsorption properties of the pristine gels and, also, that the addition of an external magnetic field additionally modifies the adsorbing character of the gels. Over the whole removal process for the three examined systems, no significant changes were detected in the maximum wavelengths of the dyes and, as can be seen in Figure 4a–c, the dye adsorption decreased with time. This general feature, observed for all dyes and adsorbents, is probably linked to the progressive saturation of the adsorbing sites of the gels over time, which eventually reaches equilibrium after approximately 400 min.

Figure 4a shows that the q_t values for MB follow the trend gel > γ-Fe$_2$O$_3$NPs@gel + mf > γ-Fe$_2$O$_3$NPs@gel, which is different from those observed for Rh6G and MO, where γ-Fe$_2$O$_3$NPs@gel exhibits the best performance in terms of dye removal (Figure 4b,c). Such a difference may be ascribed to the different structures of the dyes, which affect the gel–dye interaction at the molecular level.

Another result worth mentioning is the changes in removal efficiencies (RE%) and final adsorption capacities for different systems, summarized in Table 1. For MB, the results show that the removal efficiencies of γ-Fe$_2$O$_3$NPs@gel and γ-Fe$_2$O$_3$NPs@gel + mf are decreased by 26.3% and 10.1%, respectively, in comparison with the native gel. For Rh6G, both magnetogels and pristine gels show similar RE% values, and the combination of magnetogels with the application of a magnetic field significantly drops the adsorption efficiency by 29.1%. For MO, the magnetogel shows the highest RE% value, while the interaction of magnetogels with the magnetic field increases the dye removal by 16.7%.

Regarding the different effect of γ-Fe$_2$O$_3$NPs on dye adsorption, it is in good agreement with data from the literature on magnetogels [27,28] and is due to the ability of NPs embedded in hydrogels to affect both the cross-linking degree and porosity of the gels, influencing the surface channels and, therefore, the entry, exit, and adsorption of molecules [29]. Also, the application of an external magnetic field can further change the adsorption, because magnetogels are able to exert an on/off effect on the hydrogel pores [30]. The magnetic dipole-dipole orientation of γ-Fe$_2$O$_3$NPs towards the external magnetic field is able to affect the swelling and shrinking of γ-Fe$_2$O$_3$NPs@gels [31], most probably affecting the permeability of dyes into the gel network [32,33].

As is well known, the study of adsorption kinetics is able to provide information on the nature of the adsorption (e.g., physisorption or chemisorption) [34]. In this study, we used pseudo-first- and pseudo-second-order kinetic models [22], as described in the Supplementary Materials. The obtained kinetic parameters are summarized in Table 2, where the best correlations (in terms of R^2) of MB and Rh6G studies were mainly obtained when using the pseudo-first-order kinetic model, although not for γ-Fe$_2$O$_3$NPs@gel (Rh6G) studies. For this system, the equilibrium adsorption capacity, q_e, obtained from the pseudo-first-order relation was more similar to the experimental values reported in Figure 4 and Table 1, suggesting that the adsorption of Rh6G onto these adsorbents mainly occurs through physisorption. MB and Rh6G dyes possess cationic imine and amine groups that can favor their adsorption on the hydrogels, thanks to different electrostatic interactions that may occur, such as that between the positively charged imine nitrogen of the dyes and the negatively charged hydroxyl groups of the hydrogel. Moreover, H-bonding interactions among the amine group of the dyes and the hydroxyl groups of the hydrogels may be established [35]. Returning to the possible pseudo-second-order mechanism estimated for γ-Fe$_2$O$_3$NPs@gel (Rh6G), it is attributed to chemisorption, which may depend on the chelation between the carboxyl and amine groups of the gels (–NH, –OH, and –COOH) and the lone pair electrons of the dye molecules [36–41]. Conversely, to these cationic dyes, MO (as an anionic dye) shows much lower RE% and q_e values, which may be related to its negative charge significantly limiting its electrostatic interaction with the gels, even though it was used at higher initial concentrations (10 ppm) than MB and Rh6G (5 ppm). For MO, two adsorbents show physisorption (gel and γ-Fe$_2$O$_3$NPs@gel + mf), controlled by physical forces like dipole–dipole interactions, hydrogen bonds, van der Waals forces, and hydrophobic interactions. Based on these results, the prepared adsorbent is more suitable for the removal of cationic dyes, rather than anionic ones.

2.4.2. Experimental Set-Up Employing Syringes

The absorbing abilities of native gel, γ-Fe$_2$O$_3$NPs@gel, and γ-Fe$_2$O$_3$NPs@gel + mf towards the selected dyes were tested with the experimental set-up employing syringes (Figure 5). Three different conditions were used, including (1) gravitational passage of the dye through the gels inside the syringe (Figure S2a); (2) gravitational passage with the application of filters attached to the outlet of the syringe (Figure S2b); and (3) the application of a constant pressure to increase the flow rate of the dye inside the syringe, combined with the use of a filter at the outlet of the syringe (Figure S2c). In terms of time and reproducibility, the third set-up was selected as the optimized condition for studying the absorbing abilities of the gels. It must be taken into account that the filters absorb part of the dye; therefore, this filter absorption was quantified for each dye and then subtracted from the amount of dye removed by each hydrogel system (see Figures S3–S6).

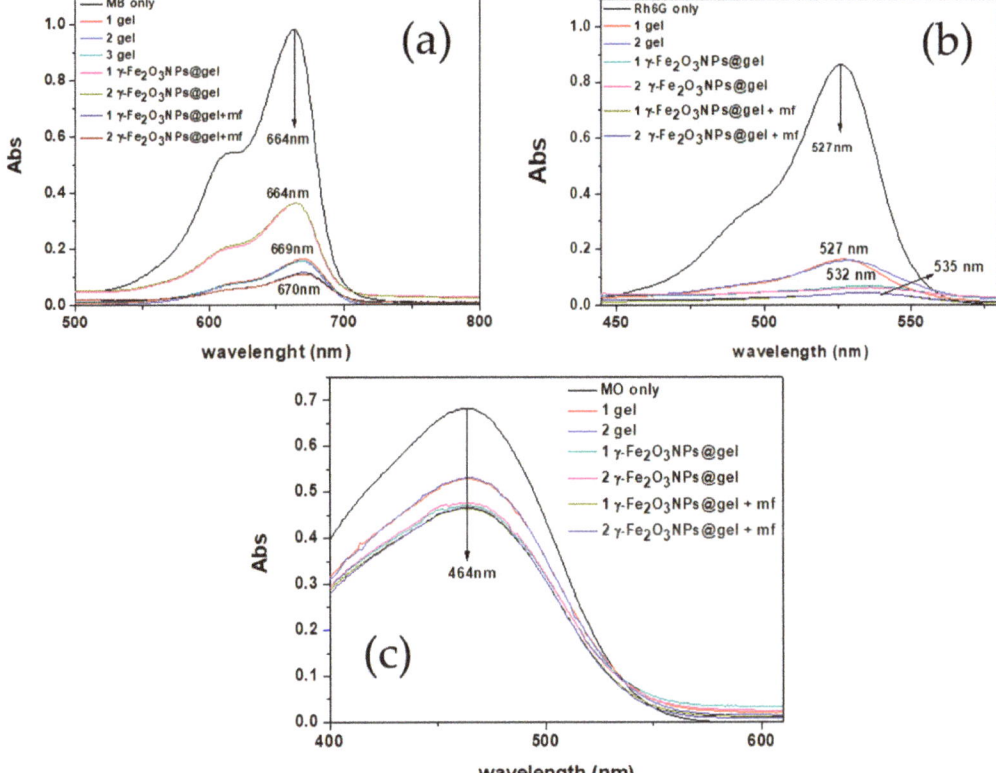

Figure 5. Absorbing abilities of gel, γ-Fe$_2$O$_3$NPs@gel, and γ-Fe$_2$O$_3$NPs@gel + mf for the removal of (**a**) MB (C_0 = 5 ppm, pH = 7.4, 1 mg of adsorbent, flow rate = 0.2 mL/min); (**b**) Rh6G (C_0 = 5 ppm, pH = 7.2, 1 mg of adsorbent, flow rate = 0.2 mL/min); (**c**) MO (C_0 = 10 ppm, pH = 7.5, 1 mg of adsorbent, flow rate = 0.2 mL/min).

For the cationic dyes, MB and Rh6G, significant adsorptions (>60%) were observed for all adsorbents (gel, γ-Fe$_2$O$_3$NPs@gel and γ-Fe$_2$O$_3$NPs@gel + mf) (Table 3). For MB, the pristine gel showed a 20.3% higher RE% than the magnetogel; however, when the external magnet was used, the γ-Fe$_2$O$_3$NPs@gel + mf system demonstrated a 4.5% higher RE% than the native hydrogel. For both the γ-Fe$_2$O$_3$NPs@gel + mf and pristine gel systems, a red-shift was also detected, which may be attributed to the decrease in the dye's concentration after passing through the gels. The UV-Vis spectrum of the initial MB solution shows a monomeric structure for the dye that can exist in two monomeric mesomers (I and II), based on data from the literature (see Figure 6) [42].

Table 3. Adsorption efficiencies of different hydrogel systems for the removal of MB, Rh6G, or MO from an aqueous phase using the experimental set-up employing syringes.

	Adsorbent	RE% of Cuvette Set-Up	RE% of Syringe Set-Up
MB	gel	30.8 ± 1	80.4 ± 1
	γ-Fe$_2$O$_3$NPs@gel	22.7 ± 2	60.1 ± 2
	γ-Fe$_2$O$_3$NPs@gel + mf	27.7 ± 1	84.9 ± 0
R6G	gel	45.7 ± 1	83.2 ± 1
	γ-Fe$_2$O$_3$NPs@gel	47.5 ± 1	95.0 ± 0
	γ-Fe$_2$O$_3$NPs@gel + mf	32.4 ± 2	97.6 ± 0
MO	gel	13.3 ± 1	21.8 ± 7
	γ-Fe$_2$O$_3$NPs@gel	14.1 ± 0	30.0 ± 6
	γ-Fe$_2$O$_3$NPs@gel + mf	13.4 ± 0	31.3 ± 6

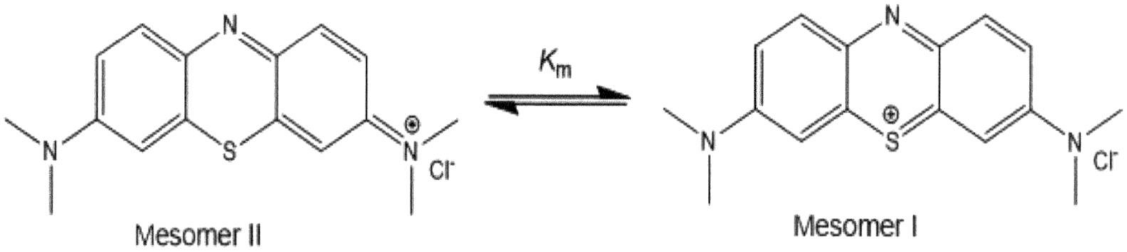

Figure 6. Alternative mesomeric structures for MB [41].

Based on Fernández-Pérez's work on the mesomeric structure of MB in water, we propose that, after the removal of the dye by the native gel or γ-Fe$_2$O$_3$NPs@gel + mf, the purified solutions contain the mesomeric II structure of MB, which has different UV-Vis adsorption with the observed red-shift, compared to the initial MB solution.

For Rh6G, more than 95% of the dye was removed by both γ-Fe$_2$O$_3$NPs@gel and γ-Fe$_2$O$_3$NPs@gel + mf and, for the observed red-shift of the purified solutions, a monomer/dimer change in Rh6G may be proposed. In fact, the initial solution of Rh6G shows a combination of monomer/dimer species for the dye. After dilution, the concentration of monomers becomes significant and a spectral change is observed. Another important factor is the possible release of ions (e.g., Na$^+$, Cl$^-$) from the hydrogel matrix to the filtered solutions, which can also affect the spectral pattern on the dyes.

Conversely, regarding the cationic dyes, for MO, no significant spectral changes were detected after the filtration; the RE% values for MO are much lower (20–30%) compared to those for MB and Rh6G, probably due to the different charges in the dyes. Regarding the amount of MO adsorbed by the three hydrogel systems, γ-Fe$_2$O$_3$NPs@gel and γ-Fe$_2$O$_3$NPs@gel + mf show the same adsorption efficiencies (approximately 30%), compared to the lower amounts observed for the pristine gel (approximately 20%).

In Table 4, the adsorption capacities of γ-Fe2O3NPs@gel + mf (in the experimental set-up employing syringes) towards cationic dyes were compared with those of similar adsorbents described in the literature. The observed lower adsorption capacities for the magnetogels described in this study may be improved by modifying the formulation of the magnetogels, e.g., using cross-linkers that could also improve the mechanical stability of the magnetogels and, therefore, their reusability as adsorbents.

Table 4. Comparison of the adsorption capacities of γ-Fe2O3NPs@gel + mf with similar adsorbents for the removal of cationic dyes.

Adsorbent	Cationic Dyes	C_0 (mg mL^{-1})	Adsorption Capacities of Syringe Set-Up (mg g^{-1})	Reference
γ-Fe$_2$O$_3$NPs@gel + mf	MB	0.005	0.9	this work
γ-Fe$_2$O$_3$NPs@gel + mf	Rh6G	0.005	1.1	this work
poly(acrylic acid-acrylamide-butyl methacrylate) magnetic hydrogel	MB	50–100	12.6	[43]
Fe$_3$O$_4$/poly(2-hydroxyethyl methacrylate-co-itaconic acid) magnetic hydrogels	MB	0.05–0.2	174.9	[44]
poly(2-(2-methoxyethoxy) ethyl methacrylate-co-oligo (ethylene glycol) methacrylate-co-acrylic acid) (PMOA) hydrogel-magnetic attapulgite/Fe$_3$O$_4$	RhB (rhodamine B)	0.001	1.65	[45]

By comparing the two experimental setups (cuvettes and syringes, see Table 3), it can be seen that the adsorption using cuvettes is much lower, as it occurs only at the interface between the gel and solution, while in the case of syringes, the solution passes through the mass of the gel, allowing a greater interaction with it and, consequently, a greater adsorption of the dyes. More importantly, in each methodology, the trends of the RE% values of the adsorbents (Figures S7–S9) are changed.

3. Conclusions

Composite hydrogels have interesting adsorption properties and, in this study, we reported that the incorporation of magnetic NPs inside peptide-based hydrogels can be a valuable approach to additionally increase the adsorption capacities of hydrogels. In particular, the γ-Fe$_2$O$_3$NPs@gel + mf system was able to absorb up to 84% of MB and 97% of Rh6G from aqueous solutions containing 5 ppm of the dyes. However, for the anionic dye MO, the gels showed much lower adsorption capacities (lower than 30%) due to the dominant electrostatic repulsions between MO and the gel components.

The kinetic models of the cationic dyes mainly showed a physisorption mechanism for these adsorbents, while, for MO, both chemisorption and physisorption were observed, depending on the type of adsorbent. The results presented here support the application of peptide-based magnetogels in wastewater treatment. Future work should focus on scaling up materials' preparation and testing them in practical scenarios.

4. Materials and Methods

4.1. Materials

L-Phenylalanyl-L-phenylalanine and N-(9-Fluorenylmethoxycarbonyl)-L-phenylalanine were obtained from Bachem GmbH (Weil am Rhein, Germany). All other chemicals, including FeCl$_2$·4H$_2$O, FeCl$_3$, and *Pseudomonas fluorescens* Lipase ($\geq$20,000 U/mg) were purchased from Sigma-Aldrich (Saint Louis, MO, USA). All chemicals were employed without further purification. Ultra-pure water was prepared with a Zeneer Power I Scholar-UV (Full Tech Instruments, Rome, Italy) apparatus. The external magnetic field was generated by using a neodymium-based magnet (volume = 39.270 cm^3, magnetization quality = N5, magnetic strength = 1.42–1.47 T).

4.2. Synthesis of γ-Fe$_2$O$_3$NPs@gel Magnetogels

Magnetogels were prepared with a two-step procedure, first preparing PAA-stabilized γ-Fe$_2$O$_3$NPs and then dispersing them in the aqueous solution of hydrogel precursors before gelation. The details of the synthetic procedures have been described previously [23].

4.3. SEM Analysis

SEM analyses were conducted with an Auriga field emission scanning electron microscope (Zeiss, Oberkochen, Germany), as previously described [23].

4.4. Rheological Measurements

The dynamo-mechanical analyses (mechanical spectroscopy) were studied for four samples including hydrogel and three magnetogels containing different concentrations of NPs. Rheological studies were carried out with an MCR 302 rotational rheometer (Anton Paar, Turin, Italy), as previously described [23].

4.5. Swelling Ability

Three mL of phosphate buffer solution (PBS, pH = 7.4) was placed on top of each hydrogel sample, followed by incubation at 30 °C in a thermostatic bath. After 24 h, the supernatants were removed and the samples were freeze-dried. The swelling degree was evaluated by using the following equation:

$$q = (W_s - W_d)/W_d \tag{1}$$

where q = swelling degree, W_s = weight of the gel after PBS removal, and W_d = weight of the freeze-dried gel.

4.6. Adsorption Experiments

For the cuvette experiments, gel and magnetogel samples were prepared in glass cuvettes. Then, 2 mL of aqueous dye solution (MB, Rh6G or MO) was cast on top of the gels. The concentration of MB and Rh6G was 5 ppm, while the MO concentration was 10 ppm.

UV-Vis spectroscopy was employed to monitor the absorbances of the solutions every 15 min. The removal efficiencies (RE) were calculated using Equation (2):

$$RE(\%) = (C_0 - C_f)/C_0 \times 100 \tag{2}$$

where C_0 = initial dye concentration, C_f = dye concentration in the eluted solution.

Calibration curves were prepared for each dye and used for the calculations. The adsorption capacities (q_t, mg g^{-1}) of the adsorbents were calculated using Equation (3) [23]:

$$q_t = (C_0 - C_t)/m \times V \tag{3}$$

in which m (g) is the dried hydrogel mass, C_0 and C_t (mg L^{-1}) are the initial and equilibrium dye concentrations, and V (L) is the solution volume, respectively.

The kinetics of dye adsorption were analyzed with non-linear pseudo-first- and pseudo-second-order kinetic models (Equations (4) and (5)) [22].

$$\log(q_e - q_t) = \log q_e - k_1 t/2.303 \tag{4}$$

$$t/q_t = 1/k_2 q_e^2 + t/q_e \tag{5}$$

Supplementary Materials: The following supporting information can be downloaded at: https://www.mdpi.com/article/10.3390/gels10050287/s1, Materials and Methods; Figure S1: (a) Calibration curve of methylene blue (MB); (b) calibration curve of rhodamine 6G (R6G); (c) calibration curve of methyl orange (MO).; Figure S2: (a) Gravitational passage of MB through the gels inside the syringe; (b) gravitational passages with applying the filter in the outlet of the syringe; (c) gravitational passage with applying the filter and a constant pressure to increase the flow rate of the dye inside the syringe.; Figure S3: (a) Absorption of MB by the filter for different cycles; (b) amount of MB adsorbed by the filter in each cycle; (c) visual observation of the MB adsorption by the filter.; Figure S4: (a) Absorption of Rh6G by the filter for different cycles; (b) amount of Rh6G adsorbed by the filter in each cycle;

(c) visual observation of the Rh6G adsorption by the filter.; Figure S5: (a) Absorption of MO by the filter for different cycles; (b) amount of MO adsorbed by the filter in each cycle; (c) visual observation of the MO adsorption by the filter. Figure S6: (a) Separation of the dyes by gel, γ-Fe$_2$O$_3$NPs@gel, and (γ-Fe$_2$O$_3$NPs@gel + mf) using the syringe with filter and external pressure for (a) MB; (b) Rh6G; (c) and MO.; Figure S7: Comparison of the cuvette and syringe methodology for the MB adsorption by (a) gel, (b) γ-Fe$_2$O$_3$NPs@gel, and (c) γ-Fe$_2$O$_3$NPs@gel + mf.; Figure S8: Comparison of the cuvette and syringe methodology for the Rh6G adsorption by (a) gel, (b) γ-Fe$_2$O$_3$NPs@gel, and (c) γ-Fe$_2$O$_3$NPs@gel + mf.; Figure S9: Comparison of the cuvette and syringe methodology for the MO adsorption MO by (a) gel, (b) γ-Fe$_2$O$_3$NPs@gel, and (c) γ-Fe$_2$O$_3$NPs@gel + mf.

Author Contributions: Conceptualization: F.H.H. and R.B.; Methodology: L.C., R.B. and F.H.H.; Investigation: P.S.P., F.H.H. and R.B.; Data curation: L.C. and C.P.; Writing—original draft preparation: F.H.H. and R.B.; Writing—review and editing, L.C. and C.P. All authors have read and agreed to the published version of the manuscript.

Funding: This research received no external funding.

Institutional Review Board Statement: Not applicable.

Informed Consent Statement: Not applicable.

Data Availability Statement: The data presented in this study are available in the article.

Conflicts of Interest: The authors declare no conflicts of interest.

References

1. Waghchaure, R.H.; Adole, V.A.; Jagdale, B.S. Photocatalytic Degradation of Methylene Blue, Rhodamine B, Methyl Orange and Eriochrome Black T Dyes by Modified ZnO Nanocatalysts: A Concise Review. *Inorg. Chem. Commun.* **2022**, *143*, 109764. [CrossRef]
2. Salahuddin, B.; Aziz, S.; Gao, S.; Hossain, M.S.A.; Billah, M.; Zhu, Z.; Amiralian, N. Magnetic Hydrogel Composite for Wastewater Treatment. *Polymers* **2022**, *14*, 5074. [CrossRef]
3. Chiam, S.L.; Pung, S.Y.; Yeoh, F.Y. Recent Developments in MnO2-Based Photocatalysts for Organic Dye Removal: A Review. *Environ. Sci. Pollut. Res.* **2020**, *27*, 5759–5778. [CrossRef]
4. Domacena, A.M.G.; Aquino, C.L.E.; Balela, M.D.L. Photo-Fenton Degradation of Methyl Orange Using Hematite (α-Fe$_2$O$_3$) of Various Morphologies. *Mater. Today Proc.* **2020**, *22*, 248–254. [CrossRef]
5. Nyankson, E.; Kumar, R.V. Removal of Water-Soluble Dyes and Pharmaceutical Wastes by Combining the Photocatalytic Properties of Ag3PO4 with the Adsorption Properties of Halloysite Nanotubes. *Mater. Today Adv.* **2019**, *4*, 100025. [CrossRef]
6. Alharthi, F.A.; Al-Zaqri, N.; El Marghany, A.; Alghamdi, A.A.; Alorabi, A.Q.; Baghdadi, N.; AL-Shehri, H.S.; Wahab, R.; Ahmad, N. Synthesis of Nanocauliflower ZnO Photocatalyst by Potato Waste and Its Photocatalytic Efficiency against Dye. *J. Mater. Sci. Mater. Electron.* **2020**, *31*, 11538–11547. [CrossRef]
7. Natarajan, S.; Bajaj, H.C.; Tayade, R.J. Recent Advances Based on the Synergetic Effect of Adsorption for Removal of Dyes from Waste Water Using Photocatalytic Process. *J. Environ. Sci.* **2018**, *65*, 201–222. [CrossRef] [PubMed]
8. Kumar, S.; Kaushik, R.D.; Purohit, L.P. Novel ZnO Tetrapod-Reduced Graphene Oxide Nanocomposites for Enhanced Photocatalytic Degradation of Phenolic Compounds and MB Dye. *J. Mol. Liq.* **2021**, *327*, 114814. [CrossRef]
9. Hasanpour, M.; Hatami, M. Photocatalytic Performance of Aerogels for Organic Dyes Removal from Wastewaters: Review Study. *J. Mol. Liq.* **2020**, *309*, 113094. [CrossRef]
10. Nawaz, A.; Khan, A.; Ali, N.; Ali, N.; Bilal, M. Fabrication and Characterization of New Ternary Ferrites-Chitosan Nanocomposite for Solar-Light Driven Photocatalytic Degradation of a Model Textile Dye. *Environ. Technol. Innov.* **2020**, *20*, 101079. [CrossRef]
11. Berradi, M.; Hsissou, R.; Khudhair, M.; Assouag, M.; Cherkaoui, O.; El Bachiri, A.; El Harfi, A. Textile Finishing Dyes and Their Impact on Aquatic Environs. *Heliyon* **2019**, *5*, e02711. [CrossRef] [PubMed]
12. Alakhras, F.; Alhajri, E.; Haounati, R.; Ouachtak, H.; Addi, A.A.; Saleh, T.A. A Comparative Study of Photocatalytic Degradation of Rhodamine B Using Natural-Based Zeolite Composites. *Surf. Interfaces* **2020**, *20*, 100611. [CrossRef]
13. Fu, F.; Wang, Q. Removal of Heavy Metal Ions from Wastewaters: A Review. *J. Environ. Manag.* **2011**, *92*, 407–418. [CrossRef] [PubMed]
14. Moosavi, S.; Lai, C.W.; Gan, S.; Zamiri, G.; Akbarzadeh Pivehzhani, O.; Johan, M.R. Application of Efficient Magnetic Particles and Activated Carbon for Dye Removal from Wastewater. *ACS Omega* **2020**, *5*, 20684–20697. [CrossRef] [PubMed]
15. De Gisi, S.; Lofrano, G.; Grassi, M.; Notarnicola, M. Characteristics and Adsorption Capacities of Low-Cost Sorbents for Wastewater Treatment: A Review. *Sustain. Mater. Technol.* **2016**, *9*, 10–40. [CrossRef]
16. Mahdi, A.E.; Ali, N.S.; Kalash, K.R.; Salih, I.K.; Abdulrahman, M.A.; Albayati, T.M. Investigation of Equilibrium, Isotherm, and Mechanism for the Efficient Removal of 3-Nitroaniline Dye from Wastewater Using Mesoporous Material MCM-48. *Prog. Color Color Coat.* **2023**, *16*, 387–398.

17. Rafieian, S.; Mirzadeh, H.; Mahdavi, H.; Masoumi, M.E. A Review on Nanocomposite Hydrogels and Their Biomedical Applications. *Sci. Eng. Compos. Mater.* **2019**, *26*, 154–174. [CrossRef]
18. Chronopoulou, L.; Binaymotlagh, R.; Cerra, S.; Hajareh Haghighi, F.; Di Domenico, E.G.; Sivori, F.; Fratoddi, I.; Mignardi, S.; Palocci, C. Preparation of Hydrogel Composites Using a Sustainable Approach for In Situ Silver Nanoparticles Formation. *Materials* **2023**, *16*, 2134. [CrossRef] [PubMed]
19. Binaymotlagh, R.; Hajareh Haghighi, F.; Di Domenico, E.G.; Sivori, F.; Truglio, M.; Del Giudice, A.; Fratoddi, I.; Chronopoulou, L.; Palocci, C. Biosynthesis of Peptide Hydrogel–Titania Nanoparticle Composites with Antibacterial Properties. *Gels* **2023**, *9*, 940. [CrossRef]
20. Fortunato, A.; Mba, M. A Peptide-Based Hydrogel for Adsorption of Dyes and Pharmaceuticals in Water Remediation. *Gels* **2022**, *8*, 672. [CrossRef]
21. Veloso, S.R.S.; Ferreira, P.M.T.; Martins, J.A.; Coutinho, P.J.G.; Castanheira, E.M.S. Magnetogels: Prospects and Main Challenges in Biomedical Applications. *Pharmaceutics* **2018**, *10*, 145. [CrossRef] [PubMed]
22. de Melo, F.M.; Grasseschi, D.; Brandão, B.B.N.S.; Fu, Y.; Toma, H.E. Superparamagnetic Maghemite-Based CdTe Quantum Dots as Efficient Hybrid Nanoprobes for Water-Bath Magnetic Particle Inspection. *ACS Appl. Nano Mater.* **2018**, *1*, 2858–2868. [CrossRef]
23. Hajareh Haghighi, F.; Binaymotlagh, R.; Chronopoulou, L.; Cerra, S.; Marrani, A.G.; Amato, F.; Palocci, C.; Fratoddi, I. Self-Assembling Peptide-Based Magnetogels for the Removal of Heavy Metals from Water. *Gels* **2023**, *9*, 621. [CrossRef] [PubMed]
24. Chronopoulou, L.; Lorenzoni, S.; Masci, G.; Dentini, M.; Togna, A.R.; Togna, G.; Bordi, F.; Palocci, C. Lipase-Supported Synthesis of Peptidic Hydrogels. *Soft Matter* **2010**, *6*, 2525–2532. [CrossRef]
25. Sáenz-Trevizo, A.; Pizá-Ruiz, P.; Chávez-Flores, D.; Ogaz-Parada, J.; Amézaga-Madrid, P.; Vega-Ríos, A.; Miki-Yoshida, M. On the Discoloration of Methylene Blue by Visible Light. *J. Fluoresc.* **2019**, *29*, 15–25. [CrossRef] [PubMed]
26. Purcar, V.; Donescu, D.; Petcu, C.; Vasilescu, M. Nanostructured Hybrid Systems with Rhodamine 6G. *J. Dispers. Sci. Technol.* **2008**, *29*, 1233–1239. [CrossRef]
27. Ruiz, C.; Vera, M.; Rivas, B.L.; Sánchez, S.A.; Urbano, B.F. Magnetic methacrylated gelatin-g-polyelectrolyte for methylene blue sorption. *RSC Adv.* **2020**, *10*, 43799–43810. [CrossRef] [PubMed]
28. Yao, G.; Bi, W.; Liu, H. pH-responsive magnetic graphene oxide/poly(NVI-co-AA) hydrogel as an easily recyclable adsorbent for cationic and anionic dyes. *Colloids Surf. A Physicochem. Eng. Asp.* **2020**, *588*, 124393. [CrossRef]
29. Gang, F.; Jiang, L.; Xiao, Y.; Zhang, J.; Sun, X. Multi-functional magnetic hydrogel: Design strategies and applications. *Nano Select* **2021**, *2*, 2291–2307. [CrossRef]
30. Veloso, S.R.S.; Andrade, R.G.D.; Castanheira, E.M.S. Review on the advancements of magnetic gels: Towards multifunctional magnetic liposome-hydrogel composites for biomedical applications. *Adv. Colloid Interface Sci.* **2021**, *288*, 102351. [CrossRef]
31. Van Berkum, S.; Biewenga, P.D.; Verkleij, S.P.; Van Zon, J.B.A.; Boere, K.W.M.; Pal, A.; Philipse, A.P.; Erné, B.H. Swelling Enhanced Remanent Magnetization of Hydrogels Cross-Linked with Magnetic Nanoparticles. *Langmuir* **2014**, *31*, 442–450. [CrossRef] [PubMed]
32. Saadli, M.; Braunmiller, D.L.; Mourran, A.; Crassous, J.J. Thermally and Magnetically Programmable Hydrogel Microactuators. *Small* **2023**, *19*, 2207035. [CrossRef] [PubMed]
33. Nagireddy, N.R.; Yallapu, M.M.; Kokkarachedu, V.; Sakey, R.; Kanikireddy, V.; Pattayil Alias, J.; Konduru, M.R. Preparation and Characterization of Magnetic Nanoparticles Embedded in Hydrogels for Protein Purification and Metal Extraction. *J. Polym. Res.* **2011**, *18*, 2285–2294. [CrossRef]
34. Sharma, G.; García-Peñas, A.; Verma, Y.; Kumar, A.; Dhiman, P.; Stadler, F.J. Tailoring Homogeneous Hydrogel Nanospheres by Facile Ultra-Sonication Assisted Cross-Linked Copolymerization for Rhodamine B Dye Adsorption. *Gels* **2023**, *9*, 770. [CrossRef] [PubMed]
35. Rana, H.; Anamika, A.; Sareen, D.; Goswami, S. Nanocellulose-Based Ecofriendly Nanocomposite for Effective wastewater Remediation: A study on its process optimization, improved swelling, adsorption, and thermal and mechanical behavior. *ACS Omega* **2024**, *9*, 8904–8922. [CrossRef]
36. Gao, H.; Jiang, J.; Huang, Y.; Wang, H.; Sun, J.; Jin, Z.; Wang, J.; Zhang, J. Synthesis of hydrogels for adsorption of anionic and cationic dyes in water: Ionic liquid as a crosslinking agent. *SN Appl. Sci.* **2022**, *4*, 118. [CrossRef]
37. Chen, T.; Liu, H.; Gao, J.; Hu, G.; Zhao, Y.; Tang, X.; Han, X. Efficient removal of methylene blue by Bio-Based sodium Alginate/Lignin composite hydrogel beads. *Polymers* **2022**, *14*, 2917. [CrossRef] [PubMed]
38. Wang, W.; Bai, H.; Zhao, Y.; Kang, S.; Yi, H.; Zhang, T.; Song, S. Synthesis of chitosan cross-linked 3D network-structured hydrogel for methylene blue removal. *Int. J. Biol. Macromol.* **2019**, *141*, 98–107. [CrossRef] [PubMed]
39. Wang, J.; Meng, X.; Yuan, Z.; Tian, Y.; Bai, Y.; Jin, Z. Acrylated Composite Hydrogel Preparation and Adsorption Kinetics of Methylene Blue. *Molecules* **2017**, *22*, 1824. [CrossRef]
40. Li, Y.; Liu, L.; Huang, W.; Xie, J.; Song, Z.; Guo, S.; Wang, E. Preparation of peanut shell cellulose Double-Network hydrogel and its adsorption capacity for methylene blue. *J. Renew. Mater.* **2023**, *11*, 3001–3023. [CrossRef]
41. Godiya, C.B.; Xiao, Y.; Lu, X. Amine Functionalized Sodium Alginate Hydrogel for Efficient and Rapid Removal of Methyl Blue in Water. *Int. J. Biol. Macromol.* **2020**, *144*, 671–681. [CrossRef] [PubMed]
42. Fernández-Pérez, A.; Marbán, G. Visible Light Spectroscopic Analysis of Methylene Blue in Water; What Comes after Dimer? *ACS Omega* **2020**, *5*, 29801–29815. [CrossRef] [PubMed]

43. Li, S.; Liu, X.; Huang, W.; Li, W.; Xia, X.; Yan, S.; Yu, J. Magnetically assisted removal and separation of cationic dyes from aqueous solution by magnetic nanocomposite hydrogels. *Polym. Adv. Technol.* **2011**, *22*, 2439–2447. [CrossRef]
44. Ludeña, M.A.; Meza, F.d.L.; Huamán, R.I.; Lechuga, A.M.; Valderrama, A.C. Preparation and Characterization of Fe3O4/Poly(HEMA-co-IA) Magnetic Hydrogels for Removal of Methylene Blue from Aqueous Solution. *Gels* **2024**, *10*, 15. [CrossRef]
45. Yuan, Z.; Wang, Y.; Han, X.; Chen, D. The adsorption behaviors of the multiple stimulus-responsive poly(ethylene glycol)-based hydrogels for removal of RhB dye. *J. Appl. Polym. Sci.* **2015**, *132*, 42244. [CrossRef]

Disclaimer/Publisher's Note: The statements, opinions and data contained in all publications are solely those of the individual author(s) and contributor(s) and not of MDPI and/or the editor(s). MDPI and/or the editor(s) disclaim responsibility for any injury to people or property resulting from any ideas, methods, instructions or products referred to in the content.

Article

Silicon-Doped Carbon Dots Crosslinked Carboxymethyl Cellulose Gel: Detection and Adsorption of Fe³⁺

Zhengdong Zhao [1,2,†], Yichang Jing [1,2,†], Yuan Shen [1,2], Yang Liu [1,2], Jiaqi Wang [1,2], Mingjian Ma [1,2], Jiangbo Pan [1,2], Di Wang [1,2,*], Chengyu Wang [1,2] and Jian Li [1,2]

[1] Key Laboratory of Bio-Based Material Science and Technology, Ministry of Education, Northeast Forestry University, Harbin 150040, China; zhaozd6866@163.com (Z.Z.); yichangjing0302@163.com (Y.J.); jlaushawn@yeah.net (Y.S.); hljqaliuyang@163.com (Y.L.); w1531565456@163.com (J.W.); 18847042762@163.com (M.M.); pjb0725@163.com (J.P.); wangcy@nefu.edu.cn (C.W.); nefujianli@163.com (J.L.)
[2] College of Material Science and Engineering, Northeast Forestry University, Harbin 150040, China
* Correspondence: diwang1030@nefu.edu.cn
† These authors contributed equally to this work.

Citation: Zhao, Z.; Jing, Y.; Shen, Y.; Liu, Y.; Wang, J.; Ma, M.; Pan, J.; Wang, D.; Wang, C.; Li, J. Silicon-Doped Carbon Dots Crosslinked Carboxymethyl Cellulose Gel: Detection and Adsorption of Fe³⁺. *Gels* **2024**, *10*, 285.
https://doi.org/10.3390/gels10050285

Academic Editor: Georgios Bokias

Received: 5 April 2024
Revised: 20 April 2024
Accepted: 20 April 2024
Published: 23 April 2024

Copyright: © 2024 by the authors. Licensee MDPI, Basel, Switzerland. This article is an open access article distributed under the terms and conditions of the Creative Commons Attribution (CC BY) license (https://creativecommons.org/licenses/by/4.0/).

Abstract: The excessive emission of iron will pollute the environment and harm human health, so the fluorescence detection and adsorption of Fe^{3+} are of great significance. In the field of water treatment, cellulose-based gels have attracted wide attention due to their excellent properties and environmental friendliness. If carbon dots are used as a crosslinking agent to form a gel with cellulose, it can not only improve mechanical properties but also show good biocompatibility, reactivity, and fluorescence properties. In this study, silicon-doped carbon dots/carboxymethyl cellulose gel (DCG) was successfully prepared by chemically crosslinking biomass-derived silicon-doped carbon dots with carboxymethyl cellulose. The abundant crosslinking points endow the gel with excellent mechanical properties, with a compressive strength reaching 294 kPa. In the experiment on adsorbing Fe^{3+}, the theoretical adsorption capacity reached 125.30 mg/g. The introduction of silicon-doped carbon dots confers the gel with excellent fluorescence properties and a good selective response to Fe^{3+}. It exhibits a good linear relationship within the concentration range of 0–100 mg/L, with a detection limit of 0.6595 mg/L. DCG appears to be a good application prospect in the adsorption and detection of Fe^{3+}.

Keywords: silicon-doped carbon dots crosslinked gel; carboxymethyl cellulose; trivalent iron detection; trivalent iron adsorption

1. Introduction

In recent years, water pollution caused by heavy metals such as Fe^{3+} has become a serious concern [1]. Excessive accumulation of Fe^{3+} in the human body can pose significant health hazards, thus necessitating its removal and detection in water [2]. Currently, the methods for treating metal ion pollution include chemical precipitation, electrolysis, ion exchange, membrane separation, etc. However, these methods are not only expensive but can also cause secondary pollution [3]. Among the numerous reported methods for removing and detecting metal ion pollution, the adsorption method has the advantages of simplicity, high efficiency, low cost, recyclability of adsorbents, and environmental friendliness [4], while fluorescence detection has the advantages of low cost and easy operation [5]. Both of them have broad application prospects and are worthy of attention [6]. Cellulose-based materials have been widely studied for heavy metal adsorption due to their excellent physicochemical properties, mechanical performance, and renewability. Gel materials with cellulose and its derivatives as the framework have attracted significant attention due to their high surface area and porous structure [7,8]. Cellulose derivatives such as carboxymethyl cellulose, cellulose acetate, and ethyl cellulose contain oxygen-containing functional groups that can form gel membranes through physical crosslinking via van der Waals forces and hydrogen bonds [9,10]. However, gels relying solely on physical crosslinking exhibit poor mechanical properties, limiting their application range. Chemically crosslinked gels, on the other hand, possess stronger stability and mechanical properties, thus finding wider applications [11]. Nevertheless, commonly used crosslinking agents, such as glutaraldehyde and epichlorohydrin, are toxic and incompatible with environmentally friendly cellulose-based materials [7,12]. Therefore, there is a need to explore novel and green crosslinking agents. Additionally, the low photoluminescence properties of cellulose and its derivatives often do not meet the requirements for fluorescent sensing, necessitating the introduction of emission sources [13–15].

Carbon dots, a new class of luminescent materials, possess strong photoluminescence and stability, making them widely applicable in the field of fluorescent sensing. Furthermore, the rich active groups on the surface of carbon dots allow them to serve as clean crosslinking agents for gel formation with framework materials. Therefore, combining carbon dots with cellulose derivatives can not only form gels but also improve the optical properties of the cellulose derivatives [16–18]. Additionally, the excellent mechanical properties of cellulose derivatives can address the challenges associated with the difficulty in processing and shaping carbon dots due to their size, achieving a win–win situation [19]. Among various carbon dots precursors, carboxymethyl chitosan, a derivative of natural polysaccharides, is chosen for synthesizing carbon dots due to its widespread availability and renewability. However, carbon dots prepared from polysaccharides as carbon sources typically exhibit poor luminescent properties. To enhance the luminescent performance of carbon dots, amino silane doping can be employed to adjust the band structure of carbon dots, leading to the synthesis of silicon-doped carbon dots with strong luminescent properties. Simultaneously, the introduction of additional amino groups increases the reactive sites [20].

In this work, silicon-doped carbon dots (Si-CDs) were prepared through a one-step hydrothermal method and subsequently chemically crosslinked with carboxymethyl cellulose through amide bonds to form a gel material (DCG). The numerous hydrophilic active groups, including hydroxyl, amino, and carboxyl, present within DCG offer numerous chelate sites for Fe^{3+}, conferring upon DCG the capability to chemically adsorb Fe^{3+} in aqueous solutions. Moreover, Si-CDs exhibit excellent fluorescent properties and specific responsiveness to Fe^{3+}. Therefore, DCG demonstrates promising application potential in the adsorption and detection of Fe^{3+}.

2. Results and Discussion

2.1. Si-CD Structural Analysis

FTIR spectra were used to detect the occurrence of the reaction, and the FTIR spectra of CMCS, APTES, and Si-CDs are shown in Figure 1a. The pure carboxymethyl chitosan powder shows stretching vibrational absorption bands of hydroxyl and amino groups at 3410 cm^{-1}. The peaks near 2900 and 2870 cm^{-1} are attributed to the stretching vibration of C-H. The peak at 1052 cm^{-1} is attributed to the bending vibration of C-H of the pyranose ring. The peaks at 1412 and 1587 cm^{-1} are attributed to the asymmetric stretching vibration peaks and symmetric stretching vibrational peaks, which are associated with the presence of -COONa and -COOH [21]. The peak at 1310 cm^{-1} is attributed to the asymmetric vibrational peak of C-N [22]. Comparing the infrared absorption curves of CMCS, new peaks appeared in the curves of Si-CDs, namely, the asymmetric stretching vibration and symmetric stretching vibration peaks of Si-O-Si at 1002 and 690 cm^{-1}, the peak of Si-O-C at 1188 cm^{-1}, and the bending vibration peak of Si-OH at 923 cm^{-1}. The peaks at 1002 and 690 cm^{-1} are the asymmetric stretching vibration and symmetric stretching vibration peaks of Si-O-Si, which proved that Si-OH produced by APTES hydrolysis is dehydrated and condensed to form siloxane structure again [23]. The disappearance of the -CH_3 peak at 2974 cm^{-1} present in APTES proves that the hydrolysis of APTES is complete [24]. The peaks related to -COO- are retained in Si-CDs, indicating that Si-CDs have -COOH and -COONa. Moreover, 1310 cm^{-1} asymmetric C-N vibrational peaks still exist, and more obvious bending vibrational peaks of -NH$_2$ are produced at 3352 and 3284 cm^{-1}, which proves that the amino group of APTES is retained on the surface of Si-CDs after the reaction [24].

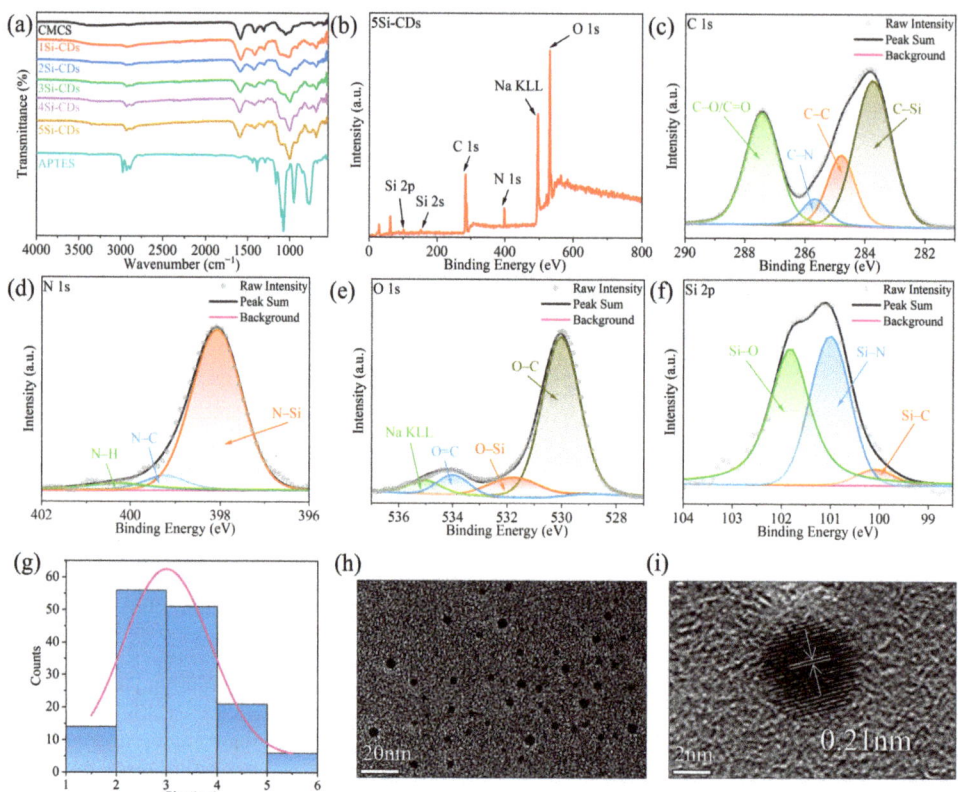

Figure 1. (**a**) FTIR spectra of CMCS, Si-CDs, and APTES. XPS spectra of 5Si-CDs: (**b**) full spectrum, (**c**) C 1s, (**d**) N 1s, (**e**) O 1s, and (**f**) Si 2p. (**g**) Size distribution histogram, (**h**) TEM image, and (**i**) HRTEM image of 5Si-CDs. Arrows and lines indicate one of the lattice spacings.

In order to further investigate the molecular structure and functional group composition of Si-CDs, XPS tests were carried out with 5Si-CDs, and peak fitting of high-resolution XPS spectra was performed. The full XPS spectra in Figure 1b shows that the Si-CDs contain the elements C, N, O, Si, and Na. In the C 1s spectra in Figure 1c, C-Si, C-C, C-N, and C-O/C=O correspond to 283.74, 284.80, 285.68, and 287.43 eV [25,26]. In the N 1s spectra in Figure 1d, N-Si, N-C, and N-H correspond to 398.08, 399.23, and 400.30 eV [27–29]. In the O 1s spectra in Figure 1e, O-C, O-Si, and O=C correspond to 530.04, 531.73, and 533.99 eV [30]. In the Si 2p spectra in Figure 1f, Si-C, Si-N, and Si-O correspond to 100.09, 101.01, and 101.83 eV [29]. Comparing the XPS full spectra of CMCS in Figure S1 with the C 1s, N 1s, and O 1s spectra, the appearance of N-Si, O-Si, and Si elemental peaks proves the successful introduction of APTES. The results of the XPS spectra are in agreement with those of the FTIR spectra, which proves that the surface of Si-CDs possesses an abundance of reactive groups, such as hydroxyl, amino, and carboxyl groups.

The corresponding histogram of size distribution in Figures 1g and S2 shows that the particle size is in line with the particle size range of carbon dots [31]. The TEM image of Si-CDs in Figures 1h and S2 shows that they are spherical, without any aggregation, and well dispersed in water. The HRTEM image with higher resolution in Figures 1i and S2 shows diffraction streaks of Si-CDs with a spacing of 0.21 nm, corresponding to the (100) crystallographic plane of graphitic carbon [32], which suggests that the core of Si-CDs is mainly carbon. Combined with the results of HRTEM, FTIR, and XPS, it can be hypothesized that the Si-CDs are a structure of the Si-O-Si skeleton encapsulating a graphitic carbon core [24,33].

2.2. DCG Structural Analysis

The FTIR spectra of CMC and DCG are shown in Figure 2a. In the infrared spectra of pure CMC powder, the telescopic vibrational peak of -OH is at 3285 cm^{-1} [34]. The peaks at 2917 and 2871 cm^{-1} are attributed to the asymmetric and symmetric telescopic vibrational absorption peaks of the C-H bond in -CH$_2$- and -CH$_3$ [35]. The peaks at 1590 and 1414 cm^{-1} are attributed to the antisymmetric -COO- and symmetric telescopic vibrational absorption peaks [36]. The telescopic vibrational absorption peak of the glycosidic bond on the cellulose pyran ring is at 1020 cm^{-1} [37]. In the infrared spectra of DCG, the C-H asymmetric stretching vibrational peaks and symmetric stretching vibrational peaks at 2922 and 2878 cm^{-1} are blue shifted. The intensity of the amino peaks near 3400 cm^{-1} increases, and a new stretching vibrational absorption peak of -CO-NH- appears at 1640 cm^{-1} [38]. The formation of chemical bonding connections between Si-CDs and CMCs is suggested, indicating that the amino group of Si-CDs has undergone a reaction with the carboxyl group of CMC to form an amide bond, thereby confirming the successful synthesis of DCG. Similarly, the appearance of the stretching vibration absorption peak of -CO-NH- at 1640 cm^{-1} also demonstrates the successful synthesis of SiDCG.

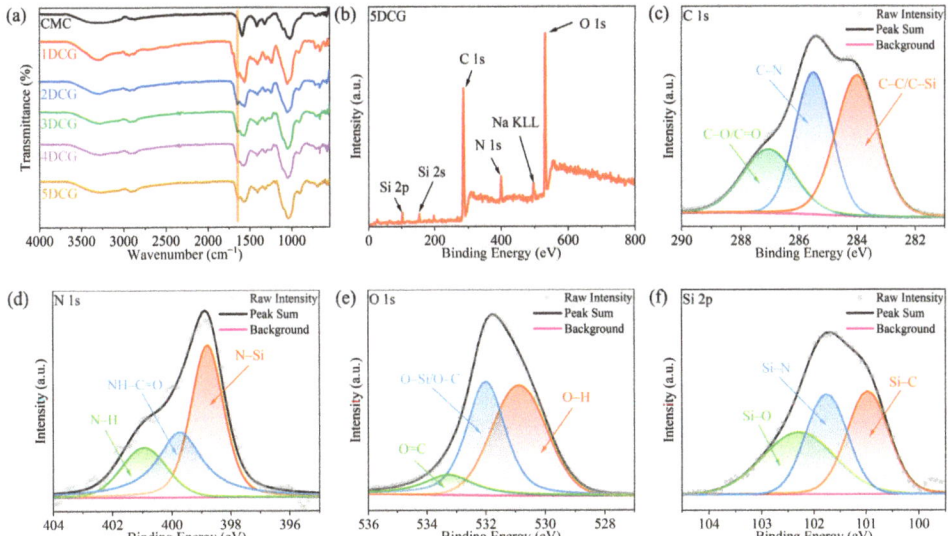

Figure 2. (**a**) FTIR spectra of CMC and DCG. XPS spectra of 5DCG: (**b**) full spectrum, (**c**) C 1s, (**d**) N 1s, (**e**) O 1s, and (**f**) Si 2p.

In order to further investigate the molecular structure and functional group composition of DCG, it was subjected to XPS tests. The full XPS spectra of DCG in Figure 2b shows that DCG contains elements such as C, N, O, and Si. In the C 1s spectra in Figure 2c, C-Si/C-C, C-N, and C-O/C=O correspond to 284.00, 285.50, and 287.05 eV [39], respectively, and the appearance of distinct C-Si and C-N peaks indicates the successful introduction of Si-CDs. In the N 1s spectra in Figure 2d, N-Si, NH-C=O, and N-H correspond to 398.80, 399.73, and 400.94 eV [29,40], respectively, and the appearance of NH-C=O peaks suggests that the Si-CDs and the CMC are chemically linked by amide bonds. In the O 1s spectra in Figure 2e, O-H, O-C/O-Si, and O=C correspond to 530.88, 532.00, and 533.27 eV, respectively. In the Si 2p spectra in Figure 2f, Si-C, Si-N, and Si-O correspond to 100.97, 101.76, and 102.30 eV, respectively [29]. Comparing the XPS full spectra, C1s spectra, and O1s spectra of CMC in Figure S3, the appearance of peaks related to N and Si proves the successful reaction of DCG with Si-CDs. The XPS results are in agreement with the FTIR results, which prove the successful synthesis of DCG.

In order to investigate the microstructural changes of CMC after crosslinking with Si-CDs, SEM tests were performed on them, taking 5DCG as an example, and the results are shown in Figure 3. The carboxymethyl cellulose gel obtained by the freeze-drying method has a porous microstructure, and it remains porous after crosslinking with Si-CDs, which does not lead to the collapse of the gel structure or the closure of the micropores, and the lamellae become more flat [16]. In order to investigate the crystal structures of CMC and DCG, XRD was performed, and the results are shown in Figure S4, where the peak around 22° corresponds to the (110) crystal plane of graphitized carbon. Overall, both CMC and DCG are amorphous structures [41].

Figure 3. SEM image of (**a**) freeze-dried CMC and (**b**) 5DCG.

2.3. Mechanical Properties and Thermal Stability Analysis of DCG

In order to investigate the mechanical properties of DCGs (1DCG, 2DCG, 3DCG, 4DCG, and 5DCG) obtained by crosslinking silicon-doped carbon dots with different molar ratios of APTES with carboxymethyl cellulose, the compressive stress-strain curves were determined by a universal mechanical testing machine, and the results are shown in Figure 4a. The calculated Young's modulus was shown in Table 1. With the increase in the proportion of APTES in Si-CDs, the elastic deformation of the gels due to the external force gradually becomes smaller, and they are able to withstand greater stress without significant deformation. The results of Young's modulus calculations are shown in the table. APTES enhances the content of amino groups on the surface of Si-CDs and increases the crosslinking point with carboxymethyl cellulose, and the Young's modulus of DCG gradually increases from 35.3233 to 294 kPa. High Young's modulus can better maintain the original structure in the process of application. Therefore, 5DCG has better stability in application due to its good mechanical properties [42].

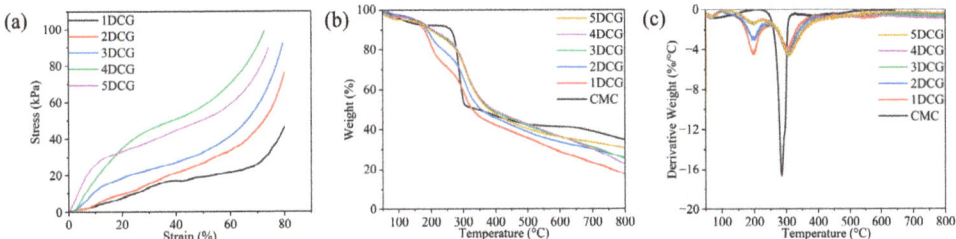

Figure 4. (a) Stress–strain curve of DCG. (b) TG and (c) DTG curves of CMC and DCG.

Table 1. Young's modulus of DCG.

Samples	1DCG	2DCG	3DCG	4DCG	5DCG
Young's modulus (kPa)	35.3233	36.7557	93.6412	156	294

In order to investigate the thermal stability and thermal decomposition of CMC and DCG, thermogravimetric analysis was carried out, and the results are shown in Figure 4b,c. The weight loss of CMC is divided into three stages: the first stage from 50 to 215 °C is caused by the evaporation of free and bound water adsorbed by CMC, and the weight loss is about 8%; the second stage from 315 to 330 °C is caused by the degradation and carbonization of the sugar chains of CMC; and the third stage from 330 to 800 °C is caused by the degradation and carbonization of the sugar chains of CMC and the sugar-containing chains of CMC, and the weight loss is about 42%. The weight loss in the second stage is caused by the degradation and carbonization of the sugar chains of CMC, and the weight loss in this stage is about 42%. The weight loss in the third stage from 330 to 800 °C is caused by the gradual and complete carbonization of the sugar chains and oxygen-containing groups of CMC, and the weight loss in this stage is about 7% [43]. The thermal decomposition behaviors of 1DCG, 2DCG, 3DCG, 4DCG, and 5DCG are similar, so the 5DCG is discussed in detail as an example. The weight loss of 5DCG is mainly divided into three stages. The weight loss in the first stage from 50 to 110 °C is caused by the successive evaporation of free and bound water physically adsorbed by the DCG. The weight loss in this stage is about 5%, which is lower than the percentage of weight loss of CMC. The reason for this change is the reduction of carboxyl groups after crosslinking CMC with Si-CDs, which reduces the percentage of bound water. The weight loss in the second stage from 110 to 230 °C is caused by the breakage of amide bonds at the crosslinking point, and the weight loss in this stage is about 8% [44]. The weight loss in the third stage from 230 to 800 °C is caused by the gradual and complete degradation and carbonization of the sugar chain and the oxygen-containing groups, and the weight loss in this stage is about 57% [39].

2.4. Analysis of the Fluorescence Properties of DCG

The fluorescence excitation spectra of 5DCG, 5Si-CDs, and CMC are shown in Figure 5a. The optimal excitation wavelength of 5DCG is approximately 360 nm; therefore, a 360 nm light source was used for the subsequent excitation of fluorescence emission spectra. The photographs of Si-CDs and SiQDs under daylight and 365 nm UV light are shown in Figure 5b. Under daylight, 1Si-CDs and 2Si-CDs powders appear yellowish-brown, while 3Si-CDs, 4Si-CDs, and 5Si-CDs powders appear yellow, and the SiQD powder appears pure white. The higher the proportion of CMCS, the deeper the color of the powder. Under 365 nm UV light, all Si-CDs exhibit visible blue fluorescence emission, and the higher the proportion of APTES, the higher the brightness of the blue fluorescence of Si-CD powder. To investigate the fluorescence properties of solid powders, fluorescence spectra under 360 nm excitation light were tested, and the results are shown in Figure 5d. The fluorescence intensity statistics of the emission peaks are shown in Figure 5e. The wavelength of the emission peak is approximately 445 nm, and its intensity order is consistent with visual observation. The main and top view photographs of the DCG and SiDCG in daylight and under UV light at 365 nm are shown in Figure 5c. Under daylight, 1DCG, 2DCG, 3DCG, 4DCG, and 5DCG gradually change from yellow to white. Under 365 nm UV light, the DCG all have bright blue fluorescence emission with increasing brightness, while the brightness of SiDCG is relatively dim. In order to investigate the fluorescence properties of DCG and SiDCG, the fluorescence spectra under 360 nm excitation light were tested, and the results are shown in Figure 5f. The fluorescence emission peak of DCG is around 445 nm. The statistics of fluorescence intensity at the emission peak are shown in Figure 5g, and the fluorescence intensities of 1DCG to 5DCG increase. Comparing the fluorescence spectra of DCG and CMC, the introduction of Si-CDs greatly enhances the fluorescence performance of CMC. The introduction of Si-CDs not only acts as a crosslinking agent but also brings excellent fluorescence performance. As the mechanical and fluorescence properties of 5DCG are optimal, 5DCG was used in the following study, referred to as DCG for brevity.

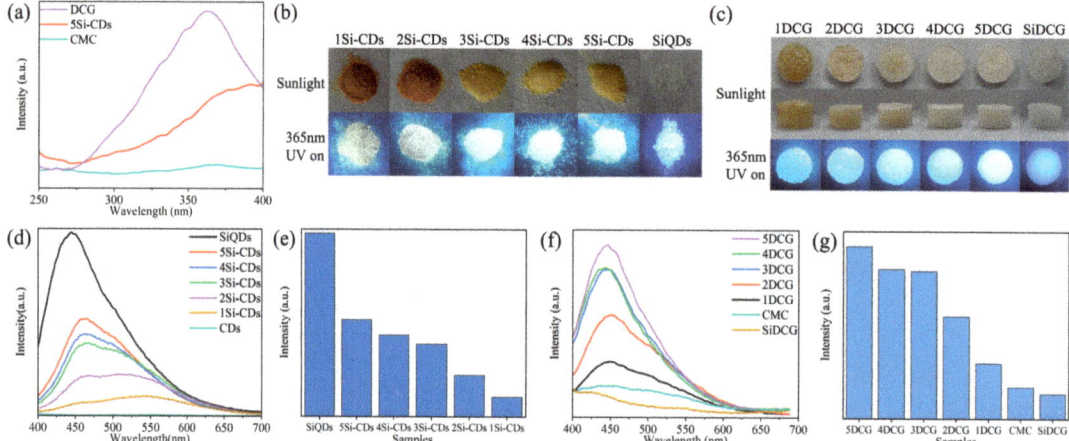

Figure 5. (a) The fluorescence excitation spectra of 5DCG, 5Si-CDs, and CMC (λ_{em} = 445 nm). (b) Photographs under daylight and 365 nm UV light, (d) fluorescence emission spectra under excitation light at 360 nm, and (e) histogram of fluorescence emission peak intensity of Si-CDs and SiQDs powders. (c) Photographs under daylight and 365 nm UV light, (f) histogram of fluorescence emission peak intensity, and (g) fluorescence emission spectra under excitation light at 360 nm of DCG and SiDCG.

2.5. DCG Detection of Fe^{3+} Concentration

To demonstrate the selectivity of DCG in detecting metal ions, the influence of several common metal cations on the fluorescence emission of DCG was explored. To eliminate the interference of anions in the experiment, chloride salts were used. As shown in Figure 6a, several ions (Ca^{2+}, Al^{3+}, Ba^{2+}, Zn^{2+}, Mg^{2+}, Ni^{2+}, Cr^{3+}, Mn^{2+}, Cu^{2+}, Co^{2+}, and Fe^{3+}) do not shift the fluorescence emission peak of DCG but only affect its intensity. As shown in Figure 6b, Ca^{2+}, Al^{3+}, Ba^{2+}, Zn^{2+}, and Mg^{2+} cause aggregation of Si-CDs, slightly enhancing the fluorescence emission intensity of DCG, while Ni^{2+}, Cr^{3+}, Mn^{2+}, Cu^{2+}, Co^{2+}, and Fe^{3+} quench the fluorescence emission of DCG. Among them, Fe^{3+} exhibits the most significant quenching effect [45,46], indicating that DCG could be used for specific detection of Fe^{3+}. As shown in Figure 6c, changes in the concentration of Fe^{3+} do not shift the fluorescence emission peak of DCG. Therefore, the intensity of the fluorescence emission peak of DCG can be used to represent the concentration of Fe^{3+}. As shown in Figure 6d, there is a good linear relationship between the concentration of Fe^{3+} and the change in fluorescence intensity of DCG in the range of 0–100 mg/L, with a linear correlation coefficient of

$R^2 = 0.999$. It highly corresponds to the single quenching mechanism of the Stern–Volmer equation (static quenching or dynamic quenching) [47]. The calculated detection limit is 0.6595 mg/L [48], which is much lower than China's wastewater quality standards for discharge to municipal sewers (GB/T 31962–2015) of 10.0 mg/L [49], thus making it capable of meeting general testing needs.

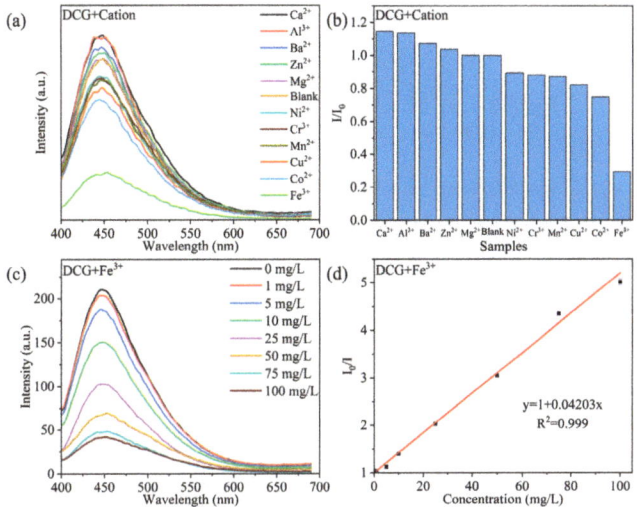

Figure 6. Ion screening and Fe^{3+} detection. (**a**) Effect of different metal cations on the fluorescence emission spectra of DCG (λ_{ex} = 360 nm). (**b**) Change of fluorescence emission peak intensity (I: fluorescence emission peak intensity after addition of metal cation, I_0: initial fluorescence intensity of DCG). (**c**) Fluorescence emission spectra of DCG after addition of different concentrations of Fe^{3+} (λ_{ex} = 360 nm). (**d**) Relationship between I_0/I and Fe^{3+} concentration and linear fitting.

2.6. Analysis of Fe^{3+} Adsorption Properties of DCG

Adsorption kinetics is an important physical quantity to study the adsorption rate. In order to study and analyze the adsorption kinetics of DCG and to determine the rate-controlling steps in the adsorption process, the pseudo-first-order kinetic model and the pseudo-second-order kinetic model were used in this study to simulate the adsorption of Fe^{3+} [44]. The pseudo-first-order kinetic and pseudo-second-order kinetic equations are shown in Equations (1) and (2), respectively. The fitting results are shown in Figure 7a,b, and the related parameters are shown in Table 2. The correlation coefficients (R^2) of the pseudo-first-order kinetic equation and pseudo-second-order kinetic equation for the adsorption of ferric ions by the Gs1el are 0.979 and 0.999, respectively. The pseudo-first-order kinetic adsorption rate constant (k_1) for the adsorption of Fe^{3+} by the DCG is 0.48252 h^{-1}, and the pseudo-second-order kinetic adsorption rate constant (k_2) is 0.01790 $g \cdot mg^{-1} \cdot h^{-1}$. The correlation coefficients show that the adsorption of Fe^{3+} by DCG is more in accordance with the pseudo-second-order kinetic equation, indicating that the adsorption of Fe^{3+} by DCG is mainly carried out by chemisorption, which is the main rate-controlling step in the adsorption process. In the 100 mg/L Fe^{3+} solution, the theoretical equilibrium adsorption capacity of DCG for Fe^{3+} according to the pseudo-second-order kinetic equation is 30.2633 mg/g.

$$\ln(q_e - q_t) = \ln q_e - k_1 t, \qquad (1)$$

$$\frac{t}{q_t} = \frac{1}{k_2 q_e^2} + \frac{t}{q_e}, \qquad (2)$$

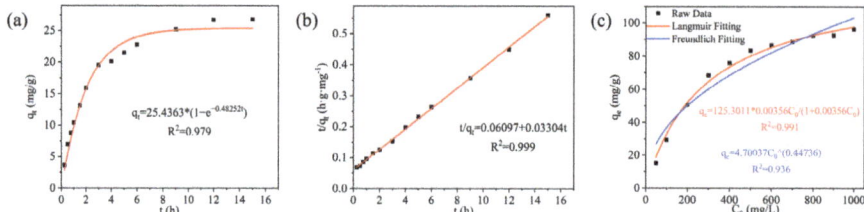

Figure 7. The performance of DCG in adsorbing Fe^{3+}. (**a**) Pseudo-first-order kinetic fitting, (**b**) pseudo-second-order kinetic fitting, and (**c**) adsorption isotherm fitting.

Table 2. Kinetic parameters of Fe^{3+} adsorption by DCG.

	Pseudo-First-Order Model			Pseudo-Second-Order Model		
	q_e, cal (mg/g)	k_1 (h^{-1})	R^2	q_e, cal (mg/g)	k_2 ($g \cdot mg^{-1} \cdot h^{-1}$)	R^2
Parameters Values	25.4363	0.48252	0.979	30.2633	0.01790	0.999

To investigate the interaction between DCG and Fe^{3+}, this study employed the Langmuir and Freundlich adsorption isotherm models to fit and analyze the experimental data for equilibrium adsorption isotherms. The Langmuir and Freundlich model equations are presented in Equations (3) and (4), respectively. The fitting results are shown in Figure 7c, and the related parameters are shown in Table 3. Compared to the fitting equation of the Freundlich model, the Langmuir model demonstrates a higher correlation coefficient (R^2) (greater than 0.99), indicating that the adsorption of iron ions from the solution by the gel aligns with the Langmuir model, which can be considered monolayer adsorption [50]. The theoretical equilibrium adsorption capacity of DCG for Fe^{3+}, according to the Langmuir model equation, is 125.3011 mg/g.

$$\frac{C_e}{q_e} = \frac{C_e}{q_e} + \frac{1}{q \times K_L}, \quad (3)$$

$$\ln q_e = \ln K_F + \frac{1}{n} \ln C_e, \quad (4)$$

Table 3. Isotherm parameters of Fe^{3+} adsorption by DCG.

	Langmuir Isotherm			Freundlich Isotherm		
	q_m, cal (mg/g)	k_L (L/mg)	R^2	k_F (L/mg)	n	R^2
Parameters Values	125.3011	0.00356	0.991	4.7004	2.2353	0.936

A comparison of the Fe^{3+} adsorption capacity and detection limits of the DCG with other cellulose-based materials is summarized in Table 4. DCG exhibits significant dual functionality, characterized by a high adsorption capacity of 125.3011 mg/L and a low detection limit for Fe^{3+} of 0.6595 mg/L. Its exceptional performance suggests a promising application in the adsorption and monitoring of Fe^{3+} in environmental pollution scenarios.

Table 4. Comparison of Fe^{3+} adsorption properties and detection limits between DCG and other materials.

Materials	q_{max} (mg/g)	LOD (mg/L)	Ref.
CMC-St/Al$_2$O$_3$	29.26	/	[51]
CMC-g-AMPS	33.65	/	[52]
FNH	98.3	62.5	[53]
CNC-g-PCysMA	60.0	/	[54]
TO-CNF	70.0	/	[54]
CP3	0.475	0.0269	[55]
DCG	125.30	0.6595	This study

3. Conclusions

In summary, a gel DCG with a three-dimensional network structure was successfully synthesized by crosslinking Si-CDs with CMC. The crosslinking structure of the gel DCG brought about by Si-CDs exhibits excellent mechanical properties, with the compressive strength increasing as the APTES doping in Si-CDs increases, reaching a maximum of 294 kPa. Si-CDs confers excellent fluorescence properties on the gel, enabling specific responsiveness to Fe^{3+}. DCG demonstrates a superior responsivity towards Fe^{3+} compared to other metal ions, thus possessing the potential for application in the detection of authentic Fe^{3+}-containing wastewater samples with complex metal ion compositions. A good linear relationship was observed within the concentration range of 0–100 mg/L, with a detection limit of 0.6595 mg/L. It highly corresponds to the single quenching mechanism of the Stern–Volmer equation. Additionally, the adsorption of Fe^{3+} by DCG is a chemical and monolayer adsorption process, with a calculated theoretical maximum adsorption capacity of 125.30 mg/g. Therefore, DCG exhibits potential applications in the adsorption and detection of Fe^{3+}.

4. Materials and Methods

4.1. Materials

Carboxymethyl chitosan (CMCS, MW: 100,000–200,000, DS ≥ 80%), carboxymethyl cellulose (CMC, MW = 250,000, DS = 1.2, μ = 400–800 mPa·s), (3-Aminopropyl)triethoxysilane (APTES, purity 98.0%), N-Hydroxysuccinimide (NHS, purity 98.0%), and zinc chloride (ZnCl$_2$) were purchased from Shanghai Maclean Biochemical Technology Co. (Shanghai, China). N-(3-Dimethylaminopropyl)-N′-ethylcarbodiimide hydrochloride (EDC, purity 98.0%) was purchased from Aladdin Reagent Co. (Shanghai, China). Ferric chloride (FeCl$_3$) was purchased from Tianjin Hengxing Chemical Preparation Co. (Tianjin, China). Calcium chloride (CaCl$_2$) was purchased from Tianjin Reference Chemical Reagent Co. (Tianjin, China). Barium chloride (BaCl$_2$) was purchased from the Tianjin Dongli district Tianda chemical reagent factory (Tianjin, China). Magnesium chloride (MgCl$_2$) was purchased from Tianjin Ruijinte Chemical Co. (Tianjin, China). Manganese chloride (MnCl$_2$) was purchased from Tianjin Xinbote Chemical Co. (Tianjin, China). Copper chloride (CuCl$_2$), cobalt chloride (CoCl$_2$), nickel chloride (NiCl$_2$), aluminum chloride (AlCl$_3$), and chromic chloride (CrCl$_3$) were purchased from Tianjin Fuchen Chemical Reagent Co. (Tianjin, China). Anhydrous ethanol and 5-sulfosalicylic acid were purchased from Tianjin Tianli Chemical Reagent Co. (Tianjin, China). These chemicals were analytically pure and were not further purified prior to use. Deionized water produced by Clever-Q30 UT (Shanghai Kehuai Instruments Co., Shanghai, China) was used in this study.

4.2. Preparation of the Si-CDs and SiQDs

First, 0.5 g of CMCS powder was dissolved in 50 mL of deionized water, and APTES (1:1, 2:1, 3:1, 4:1, and 5:1 molar ratio to CMCS structural units) was added and stirred at room temperature for 10 min. Then, the reaction was transferred to a 100 mL PTFE-lined stainless-steel autoclave and placed in a blower oven at a constant temperature of

180 °C for 6 h [29]. After being cooled to room temperature, the reaction was filtered using a 0.22 μm micropore membrane and then dialyzed with a 1000 Da dialysis bag for 3 days. The dialysate was concentrated by a rotary evaporator and freeze-dried to obtain a yellow powder. 1Si-CDs, 2Si-CDs, 3Si-CDs, 4Si-CDs, and 5Si-CDs with blue fluorescence emission are collectively referred to here as Si-CDs. Next, 3 g of APTES was dissolved in 50 mL of deionized water, and SiQDs [24] were prepared by the same method for performance comparison.

4.3. Preparation of the DCG and SiDCG

First, 0.1 g of CMC powder was dissolved in 7 mL of deionized water, followed by the addition of 1 mL of a 50 mg/mL aqueous solution of 1Si-CDs, 2Si-CDs, 3Si-CDs, 4Si-CDs, and 5Si-CDs. This was followed by stirring for 10 min and then sonication for 10 min to ensure Si-CDs were uniformly dispersed between the CMC chains. Then, 1 mL of freshly prepared 0.5 mmol/L EDC aqueous solution was added under stirring, and 1 mL of freshly prepared 0.5 mmol/L NHS aqueous solution was added after stirring for another 5 min. After 24 h of reaction at room temperature, the unreacted material was removed by dialysis in a 50% ethanol aqueous solution, and freeze-drying yielded yellow to white 1DCG, 2DCG, 3DCG, 4DCG, and 5DCG with blue fluorescence emission. They are collectively referred to here as DCG. The gel obtained by crosslinking SiQDs with CMC is referred to as SiDCG.

4.4. Structure Characterizations

Fourier transform infrared spectroscopy (FTIR) images were obtained using the PerkinElmer Frontier spectrometer, produced by PerkinElmer, Waltham, MA, USA, with a measurement range of 4000–550 and a resolution of 2 cm^{-1}. The UV–visible diffuse reflection spectra and UV–visible absorption spectra were measured using the UV–visible spectrophotometer TU-1950, produced by PERSEE, Beijing, China. X-ray photoelectron spectroscopy (XPS) images were obtained using K-alpha, an X-ray photoelectron spectrometer produced by Thermo Fisher Scientific, Waltham, MA, USA. Transmission electron microscope (TEM) images were obtained with the transmission electron microscope JEM-2100, produced by Japan Electronics Co., Ltd, Tokyo Metropolis, Japan. Fluorescence spectra were recorded with the LS-55, a fluorescence spectrophotometer produced by PerkinElmer, USA. Thermogravimetric analysis (TGA) was carried out using the STA 6000-SQ8 analyzer, produced by PerkinElmer in the USA, heated from 40 to 800 °C in a nitrogen atmosphere at a rate of 10 °C/min. X-ray diffraction (XRD) patterns were measured using the X-ray diffractometer X'Pert^3 Powder, produced by PANalytical, Almelo, Netherlands. The surface morphology of the gel was observed using a TM3030 scanning electron microscope manufactured by Hitachi, Tokyo, Japan, with a test accelerating voltage of 5 kV. The mechanical properties were obtained by the universal mechanical testing machine INSTRON5942, produced by INSTRON, Boston, MA, USA.

4.5. Ion Screening and Fe^{3+} Detection Experiments

Ion screening: An aqueous solution of 1 mol/L metal salt was configured, and 0.5 mL was taken and added dropwise to the surface of DCG. Then, the fluorescence emission spectrum of DCG was measured by a fluorescence spectrophotometer.

Fe^{3+} detection: An aqueous solution of 1–100 mg/L $FeCl_3$ was configured, and 0.5 mL was taken and added dropwise to the surface of DCG. Then, the fluorescence emission spectrum of DCG was measured by a fluorescence spectrophotometer.

4.6. Adsorption Performance Experiment

Adsorption kinetics: 0.1 g of DCG was immersed in 50 mL of $FeCl_3$ aqueous solution (100 mg/L) and placed in a water bath thermostatic oscillator at 25 °C. The amount of adsorption was analyzed by a UV–visible spectrophotometer using sulfosalicylic acid as the chromogenic agent [56]. The current Fe^{3+} adsorption capacity (q_t) of DCG was calculated using Equation (5):

$$q_t = \frac{(C_0 - C_t)V}{m}, \tag{5}$$

Adsorption thermodynamics: 0.1 g of DCG was immersed in 50 mL of $FeCl_3$ aqueous solution (100 mg/L) and placed in a constant-temperature water bath shaker at 25 °C. The adsorption amount was analyzed by the same method as above after 72 h. The approximate equilibrium adsorption capacity (q_e) was calculated using Equation (6):

$$q_e = \frac{(C_0 - C_e)V}{m}, \tag{6}$$

Supplementary Materials: The following supporting information can be downloaded at https://www.mdpi.com/article/10.3390/gels10050285/s1, Figure S1: XPS spectrum of CMCS. Figure S2: Histograms of particle size distribution, TEM images, and HRTEM images of 1Si-CDs and 3Si-CDs. Figure S3: XPS spectra of CMC [57]. Figure S4: XRD spectra of CMC and DCG. Figure S5: Fluorescence emission lifetime decay diagram and fitting curve of Si-CDs [58]. Table S1–S4: Peak fitting data of XPS for 5Si-CDs, CMCS, 5DCG, and CMC.

Author Contributions: Conceptualization, Z.Z. and Y.J.; writing—original draft preparation, Z.Z. and Y.J.; validation, Y.S.; writing—review and editing, D.W., Z.Z. and Y.J.; supervision, D.W., Y.L., J.W., M.M., J.P., C.W. and J.L.; project administration, D.W.; funding acquisition, D.W. All authors have read and agreed to the published version of the manuscript.

Funding: The research was supported by financial support from the National Natural Science Foundation of China (GN: 31770593) and the Fundamental Research Funds for the Central Universities (GN: 2572023AW50).

Data Availability Statement: Data are contained within the article.

Acknowledgments: The authors would also like to acknowledge the technical support from the Analysis and Testing Center of Northeast Forestry University.

Conflicts of Interest: The authors declare no conflicts of interest.

References

1. Cheng, Z.H.; Fu, F.L.; Dionysiou, D.D.; Tang, B. Adsorption, oxidation, and reduction behavior of arsenic in the removal of aqueous As(III) by mesoporous Fe/Al bimetallic particles. *Water Res.* **2016**, *96*, 22–31. [CrossRef]
2. Atchudan, R.; Edison, T.; Aseer, K.R.; Perumal, S.; Karthik, N.; Lee, Y.R. Highly fluorescent nitrogen-doped carbon dots derived from *Phyllanthus acidus* utilized as a fluorescent probe for label-free selective detection of Fe^{3+} ions, live cell imaging and fluorescent ink. *Biosens. Bioelectron.* **2018**, *99*, 303–311. [CrossRef]
3. Zhang, X.; Zhang, H.; Wang, B.; Zeng, X.; Wang, J.; Ren, B.; Yang, X.; Bai, X. Preparation of non-swelling hydrogels and investigation on the adsorption performance of iron ions. *J. Appl. Polym. Sci.* **2022**, *139*, e52411. [CrossRef]
4. Sun, Y.C.; Yu, F.X.; Han, C.H.; Houda, C.; Hao, M.G.; Wang, Q.Y. Research progress on adsorption of arsenic from water by modified biochar and its mechanism: A review. *Water* **2022**, *14*, 1691. [CrossRef]
5. Burratti, L.; Ciotta, E.; De Matteis, F.; Prosposito, P. Metal nanostructures for environmental pollutant detection based on fluorescence. *Nanomaterials* **2021**, *11*, 276. [CrossRef]
6. Jamshaid, A.; Hamid, A.; Muhammad, N.; Naseer, A.; Ghauri, M.; Iqbal, J.; Rafiq, S.; Shah, N.S. Cellulose-based materials for the removal of heavy metals from wastewater—An overview. *ChemBioEng Rev.* **2017**, *4*, 240–256. [CrossRef]
7. Singh, A.K.; Itkor, P.; Lee, Y.S. State-of-the-art insights and potential applications of cellulose-based hydrogels in food packaging: Advances towards sustainable trends. *Gels* **2023**, *9*, 433. [CrossRef]
8. Yuan, Y.W.; Li, R.Y.; Peng, S.J. Research progress on chemical modification of waste biomass cellulose to prepare heavy metal adsorbents. *Polym. Bull.* **2023**, *80*, 11671–11700. [CrossRef]
9. Nakayama, R.; Yano, T.; Namiki, N.; Imai, M. Highly size-selective water-insoluble cross-linked carboxymethyl cellulose membranes. *J. Polym. Environ.* **2019**, *27*, 2439–2444. [CrossRef]
10. Gul, B.Y.; Pekgenc, E.; Vatanpour, V.; Koyuncu, I. A review of cellulose-based derivatives polymers in fabrication of gas separation membranes: Recent developments and challenges. *Carbohydr. Polym.* **2023**, *321*, 18. [CrossRef]
11. Xiao, Z.H.; Liu, Y.; Yang, J.S.; Jiang, H.; Tang, L.Q.; Chen, H.; Sun, T.L. Rate-dependent fracture behavior of tough polyelectrolyte complex hydrogels from biopolymers. *Mech. Mater.* **2021**, *156*, 8. [CrossRef]

12. Nath, P.C.; Debnath, S.; Sharma, M.; Sridhar, K.; Nayak, P.K.; Inbaraj, B.S. Recent advances in cellulose-based hydrogels: Food applications. *Foods* **2023**, *12*, 350. [CrossRef]
13. Wang, P.; Zheng, D.; Liu, S.; Luo, M.; Li, J.; Shen, S.; Li, S.; Zhu, L.; Chen, Z. Producing long afterglow by cellulose confinement effect: A wood-inspired design for sustainable phosphorescent materials. *Carbon* **2021**, *171*, 946–952. [CrossRef]
14. Mokhtar, O.M.; Attia, Y.A.; Wassel, A.R.; Khattab, T.A. Production of photochromic nanocomposite film via spray-coating of rare-earth strontium aluminate for anti-counterfeit applications. *Luminescence* **2021**, *36*, 1933–1944. [CrossRef]
15. Liu, J.M.; Huang, X.M.; Zhang, L.H.; Zheng, Z.Y.; Lin, X.; Zhang, X.Y.; Jiao, L.; Cui, M.L.; Jiang, S.L.; Lin, S.Q. A specific tween-80-rhodamine S-MWNTs phosphorescent reagent for the detection of trace calcitonin. *Anal. Chim. Acta* **2012**, *744*, 60–67. [CrossRef]
16. Wu, B.; Zhu, G.; Dufresne, A.; Lin, N. Fluorescent aerogels based on chemical crosslinking between nanocellulose and carbon dots for optical sensor. *ACS Appl. Mater. Interfaces* **2019**, *11*, 16048–16058. [CrossRef]
17. Chen, X.; Song, Z.; Li, S.; Thang, N.T.; Gao, X.; Gong, X.; Guo, M. Facile one-pot synthesis of self-assembled nitrogen-doped carbon dots/cellulose nanofibril hydrogel with enhanced fluorescence and mechanical properties. *Green Chem.* **2020**, *22*, 3296–3308. [CrossRef]
18. Lv, H.; Wang, S.; Wang, Z.; Meng, W.; Han, X.; Pu, J. Fluorescent cellulose-based hydrogel with carboxymethyl cellulose and carbon quantum dots for information storage and fluorescent anti-counterfeiting. *Cellulose* **2022**, *29*, 6193–6204. [CrossRef]
19. Wang, Y.J.; Lv, T.J.; Yin, K.Y.; Feng, N.; Sun, X.F.; Zhou, J.; Li, H.G. Carbon dot-based hydrogels: Preparations, properties, and applications. *Small* **2023**, *19*, 28. [CrossRef]
20. Desai, M.L.; Basu, H.; Saha, S.; Singhal, R.K.; Kailasa, S.K. Investigation of silicon doping into carbon dots for improved fluorescence properties for selective detection of Fe^{3+} ion. *Opt. Mater.* **2019**, *96*, 109374. [CrossRef]
21. Borsagli, F.G.; Borsagli, A. Chemically modified chitosan bio-sorbents for the competitive complexation of heavy metals ions: A potential model for the treatment of wastewaters and industrial spills. *J. Polym. Environ.* **2019**, *27*, 1542–1556. [CrossRef]
22. Li, Y.; Zhang, Z.; Fu, Z.; Wang, D.; Wang, C.; Li, J. Fluorescence response mechanism of green synthetic carboxymethyl chitosan-Eu^{3+} aerogel to acidic gases. *Int. J. Biol. Macromol.* **2021**, *192*, 1185–1195. [CrossRef] [PubMed]
23. Wu, Y.; Zhao, L.; Cao, X.; Zhang, Y.; Jiang, X.; Sun, Z.; Zhan, Y. Bright and multicolor emissive carbon dots/organosilicon composite for highly efficient tandem luminescent solar concentrators. *Carbon* **2023**, *207*, 77–85. [CrossRef]
24. Zhang, H.; Wang, H.; Yang, H.; Zhou, D.; Xia, Q. Luminescent, protein-binding and imaging properties of hyper-stable water-soluble silicon quantum dots. *J. Mol. Liq.* **2021**, *331*, 115769. [CrossRef]
25. Ye, H.L.; Shang, Y.; Wang, H.Y.; Ma, Y.L.; He, X.W.; Li, W.Y.; Li, Y.H.; Zhang, Y.K. Determination of Fe(III) ion and cellular bioimaging based on a novel photoluminescent silicon nanoparticles. *Talanta* **2021**, *230*, 122294. [CrossRef] [PubMed]
26. Gao, G.; Jiang, Y.-W.; Jia, H.-R.; Yang, J.; Wu, F.-G. On-off-on fluorescent nanosensor for Fe^{3+} detection and cancer/normal cell differentiation via silicon-doped carbon quantum dots. *Carbon* **2018**, *134*, 232–243. [CrossRef]
27. Zhou, J.; Zhao, R.; Liu, S.; Feng, L.; Li, W.; He, F.; Gai, S.; Yang, P. Europium doped silicon quantum dot as a novel FRET based dual detection probe: Sensitive detection of tetracycline, zinc, and cadmium. *Small Methods* **2021**, *5*, e2100812. [CrossRef]
28. Wang, X.; Yang, Y.; Huo, D.; Ji, Z.; Ma, Y.; Yang, M.; Luo, H.; Luo, X.; Hou, C.; Lv, J. A turn-on fluorescent nanoprobe based on N-doped silicon quantum dots for rapid determination of glyphosate. *Microchim. Acta* **2020**, *187*, 341. [CrossRef]
29. Cao, Q.; Luo, Y.X.; Liu, W.P.; Li, Y.S.; Gao, X.F. Enzyme-free fluorescence determination of uric acid and trace Hg(II) in serum using Si/N doped carbon dots. *Spectrochim. Acta Part A Mol. Biomol. Spectrosc.* **2021**, *263*, 120182. [CrossRef]
30. Wang, X.; Liu, Y.; Wang, Q.; Bu, T.; Sun, X.; Jia, P.; Wang, L. Nitrogen, silicon co-doped carbon dots as the fluorescence nanoprobe for trace p-nitrophenol detection based on inner filter effect. *Spectrochim. Acta Part A Mol. Biomol. Spectrosc.* **2021**, *244*, 118876. [CrossRef]
31. Li, S.; Li, L.; Tu, H.; Zhang, H.; Silvester, D.S.; Banks, C.E.; Zou, G.; Hou, H.; Ji, X. The development of carbon dots: From the perspective of materials chemistry. *Mater. Today* **2021**, *51*, 188–207. [CrossRef]
32. Hu, G.; Wang, Y.; Zhang, S.; Ding, H.; Zhou, Z.; Wei, J.; Li, X.; Xiong, H. Rational synthesis of silane-functionalized carbon dots with high-efficiency full-color solid-state fluorescence for light emitting diodes. *Carbon* **2023**, *203*, 1–10. [CrossRef]
33. Villalba-Rodríguez, A.M.; González-González, R.B.; Martínez-Ruiz, M.; Flores-Contreras, E.A.; Cárdenas-Alcaide, M.F.; Iqbal, H.M.N.; Parra-Saldívar, R. Chitosan-based carbon dots with applied aspects: New frontiers of international interest in a material of marine origin. *Mar. Drugs* **2022**, *20*, 782. [CrossRef]
34. Salehi, B.; Zhang, B.; Nowlin, K.; Wang, L.; Shahbazi, A. A multifunctional cellulose- and starch-based composite hydrogel with iron-modified biochar particles for enhancing microalgae growth. *Carbohydr. Polym.* **2024**, *327*, 121657. [CrossRef]
35. Zengin Kurt, B.; Uckaya, F.; Durmus, Z. Chitosan and carboxymethyl cellulose based magnetic nanocomposites for application of peroxidase purification. *Int. J. Biol. Macromol.* **2017**, *96*, 149–160. [CrossRef]
36. Wu, L.; Lin, X.; Zhou, X.; Luo, X. Removal of uranium and fluorine from wastewater by double-functional microsphere adsorbent of SA/CMC loaded with calcium and aluminum. *Appl. Surf. Sci.* **2016**, *384*, 466–479. [CrossRef]
37. Sun, X.; Liu, C.; Omer, A.M.; Lu, W.; Zhang, S.; Jiang, X.; Wu, H.; Yu, D.; Ouyang, X.K. pH-sensitive ZnO/carboxymethyl cellulose/chitosan bio-nanocomposite beads for colon-specific release of 5-fluorouracil. *Int. J. Biol. Macromol.* **2019**, *128*, 468–479. [CrossRef]
38. Zeng, M.; Li, T.; Liu, Y.; Lin, X.; Zu, X.; Mu, Y.; Chen, L.; Huo, Y.; Qin, Y. Cellulose-based photo-enhanced persistent room-temperature phosphorescent materials by space stacking effects. *Chem. Eng. J.* **2022**, *446*, 136935. [CrossRef]

39. Wang, J.; Du, P.; Hsu, Y.-I.; Uyama, H. Cellulose luminescent hydrogels loaded with stable carbon dots for duplicable information encryption and anti-counterfeiting. *ACS Sustain. Chem. Eng.* **2023**, *11*, 10061–10073. [CrossRef]
40. Xiao, S.J.; Wieland, M.; Brunner, S. Surface reactions of 4-aminothiophenol with heterobifunctional crosslinkers bearing both succinimidyl ester and maleimide for biomolecular immobilization. *J. Colloid Interface Sci.* **2005**, *290*, 172–183. [CrossRef]
41. Mogharbel, A.T.; Hameed, A.; Sayqal, A.A.; Katouah, H.A.; Al-Qahtani, S.D.; Saad, F.A.; El-Metwaly, N.M. Preparation of carbon dots-embedded fluorescent carboxymethyl cellulose hydrogel for anticounterfeiting applications. *Int. J. Biol. Macromol.* **2023**, *238*, 124028. [CrossRef] [PubMed]
42. Ball, V. Crosslinking of bovine gelatin gels by genipin revisited using ferrule-top micro-indentation. *Gels* **2023**, *9*, 149. [CrossRef] [PubMed]
43. Pettignano, A.; Charlot, A.; Fleury, E. Solvent-free synthesis of amidated carboxymethyl cellulose derivatives: Effect on the thermal properties. *Polymers* **2019**, *11*, 1227. [CrossRef] [PubMed]
44. Wang, W.; Yu, F.; Ba, Z.; Qian, H.; Zhao, S.; Liu, J.; Jiang, W.; Li, J.; Liang, D. In-depth sulfhydryl-modified cellulose fibers for efficient and rapid adsorption of Cr(VI). *Polymers* **2022**, *14*, 1482. [CrossRef] [PubMed]
45. Han, S.; Ni, J.; Han, Y.; Ge, M.; Zhang, C.; Jiang, G.; Peng, Z.; Cao, J.; Li, S. Biomass-based polymer nanoparticles with aggregation-induced fluorescence emission for cell imaging and detection of Fe^{3+} ions. *Front. Chem.* **2020**, *8*, 563. [CrossRef] [PubMed]
46. Zhou, J.; Ge, M.; Han, Y.; Ni, J.; Huang, X.; Han, S.; Peng, Z.; Li, Y.; Li, S. Preparation of biomass-based carbon dots with aggregation luminescence enhancement from hydrogenated rosin for biological imaging and detection of Fe^{3+}. *ACS Omega* **2020**, *5*, 11842–11848. [CrossRef]
47. Liu, M.-L.; Chen, B.-B.; Li, C.-M.; Huang, C.-Z. Carbon dots prepared for fluorescence and chemiluminescence sensing. *Sci. China Chem.* **2019**, *62*, 968–981. [CrossRef]
48. Chanmungkalakul, S.; Ervithayasuporn, V.; Hanprasit, S.; Masik, M.; Prigyai, N.; Kiatkamjornwong, S. Silsesquioxane cages as fluoride sensors. *Chem. Commun.* **2017**, *53*, 12108–12111. [CrossRef] [PubMed]
49. He, J.J.; Wan, Y.; Zhou, W.J. ZIF-8 derived Fe-N coordination moieties anchored carbon nanocubes for efficient peroxymonosulfate activation via non-radical pathways: Role of FeN_x sites. *J. Hazard. Mater.* **2021**, *405*, 15. [CrossRef]
50. Li, J.; Zuo, K.M.; Wu, W.B.; Xu, Z.Y.; Yi, Y.G.; Jing, Y.; Dai, H.Q.; Fang, G.G. Shape memory aerogels from nanocellulose and polyethyleneimine as a novel adsorbent for removal of Cu(II) and Pb(II). *Carbohydr. Polym.* **2018**, *196*, 376–384. [CrossRef]
51. Murad, G.A.; Dakroury, G.A.; Elgoud, E.M.A. Exploiting carboxymethyl cellulose-starch/alumina nano gel to eliminate Fe(III) from ore leachates of rare earth elements. *Cellulose* **2024**, *31*, 969–992. [CrossRef]
52. Panchan, N.; Niamnuy, C.; Dittanet, P.; Devahastin, S. Optimization of synthesis condition for carboxymethyl cellulose-based hydrogel from rice straw by microwave-assisted method and its application in heavy metal ions removal. *J. Chem. Technol. Biotechnol.* **2018**, *93*, 413–425. [CrossRef]
53. Yang, J.C.; Luo, Z.X.; Wang, M. Novel fluorescent nanocellulose hydrogel based on nanocellulose and carbon dots for detection and removal of heavy metal ions in water. *Foods* **2022**, *11*, 1619. [CrossRef] [PubMed]
54. Georgouvelas, D.; Abdelhamid, H.N.; Li, J.; Edlund, U.; Mathew, A.P. All-cellulose functional membranes for water treatment: Adsorption of metal ions and catalytic decolorization of dyes. *Carbohydr. Polym.* **2021**, *264*, 118044. [CrossRef] [PubMed]
55. Ma, Y.Q.; Cheng, X.J. Readily soluble cellulose-based fluorescent probes for the detection and removal of Fe^{3+} ion. *Int. J. Biol. Macromol.* **2023**, *253*, 13. [CrossRef] [PubMed]
56. Ogawa, K.Y.; Tobe, N. A spectrophotometric study of the complex formation between Iron(III) and sulfosalicylic acid. *Bull. Chem. Soc. Jpn.* **2006**, *39*, 223–227. [CrossRef]
57. Ye, J.; Wang, B.; Xiong, J.; Sun, R. Enhanced fluorescence and structural characteristics of carboxymethyl cellulose/Eu(III) nano-complex: Influence of reaction time. *Carbohydr. Polym.* **2016**, *135*, 57–63. [CrossRef]
58. Röding, M.; Bradley, S.J.; Nydén, M.; Nann, T. Fluorescence lifetime analysis of graphene quantum dots. *J. Phys. Chem. C* **2014**, *118*, 30282–30290. [CrossRef]

Disclaimer/Publisher's Note: The statements, opinions and data contained in all publications are solely those of the individual author(s) and contributor(s) and not of MDPI and/or the editor(s). MDPI and/or the editor(s) disclaim responsibility for any injury to people or property resulting from any ideas, methods, instructions or products referred to in the content.

Article

Polyhedral Oligomeric Sesquioxane Cross-Linked Chitosan-Based Multi-Effective Aerogel Preparation and Its Water-Driven Recovery Mechanism

Yang Liu [1,2,†], Mingjian Ma [1,2,†], Yuan Shen [1,2], Zhengdong Zhao [1,2], Xuefei Wang [1,2], Jiaqi Wang [1,2], Jiangbo Pan [1,2], Di Wang [1,2,*], Chengyu Wang [1,2] and Jian Li [1,2]

1. Key Laboratory of Bio-Based Material Science and Technology, Ministry of Education, Northeast Forestry University, Harbin 150040, China; hljqaliuyang@163.com (Y.L.); 18847042762@163.com (M.M.); jlaushawn@yeah.net (Y.S.); zhaozd6866@163.com (Z.Z.); fei1807352093@163.com (X.W.); w1531565456@163.com (J.W.); pjb0725@163.com (J.P.); wangcy@nefu.edu.cn (C.W.); nefujianli@163.com (J.L.)
2. College of Material Science and Engineering, Northeast Forestry University, Harbin 150040, China
* Correspondence: diwang1030@nefu.edu.cn
† These authors contributed equally to the work.

Abstract: The use of environmentally friendly and non-toxic biomass-based interfacial solar water evaporators has been widely reported as a method for water purification in recent years. However, the poor stability of the water transport layer made from biomass materials and its susceptibility to deformation when exposed to harsh environments limit its practical application. To address this issue, water-driven recovery aerogel (PCS) was prepared by cross-linking epoxy-based polyhedral oligomeric silsesquioxane (EP-POSS) epoxy groups with chitosan (CS) amino groups. The results demonstrate that PCS exhibits excellent water-driven recovery performance, regaining its original volume within a very short time (1.9 s) after strong compression ($\varepsilon > 80\%$). Moreover, PCS has a water absorption rate of 2.67 mm s^{-1} and exhibits an excellent water absorption capacity of 22.09 g g^{-1} even after ten cycles of absorption-removal. Furthermore, a photothermal evaporator (PCH) was prepared by loading the top layer with hydrothermally reacted tannins (HAs) and Zn^{2+} complexes. The results indicate that PCH achieves an impressive evaporation rate of 1.89 kg m^{-2} h^{-1} under one sun illumination. Additionally, due to the antimicrobial properties of Zn^{2+}, PCH shows inhibitory effects against Staphylococcus aureus and Escherichia coli, thereby extending the application of solar water evaporators to include antimicrobial purification in natural waters.

Keywords: chitosan-based aerogel; EP-POSS; water-driven recovery; water purification

1. Introduction

In recent years, the scarcity of freshwater resources has emerged as a global problem that cannot be ignored [1]. Conventional water purification techniques like atmospheric water harvesting [2] and seawater desalination [3] not only consume substantial energy but also inflict significant damage on the surrounding environment [4]. In comparison to traditional solar evaporation technology, interfacial solar evaporation technology constructs a Janus structure comprising a photothermic conversion layer and a water transport layer using suitable materials [5]. That can improve the evaporation rate and solar energy conversion efficiency by enhancing the light absorption capacity, reducing the thermal conductivity, and optimizing the transport of the water body and so on [6–8], which has become the most widely used new solar evaporation technology at present [9].

The primary function of the water transport layer is to timely convey bulk water to the evaporator surface while minimizing heat loss and augmenting surface warming effects. Consequently, it significantly influences the evaporation rate of interfacial solar evaporators. Currently, three-dimensional porous structures such as foam [10] and aerogel [11] are predominantly employed for constructing the water transport layer. Renewable biomass materials are extensively utilized in fabricating the water transport layer due to their environmentally friendly nature. As one of the most abundant biomass materials, chitosan (CS) is rich in hydroxyl and amino groups, which can provide a large number of reaction sites [12], and is widely used in the preparation of biomass aerogels. CS-based materials are currently prevalent in various fields including wastewater treatment [13], antimicrobial activities [14,15], food preservation [16], sound absorption [17], reinforcement filler [18,19], drug delivery [20], oil spill cleanup [21,22], healing wounds [23,24], etc. However, due to the poor mechanical properties and unstable structure of biomass aerogels [25], CS-based aerogels have disadvantages. Therefore, in practical applications, the water transport layer made of CS materials are susceptible to external interference, such as extrusion from external forces leading to evaporator deformation and a decreased water transport rate, which affects the evaporation rate. At present, only a few water-driven recovery aerogels have been developed using CS as a raw material due to the aforementioned shortcomings of CS aerogels. Therefore, it is necessary to modify or cross-link CS materials with other components to improve their limitations.

Currently, CS as a base aerogel is usually cross-linked and modified with glutaraldehyde [26], epichlorohydrin [27], and other organic cross-linking agents [28], as they react with the amino groups of the CS to improve the stability and mechanical properties of the aerogel. However, most organic cross-linking agents cause significant environmental pollution; therefore, their application in solar water evaporators may lead to secondary pollution concerns. Therefore, known for its excellent biocompatibility and non-toxicity [29], polyhedral oligomeric sesquioxane (POSS) is a promising choice. POSS is considered the smallest silica particle available [30] that can effectively reduce stress concentration within matrix materials through its unique nano-size effect while absorbing energy and providing good mechanical properties. POSS material combines the reactive properties of organic materials with the excellent physical properties of inorganic materials at a molecular level [31], which can enhance the various aspects of the materials such as heat resistance [32], flame retardancy [33], and mechanical properties [34]. In addition, the water-driven recovery of the aerogel can be achieved by the formation of hydrogen bonds through the combination of water with Si-OH in the irregular POSS. By incorporating POSS into CS-based materials, we can overcome the limitations associated with CS-based materials by leveraging the unique properties offered by POSS and endow CS-based aerogels with water-driven recovery capabilities. Through the function of water-driven recovery, it realizes the rapid recovery of the CS water transport layer after receiving the extrusion of external force, and solves the problem of the practical use of the CS water transport layer in harsh environments.

The main function of the photothermal layer, as another important component of the solar evaporator, is to absorb solar energy and convert it into the heat energy required for water evaporation. Compared to other photothermal materials, inexpensive, simple to prepare, and photothermal-convertible tannic acid–metal complexes offer a favorable option. Tannin, a natural plant polyphenol widely found in plant barks, roots, and leaves [35] contains numerous phenolic hydroxyl groups [36]. These groups enable tannin to form coordination reactions with metal ions [37], resulting in tannic metal complexes that exhibit excellent photothermic properties by absorbing sunlight energy through the conjugated structure of the tannin and transferring it to metal ions. This promotes electronic resonance and converts optical energy into heat energy [38]. And, the simple hydrothermal treatment of tannins (HAs) can increase its light absorption and improve its energy absorption capacity. However, the presence of metal ions in water pollution cannot be ignored; therefore, according to guidelines from the World Health Organization and drinking water standards issued by countries worldwide [39], Zn^{2+} has a maximum allowable concentration far exceeding that of Fe^{3+}, making it suitable for coordination with tannin to generate metal complexes for efficient photothermic conversion. Additionally, the Zn^{2+} demonstrates potent inhibitory effects against both Escherichia coli and Staphylococcus aureus, making it a popular and low-cost antimicrobial agent. The use of HA-Zn^{2+} complexes as photothermic materials enable solar evaporators to perform various functions, such as antimicrobial activity and water evaporation in natural waters.

The present study aims to enhance the mechanical properties and stability of CS-based solar water evaporators for improved adaptability in harsh environments. To achieve this, triethoxy (3-epoxypropyl oxypropyl) silane was synthesized into epoxy-based polyhedral oligomeric silsesquioxane (EP-POSS), which was then cross-linked with CS as a cross-linking agent and freeze-dried to obtain aerogel (PCS). Notably, our PCS demonstrated remarkable self-recovery ability driven by water, along with superior mechanical strength and water absorption capacity compared to pure CS aerogels. Furthermore, interfacial photothermal aerogels (PCHs) were developed by incorporating HA-Zn^{2+} complexes onto the top of the PCH for efficient solar-driven evaporation. And PCH exhibited exceptional performance in sewage purification and antimicrobial activities. Consequently, PCH holds great potential for applications in water purification under challenging environmental conditions.

2. Results and Discussion

2.1. PCS Morphology and Structure

In this study, EP-POSS was synthesized in a single step through silane hydrolysis, as shown in Figure 1, and organic–inorganic hybrid materials were successfully introduced into biomass macromolecular chains by cross-linking the epoxy group of the EP-POSS with the amino group of the CS under acidic conditions. The resulting PCS was then prepared by freeze-drying, forming a PCS gel with a three-dimensional network pore structure. This structure was achieved through the direct sublimation of frozen ice crystals during freeze-drying, resulting in an ultra-lightweight aerogel with a density of only 11.33 mg cm^{-3}. And, the method allows for the simple and convenient production of aerogels of various sizes by utilizing molds of different sizes. Additionally, the volumetric shrinkage of the aerogel after freeze-drying is negligible. On this basis, we prepared PCHs by loading HA-Zn^{2+} complexes onto the top of the PCS.

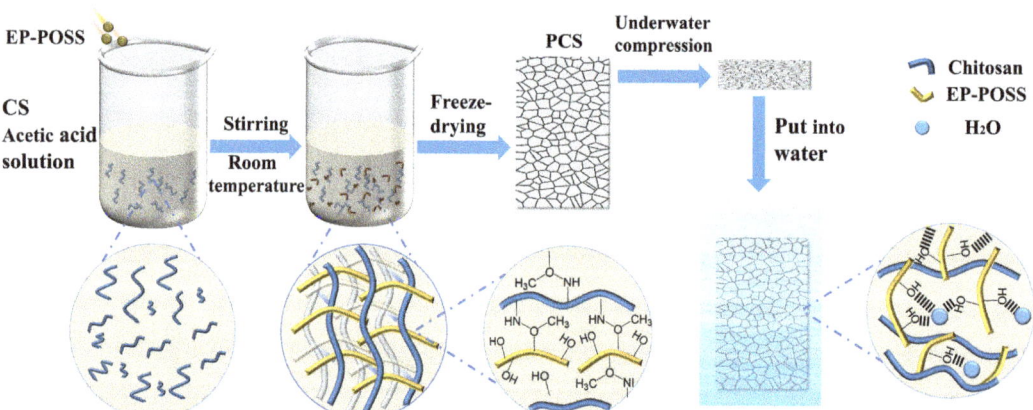

Figure 1. The preparation of the PCS by freeze-drying after the cross-linking of the amino group and epoxide group and the mechanism of the water-driven self-recovery were studied.

The FT-IR spectra of the POSS and PCS are shown in Figure 2a. The broad peak of the EP-POSS at 3100–3600 cm^{-1} is a stretching vibrational absorption peak of ν(O-H), which indicates that the hydrolysis of silanes occurs in the presence of water. The stretching vibrational absorption peak of ν(Si-O-Si) at 1083 cm^{-1} suggests that the condensation of Si-OH occurs in the presence of the acid as a catalyst so as to allow the Si-O-Si structure to be generated, and the presence of a hydroxyl peak indicated that the reaction is not carried out completely. The characteristic absorption peak of the epoxy group at 911 cm^{-1} indicated that the epoxy group does not undergo ring opening under the action of the acid. In order to further prove the molecular structure of the EP-POSS, it was subjected to FT-ICR-MS, and the mass spectral results are shown in Figure 2b. The peak located at 1337.55068 is a complete hexahedral cage of the EP-POSS, whose content is very small, and the peak located at 578.30584 is mainly attributed to $[C_{20}H_{42}Si_3O_{13} + H]^+$, as shown in Figure S1. Combining the mass spectrometry and the infrared analysis results, it can be seen that most of the EP-POSS prepared directly by silane hydrolysis are silane chains that do not form a complete cage structure, and only a small portion of them is formed. The broad peak of the PCS located at 3050–3600 cm^{-1} is the stretching vibrational absorption peak of the ν(O-H), which indicates the existence of the strong intermolecular and intramolecular hydroxyl group with hydrogen bonding. The stretching vibrational absorption peaks for ν(-CH$_2$) and ν(-CH$_3$) at 2866 cm^{-1} to 2937 cm^{-1}, the distorted vibrational absorption peaks for δ(-NH$_2$) at 1573 cm^{-1}, and the distorted vibrational absorption peaks for δ(-OH) at 1259 cm^{-1}. The stretching vibrational absorption peaks for ν(Si-O) at 800 cm^{-1}, the δ(-NH$_2$) characteristic absorption peak at 1573 cm^{-1} is weakened and shifted to a lower wave number to 1563 cm^{-1}, the δ(N-H) absorption peak [40] at 1412 cm^{-1} is shifted to 1406 cm^{-1}, and the ν(C-O) absorption band on the chitosan skeleton at 1000–1200 cm^{-1}, with the internal absorption peaks shifted to different degrees. These suggest that the amino group of CS reacts with the EP-POSS epoxy group to cross-link to form an aerogel.

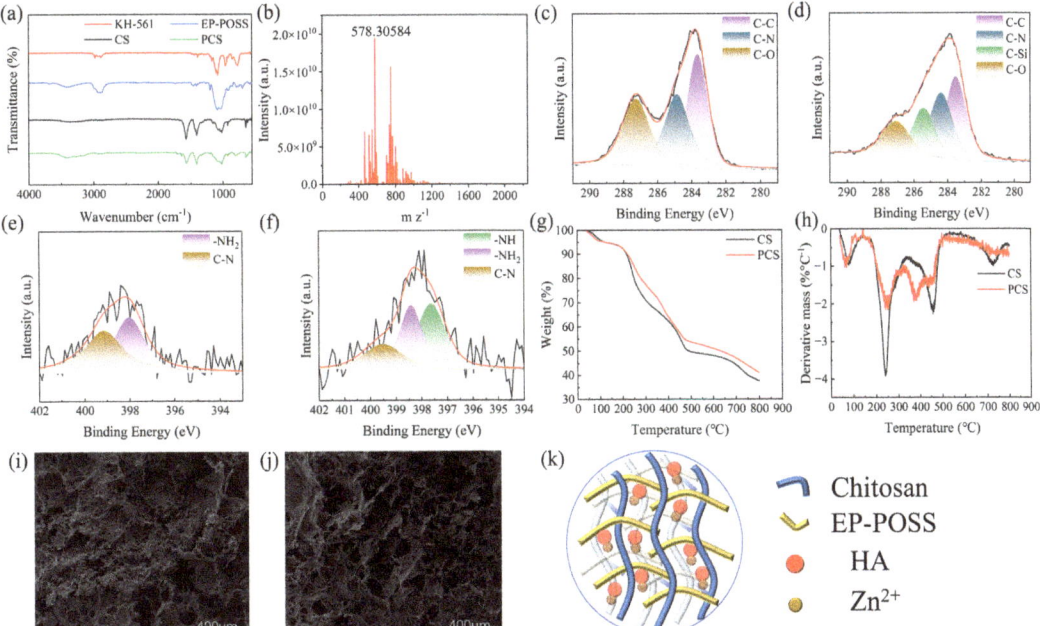

Figure 2. (a) FTIR spectra of CS, KH-561, EP-POSS, and PCS. (b) FT-ICR-MS spectra of EP-POSS. XPS spectra. (c) Pure CS aerogel C1s spectra. (d) PCS C1s spectra. (e) Pure CS aerogel N1s spectra. (f) PCS N1s spectra. (g) TG chart of pure CS aerogel and PCS. (h) DTG chart of pure CS aerogel and PCS. SEM images. (i) Pure CS aerogel. (j) PCS. (k) Schematic internal structure of PCH.

To further demonstrate the cross-linking of the EP-POSS with CS, the aerogel is subjected to XPS test, and the binding energy peaks of the aerogel are shown in Figure 2c–f. The binding energy peaks of the pure CS aerogel at 283.87 eV, 285.46 eV, and 287.27 eV in the C1s binding energy spectrum denote C-C, C-N, and C-O, respectively, while the binding energy peaks at 398.02 eV and 399.15 eV in the N1s binding energy spectrum denote is -NH$_2$ and C-N, respectively. After the addition of the EP-POSS, there is a new peak located at 285.4 eV in the C1s binding energy spectrum of the PCS, which is attributed to C-Si. Also, a new peak located at 397.62 eV in the N1s spectrum is ascribed to -NH-. Moreover, the peak of the -NH$_2$ at 398.02 eV and C-N at 399.15 eV in the binding energy spectrum of N1s after the addition of the EP-POSS are shifted to the higher binding energy at 398.41 eV, 399.5 eV, indicating that the amino group of the CS cross-links with the epoxide group of the EP-POSS and the primary amine changes to secondary amine. To sum up, it can be proved that the EP-POSS has been cross-linked with the CS through IR spectra and XPS photoelectron spectroscopy.

The thermal decomposition behavior of the PCS is shown in Figure 2g,h. It is mainly divided into three stages. The first stage is at 40–100 °C and 100–200 °C, which is mainly the evaporation of the physically adsorbed free water and internally bound water [41]. The second stage is at 200–300 °C, which contains the decomposition of the CS sugar chains and organic groups on the side face of the EP-POSS [42]. The third stage at 300–800 °C mainly consisted of the complete carbonylation decomposition of the CS [43]. The carbon residual content of the aerogel increases from 37.6% to 41.2% after the addition of the EP-POSS, and the temperature at which the maximum thermal decomposition rate is reached increases from 239.5 °C to 250.2 °C, with a decrease in the maximum thermal decomposition rate, which indicates that the thermal stability of the CS-based aerogel is significantly improved after the addition of the EP-POSS.

According to the SEM of the PCS shown in Figure 2i,j, it can be seen that the pure CS aerogel without the EP-POSS is made up of the CS fragments of different sizes stacked on top of each other, which has few pore structures and is poorly connected to each other. This is due to pure CS aerogel being physically cross-linked by the ionic bonding cross-linking method, which makes the pure CS aerogel highly susceptible to decomposition in a liquid environment [25]. After the addition of the EP-POSS, the pores of the aerogel become more abundant, forming a honeycomb-like three-dimensional pore structure ranging from tens of micrometers to several hundred micrometers. This is due to the fact that the reaction between the epoxy group of the EP-POSS and the amino group of the CS makes it a much tighter chemical cross-linking, which makes its structure more stable. The rich pore structure can form a stronger capillary force when transporting water, which is more favorable for the transport of water [44]. According to the BET test, the average adsorption pore size of the PCS is about 54.45 nm (Figure S4).

2.2. Water-Driven Recovery Properties and Mechanism of PCS

The water contact angle of the PCS is shown in Figure 3f. A contact angle of 52.4° between the PCS and the aerogel is formed at the moment of the droplet's fall, which is smaller than the 86.5° (Figure S6) of the pure CS aerogel, and the PCS is completely absorbed the water within 0.08 s (Video S1), whereas the pure CS aerogel requires 0.2 s (Video S2), which shows that the PCS has excellent hydrophilicity and strong water absorption capacity. The water-driven recovery of the PCS is excellent, as shown in Figure 3a. The PCS with a height of 25 mm was immersed in the water and then strongly squeezed ($\varepsilon > 80\%$) to a height of 5 mm PCS sheet. Then, when the pressure is released, air cannot occupy the pores of the PCS and the aerogel cannot return to its initial volume. However, after re-inserting the PCS into the water, the PCS sheet recovers 50% of its volume within 1 s and almost completely recovers its initial shape within 1.9 s (Video S3). This indicates that water enters and occupies the pores of the compressed aerogel; hence, the aerogel expands to its original volume and shape. This rapid water-driven recovery property is related to the cross-linking of the chain EP-POSS with the introduction of the hydroxyl groups [45].

To further demonstrate the relationship between this water-driven self-recovery phenomenon and the PCS hydroxyl groups, the infrared absorption spectra of the PCS and CS aerogels were analyzed by peak splitting. Figure 3b shows the FT-IR absorbance profiles of the PCS with different ratios of the EP-POSS added, from which it can be seen that the ν(O-H) peak changes significantly, from a broad peak containing two absorption peaks to a broad peak containing three absorption peaks, and the intensity of the ν(O-H) peak increases significantly with the addition of the EP-POSS. The FT-IR spectra of the CS aerogel and PCS are obtained through Gaussian fitting analysis, as shown in Figure 3c,d. After fitting the ν(O-H) peaks, the absorption peaks are formed by hydrogen-bonded hydroxyls located at 3390–3415 cm^{-1}, and free carbon hydroxyls located at 3174–3209 cm^{-1}, which can be seen in the bifurcation diagram of the pure CS aerogel. After adding the EP-POSS for cross-linking, the ν(O-H) peaks change significantly, and there appear absorption peaks at 3260–3285 cm^{-1} formed by free Si-OH after fitting the peaks [46]. The spectra show that the incompletely condensed chain EP-POSS introduces a large amount of Si-OH into the aerogel after the introduction of the chain EP-POSS cross-linking, and the formation of the hydrogen bonding between the Si-OH and CS makes the PCS structure more stable [47]. However, the addition of water causes the hydrogen bonding between Si-OH and CS to be broken after immersing the PCS in water, and the pores of the aerogel are occupied by water and the PCS became soft by the water wetting. After compressing the PCS, the residual moisture formed hydrogen bonds with the hydroxyl groups in the aerogel, and air could not disrupt this structure to occupy the pores and restore it to its original shape. When the PCS is put into the water again, due to its strong water absorption, enough water re-enters the PCS and forms hydrogen bonds with the hydroxyl groups in the PCS [48], which rapidly fills up the pore structure of the PCS and completely restores the original shape. Therefore, it can be proved that the hydroxyl group, which is introduced through the addition of the EP-POSS, combines with water to form hydrogen bonds, thereby causing the PCS to exhibit water-driven recovery behavior. It is the introduction of hydroxyl groups and the EP-POSS that allows the PCS to exhibit superior mechanical properties and recovery compared to previous water-driven recovery aerogels.

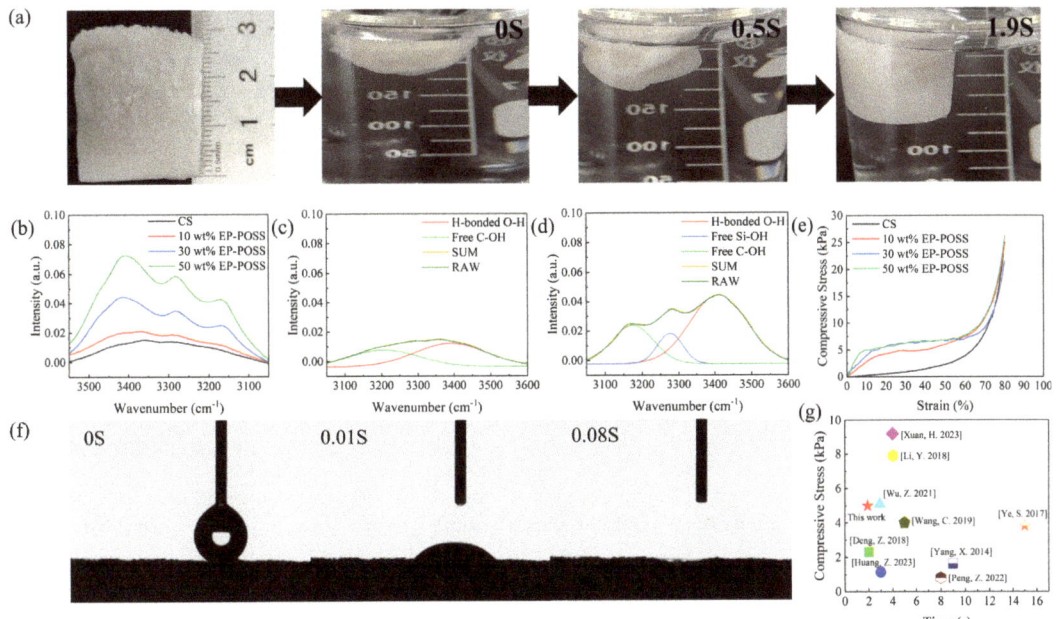

Figure 3. (a) Image of the PCS-compressed water-induced recovery process. (b) FTIR spectra of the hydroxyl absorbance of the PCS with different ratios of the EP-POSS added. Gaussian fitting. (c) Fitt-

ed spectra of the pure CS aerogel. (**d**) Fitted spectra of the PCS. (**e**) Compressive stress–strain curves of the pure CS aerogel with the addition of different ratios of the EP-POSS. (**f**) Contact angle image of the PCS aerogel. (**g**) Comparison of the mechanical properties and water-driven recovery of the aerogels in this study with previous [Ye, S. 2017] [45], [Huang, Z. 2023] [49], [Deng, Z. 2018] [50], [Xuan, H. 2023] [51], [Yang, X. 2014] [52], [Peng, Z. 2022] [53], [Li, Y. 2018] [54], [Wang, C. 2019] [55], [Wu, Z. 2021] [56].

2.3. Mechanical Properties of PCS

The compressive stress–strain curves of the PCS and pure CS aerogel are shown in Figure 3e. At up to 80% strain, it can be found that the compressive stress–strain curves can be divided into three regions, which are the elastic region ($0 < \varepsilon < 10\%$) where the PCS undergoes elastic deformation, the yield region ($10\% < \varepsilon < 60\%$) where the pore structure of the PCS changes continuously with increasing pressure and undergoes irreversible plastic deformation, and a dense region ($\varepsilon < 60\%$) where the PCS structure undergoes complete collapse failure under strong pressure. At low stresses, the 10% strain corresponds to the stresses of 2.77 kPa, 3.95 kPa, and 4.99 kPa, and the Young's modulus of 27.39 kPa, 42.42 kPa, and 56.52 kPa, which are much higher than those of the pure CS aerogel of 0.25 kPa and 2.26 kPa, respectively, when the proportion of the EP-POSS is 10 wt%, 30 wt%, and 50 wt%. The results show that the compressive strength of the PCS increases with the increase in the EP-POSS content. The main reason is that when the content of the EP-POSS increases, the degree of cross-linking of the PCS is enhanced, the hydrogen bonds formed between the Si-OH and CS are increased, and the structure of the PCS is more stable so that the aerogels' mechanical properties are improved [47]. The improved mechanical properties may also be related to the unique nano-size effect of the EP-POSS, which may positively affect the mechanical properties of the PCS [57].

2.4. Water Transport Properties of PCS

In order to investigate the cyclic usability of the PCS compression–expansion and changes in water absorption capacity, the PCS went through the water absorption expansion–dehydration compression cycle 10 times. The results are shown in Figure 4a–e. Pure CS aerogel has a maximum water absorption of 33.07 g g^{-1} when it first enters water; however, it produces hydrogen bonds in the presence of water, destroying the cross-linking of the aerogel [58], at which point the aerogel becomes transparent (Video S4). The pure CS aerogel broke after compression and drainage (Video S5), and the water absorption dropped sharply. The second time it is only 26.01 g g^{-1}, less than 80% of the first water absorption, and the tenth time it is only 14.44 g g^{-1}, less than 50% of the first water absorption. With the increase in the EP-POSS incorporation content, the amount of water absorbed by PCS after compression–absorption was gradually stabilized. The maximum water absorption capacity of the PCS containing 50 wt% EP-POSS in the first suction (Video S6) is 27.50 g g^{-1} after compression and drainage, and the second water-absorbing capacity is 27.06 g g^{-1}, reaching 98.4% of the first time. The water-absorbing capacity after ten compression and drainage cycles is still 23.09 g g^{-1}, reaching 83.9% of the first time, and there is no obvious breakage of the PCS in the whole process of suction, compression, and drainage cycles. It shows that it can still have good water transport performance and keep the evaporation continuity of the photothermal layer after several times of severe external extrusions.

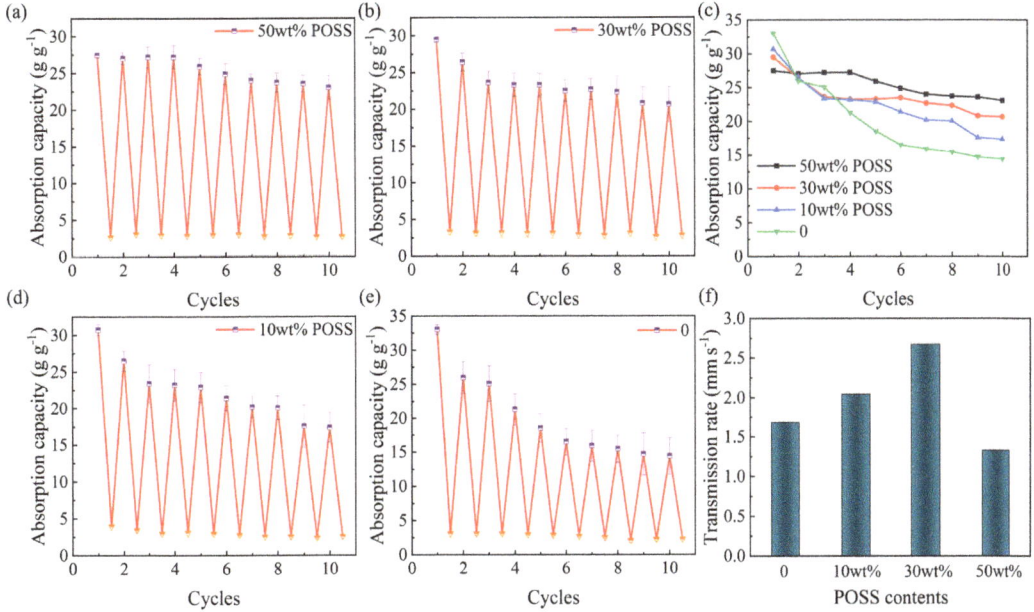

Figure 4. Compression–expansion cycle water absorption test of the pure CS aerogel and different ratios of PCS (**a**) 50 wt% POSS, (**b**) 30 wt% POSS, (**d**) 10 wt% POSS, and (**e**)pure CS aerogel. (**c**) Comparison of the compression–expansion cycle water absorption with different ratios of PCS. (**f**) The water transport rate of the pure CS aerogel and different ratios of PCS.

Figure 4f shows the water transport rate of the aerogel, and the aerogel end is vertically contacted to the water surface to record the time required for the water to rise to 20 mm. It can be seen that the water transport rate of the pure CS aerogel is 1.69 mm s^{-1}, while that of the PCS with 30 wt% POSS reaches 2.67 mm s^{-1}, which is higher than the other reported water transport layer [59,60]. The enhanced water transport rate may be due to the introduction of hydrophilic hydroxyl groups in the chain EP-POSS cross-linked with CS. However, the water transport rate of the PCS decreased to 1.33 mm s^{-1} when the content of the EP-POSS was increased to 50 wt%, which could be attributed to the introduction of the excessive Si-O-Si structure into the PCS. The Si-O-Si structure acts as a hydrophobic group [61], and the introduction of a small amount may lead to a faster upward escape of water molecules, while the introduction of too much may have a greater negative effect. To demonstrate that the PCS can be applied to a conditioned harsh environment without breakage, it was further placed in an ultrasonic cleaner with a power of 360 W (Video S7, Figure S7) to simulate the external environment. There was no significant change in the aerogel after ultrasonic vibrating for 30 min, indicating that the aerogel can be applied to be used in harsh environments.

2.5. Photothermal Evaporation Properties of PCH

The photothermal effect of aerogels is realized through the coordination of HA with metal ions. The UV-Vis spectra of the aerogels loaded with different photothermal layers are shown in Figure 5a. In the visible band, the average absorptivity of the HA-Zn^{2+} complex aerogel can reach 98.9%, which is higher than that of the aerogel with HA alone and the aerogel with tannin alone. Regarding the absorbance versus the color of the aerogels, the higher the absorbance in the visible band, the darker the color. In contrast, the average absorption rate of the photothermal layer loaded with tannin-Zn^{2+} is only 81.9%, indicating that HA-Zn^{2+} can absorb more energy than tannin-Zn^{2+}.

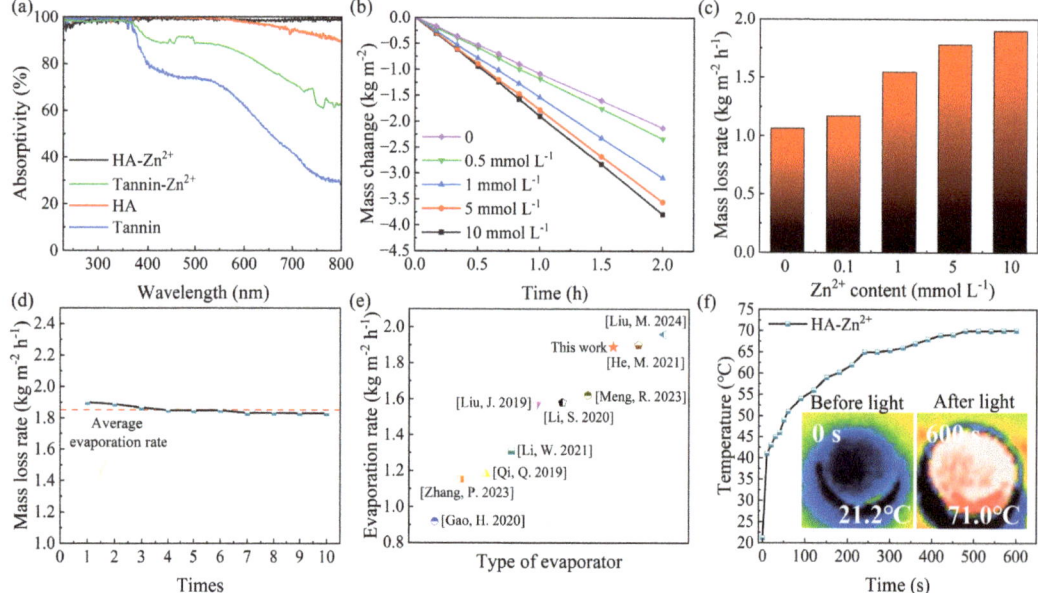

Figure 5. (a) UV-Vis spectra of different photothermal layers in wet state. (b) Loss of water mass in the light intensity of one sun. (c) The rate of evaporation in the light intensity of one sun. (d) The evaporation rate of 10 consecutive evaporations. (e) Comparison of evaporation rates between this work and other literature under the same solar irradiation condition [Gao, H. 2020] [62], [Zhang, P. 2023] [63], [He, M. 2021] [64], [Li, S. 2020] [65], [Liu, J. 2019] [66], [Liu, M. 2024] [67], [Meng, R. 2023] [68], [Qi, Q. 2019] [69], [Li, W. 2021] [70]. (f) Infrared thermal imaging and temperature change curve of the dried PCH under one solar illumination intensity.

The relationship between photothermal performance and absorbance is verified by photothermal evaporation experiments. Under the irradiation of a full-band xenon lamp simulating sunlight, Figure 5b,c show the HA-Zn^{2+} evaporator loaded with different concentrations of Zn^{2+} ion solutions undergoing continuous photothermal evaporation tests for 2 h. As the Zn^{2+} ion concentration increased, the evaporation rate of the PCH rose by 78.6%, reaching a rate of 1.89 kg m^{-2} h^{-1}. The evaporation rate is higher than most evaporators that use biomass material as a substrate (Figure 5e). However, when the concentration of the Zn^{2+} ion solution is increased from 5 mmol L^{-1} to 10 mmol L^{-1}, the increase in the evaporation rate is not significant. This may be due to the upper limit of the coordination capacity of the HA. In order to demonstrate the continuous availability of the PCH, 10 consecutive photothermal evaporation tests were performed on the PCH loaded with the highest complexing concentration, and the results are shown in Figure 5d. Benefiting from the excellent underwater stability and water transport of the PCS, the average evaporation rate of the PCH is 1.85 kg m^{-2} h^{-1} in 10 consecutive evaporation experiments, which is stabilized at 96.4% of the maximum evaporation rate after 7 evaporation experiments. The experimental results show that the PCH aerogel has the potential for long-term use.

The infrared thermal imaging of the PCH under simulated sunlight is shown in Figure 5f. Under the light intensity of one sun, the surface temperature of the PCH can rapidly increase to 51.0 °C in one minute and 65.4 °C in five minutes, and the temperature reaches a steady maximum of 70.1 °C in ten minutes. The maximum temperature of the PCS not loaded with a photothermal layer is only 31.1 °C after 10 min of irradiation. And, the maximum temperature of the PCS loaded with HA is only 59.7 °C, while the maximum temperature of the PCS loaded with tannin-Zn^{2+} is only 48.3 °C after 10 min of irradiation

(Figure S9). The results indicate that the energy absorbed by the complex can be released as a part of the thermal energy through the vibration of the metal ions by the coordination of the TA with metal ions, thus enhancing its photothermal conversion ability [38].

2.6. Photothermal Purification Performance

In order to investigate the purification ability of the PCH on wastewater, PCH was put into the concentration of 5 g L^{-1} methylene blue solution, rhodamine B solution, and their mixed solution for evaporation treatment, and the evaporated water was collected and detected by ultraviolet-visible spectroscopy to calculate the concentration of the organic dyes. The results are shown in Figure 6a, which shows that the photothermal evaporation treatment has a good purification effect on wastewater polluted with organic dyes. The concentration of methylene blue in the water after photothermal evaporation treatment is only 0.045 mg L^{-1}, the concentration of rhodamine B is 0.02485 mg L^{-1}, and the concentration of the mixed solution is 0.02219 mg L^{-1}, which is much lower than their original concentration. In order to further investigate the purification ability of the PCH, sulfuric acid solution and sodium hydroxide solution with pH values of 1 and 13, respectively, are purified by the PCH and subjected to photothermal evaporation treatment, and the results are shown in Figure 6b; also, the pH value of the solution after the photothermal evaporation treatment is detected to be neutral by pH test strips.

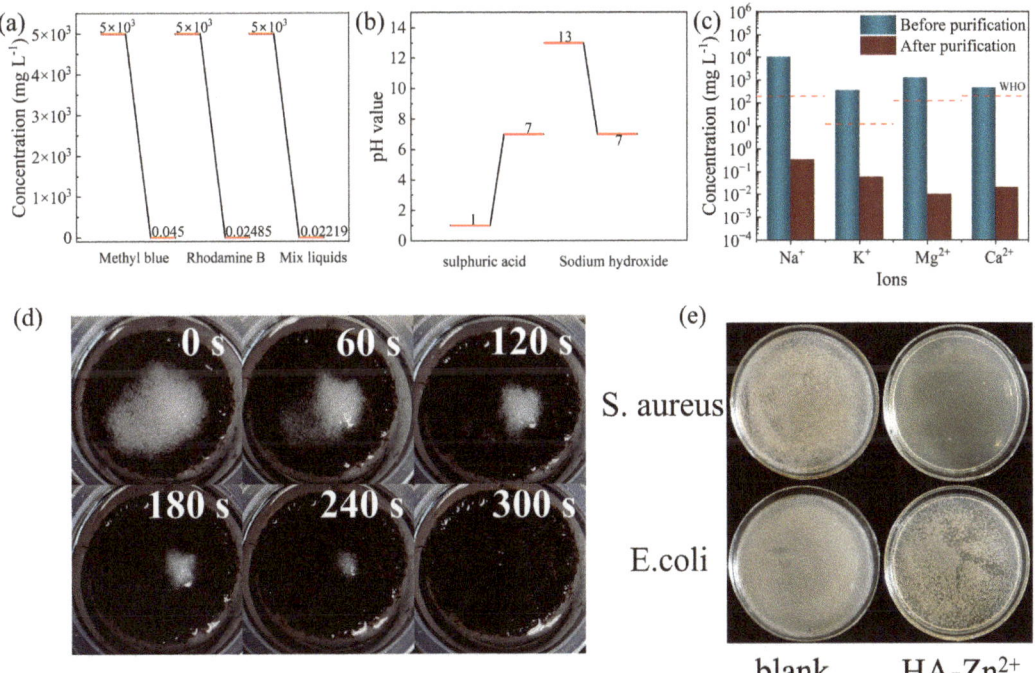

Figure 6. Evaporator water purification. (**a**) Purification of organic dyes. (**b**) Acid–base purification. (**c**) Artificial seawater. (**d**) Salt dissolution image of a PCH. (**e**) Inhibitory effect of PCH on Escherichia coli and Staphylococcus aureus.

To prove the seawater purification ability of the PCH, using artificial seawater [71] for evaporation test experimental results as shown in Figure 6c. The evaporation of water collected by the content of the four metal ions have dropped by four orders of magnitude, in line with the World Health Organization's drinking water standards. In addition, we also tested the salt removal ability of the PCH; we placed the PCH in a solution simulating seawater, and placed 1 g NaCl on the surface of the PCH for evaporation experiments. As shown in Figure 6d, NaCl dissolved within 300 s, indicating that the PCH has a good ability to remove salt.

Because Zn^{2+} itself has good antibacterial properties, Zn^{2+} produces significant toxic effects on bacteria by interfering with its carbon metabolism [72]. Co-mixing the PCH with diluted bacterial broth solution in a constant temperature incubator at 37 °C for 24 h, the bacterial mixture is diluted and applied to agar Petri dishes and incubation is continued for 12 h. As shown in Figure 6e, the inhibitory effect of the PCH loaded with Zn^{2+} complex for *S. aureus* is much more than that of the blank control group. Similarly in the experiments with *E. coli*, the PCH loaded with Zn^{2+} complexes also show significant bacterial inhibition. In addition, due to the abundance of CS as well as HA and their low price, PCH has a lower cost than other reported solar water evaporators [73], which is only 1.97 $ m^{-2} (Table S1), so that it has the basis for practical application.

3. Conclusions

In summary, in this paper, a novel CS-based aerogel was prepared by combining organic–inorganic materials with biomass materials, cross-linking EP-POSS with CS, and freeze-drying, which has excellent mechanical properties as well as superb water-absorbing capacity. The water absorption rate of PCS can reach 2.67 mm s^{-1} at 30 wt% of EP-POSS, and the water absorption capacity of PCS can reach 22.09 g g^{-1} at 50 wt% of EP-POSS. Moreover, the introduction of hydroxyl groups allows PCS to form hydrogen bonds with H_2O in water, and PCS has a rapid water-driven self-recovery ability under the effect of hydrogen bonding. Based on the PCS with 30 wt% EP-POSS added, which exhibited the fastest water transport rate, it was utilized as a solar water evaporator by depositing the HA-Zn^{2+} photothermal layer on its top layer. The tested PCH has a stable solar water evaporation capability and due to the underwater self-recovery capability of the PCS, it can remain water-transportable even after external crushing. After 10 consecutive photothermal evaporation tests, the PCH does not show any serious performance degradation, and it has excellent water purification and salt drainage capabilities. In addition, the PCH shows excellent antimicrobial performance due to the antimicrobial properties of Zn^{2+}, so that it can meet the needs of water evaporation in harsh environments.

This photothermal aerogel, exhibiting remarkable water-driven recovery capabilities, effectively harnesses its water-driven recovery behavior to adapt to diverse aquatic environments, including those containing numerous rigid objects that could potentially compromise the integrity of the evaporator. Furthermore, its resilience enables its deployment in extremely adverse climatic conditions, mitigating the risk of structural degradation. As such, this evaporator is anticipated to significantly advance the practical utilization of solar energy evaporation systems.

4. Materials and Methods

4.1. Materials

Chitosan (CS, MW = 30,000, ≥85% deacetylation), $ZnCl_2$, and NaBr (analytical purity) were purchased from Shanghai Maclean Biochemical Technology Co. (Shanghai, China). Triethoxy (3-epoxypropoxy) silane (KH-561, purity > 96.0%) and tannin (analytical purity) were purchased from Aladdin Reagent Co. (Shanghai, China). Hydrochloric acid, $MgSO_4$, and KCl (analytical purity) were purchased from Yantai Shuang Chemical Co. (Yantai, China). Tetrahydrofuran (THF, analytical purity), methanol (analytical purity), acetic acid (analytical purity), and $AlCl_3$ (analytical purity) were purchased from Tianjin Tianli Chemical Reagent Co. (Tianjin, China). $MnCl_2$ (analytical purity) was purchased from Tianjin Xinput Chemical Co. (Tianjin, China). NaCl, $CaCl_2$, and $MgCl_2$ (analytical purity) were purchased from Tianjin Hengxing Chemical Reagent Manufacturing Co. (Tianjin, China). $NaHCO_3$ (analytical purity) was purchased from Tianjin Dayong Chemical Reagent Manufacturing Co. (Tianjin, China).

4.2. Preparation of PCH

4.2.1. Preparation of the EP-POSS

EP-POSS was directly synthesized by silane hydrolysis; 10 mL of KH-561, 150 mL of methanol, 0.5 mL of concentrated hydrochloric acid, and 0.65 mL of deionized water were placed in a flask and the reaction was carried out with magnetic stirring in an oil bath at 90 °C for 10 days, and the solvent was slowly evaporated by continued magnetic stirring and heating after 10 days until the solution was a pale yellow transparent viscous liquid. Deionized water is added to make the liquid become a white suspension, and the yellowish viscous product EP-POSS can be obtained after freeze-drying.

4.2.2. Preparation of the PCS

An amount of 1.5 g of CS was dissolved in 100 mL of 1% v v^{-1} acetic acid solution, different ratios (0.3 g, 0.9 g, 1.5 g) of EP-POSS were added, and the pH was adjusted to 3 using 1 mol L^{-1} hydrochloric acid, and the reaction was performed with magnetic stirring at room temperature for 48 h. After the reaction was performed with saturated sodium bicarbonate solution, the pH was adjusted to 6.4, and the aqueous gel was obtained by ultrasonication for 5 min to remove air bubbles and water bathing at 50 °C for 2 h. The aqueous gel was frozen and then freeze-dried to obtain PCS with different ratios. Pure CS aerogel was prepared by the same method without adding POSS.

4.2.3. Construction of Photothermal Layers of HA-Zn^{2+} Complexes

A total of 4 g of tannin was dissolved in 100 mL water, hydrothermal reaction was carried out at 240 °C for 4 h, the reaction solution was filtered by 0.45 μm filter membrane, and the filtered solution was dried to obtain HA. After mixing an aqueous HA solution of 1 g L^{-1} with Zn^{2+} solutions of different concentrations (0.5 mmol L^{-1}, 1 mmol L^{-1}, 5 mmol L^{-1}, and 10 mmol L^{-1}) 1:1, the mixed solution was added to the above PCS solution, and the pH was adjusted to 6.4 by sodium bicarbonate, which was then covered on the above aqueous gel, freeze it as a whole for 24 h, and then freeze-dried to obtain PCH. The evaporator containing tannin metal complexes was prepared according to the above method, and the tannin concentration is 1 g L^{-1}.

4.3. Structure Characterizations

Fourier Transform Infrared spectrometer (FTIR, PerkinElmer Frontier spectrometer, PerkinElmer, Waltham, MA, USA, measuring range 4000–500, resolution 2 cm^{-1}), UV-visible spectrophotometer (TU-1901, Beijing Pulse Analytical, Beijing, China), X-ray photo-electron spectrometer (XPS, Thermo Kalpha, ThermoFisher Technologies, Waltham, MA, USA, using K-Alpha X-rays), and Fourier Transform Ion Cyclotron mass spectrometry (FT-ICR-MS, MALDI-SolariX FTMS, Bruker, Billerica, MA, USA, with samples dissolved in a solvent of toluene: methanol 9:1) were used for the analysis of the chemical structure and composition of the EP-POSS and PCS. Scanning electron microscope (SEM, TM3030, Hitachi, Tokyo, Japan, sample gold spraying treatment, acceleration voltage 5 kV, working distance 2–40 mm) was used for the micro-morphological analysis of the aerogels. Universal mechanical testing machine (UTM-2203, Shenzhen Sansi Zongheng Technology Co., Ltd., Shenzhen, China, compression rate 5 mm min^{-1}, room temperature, vertically compressed) was used to test the mechanical properties of the aerogels, and the compression modulus was determined by the slope of the stress–strain curve at low strain. A video optical contact angle meter (OCA20, Data-physics, Baden-Wuerttemberg, Filderstadt, Germany, 5 μL water droplets) was used to measure the aqueous contact angle of the aerogels. A Thermogravimetric analyzer was also used (TGAS, STA 6000-SQ8 PerkinElmer, Waltham, MA, USA, measuring range 40–800 °C, heating rate 10 °C min^{-1}, nitrogen protection flow rate of 20 mL min^{-1}).

4.4. Water Absorption Test

The water absorption capacity (C_0 g g^{-1}) was measured by placing the PCS with a mass of m_0 in water. The mass m_x was measured by removing the surface liquid from the PCS after placing it in water for 30 s. The formula is shown below:

$$C_0 = \frac{m_x - m_0}{m_0}$$

4.5. Evaporation Performance Test

The solar evaporation experiments were performed at an indoor temperature of 25 ± 1 °C using a xenon lamp light source (CEL-HXF300, Beijing CEC Jinyuan, Beijing, China) as the sunlight simulator and a solar power meter (SM206- SOLAR, Shenzhen Xinbao Science Instrument, Shenzhen, China) was used to determine the average light intensity on the evaporated surface as 1 kW m^{-2}. The PCH was placed in a beaker filled with water, and to avoid the influence of the heat of the light source it was placed 40 cm below the light source, and the mass change was recorded by an electronic balance (FA2104, Shanghai Shunyu Hengping, Shanghai, China). An infrared thermal imager (Uti260A, Haikangweishi, Hangzhou, Zhejiang, China) was used to take infrared pictures and measure the surface temperature. The photothermal purification experiments were carried out by a double-beam UV-visible spectrophotometer to determine the concentration of the organic dyes and determine their pH by a pH meter (PB-10, Sartorius, Gottingen, Lower Saxony, Germany).

4.6. Artificial Seawater Purification Test

Artificial seawater was formulated with the following composition: 26.5 g NaCl, 3.3 g MgSO$_4$, 2.4 g MgCl$_2$, 1.3 g CaCl$_2$, 0.7 g KCl, 0.2 g NaHCO$_3$, and 0.08 g NaBr dissolved in 1 L of water. The evaporated water was collected and tested for metal ions using an ICP-OES (Agilent 5800 ICP-OES, Santa Clara, CA, USA).

4.7. Antimicrobial Performance Test

Bacteria were scraped into sterile tubes and saline-fixed to 1.5×10^6 CFU mL^{-1} before bacterial solution was added. After diluting the bacterial solution 50-fold using beef broth, 1 mL of the mixed solution was taken and added to 9 mL of the saline and stored at 37 °C for 24 h. The mixed solution was diluted 10-fold using saline, and 20 µL was taken and spread evenly in the agar medium and incubated at 37 °C for 12 h.

Supplementary Materials: The following supporting information can be downloaded at: https://www.mdpi.com/article/10.3390/gels10040279/s1, Figure S1. EP-POSS structure inferred from FT-ICR-MS and FT-IR spectra (a) Located at 578 m/z and (b)750 m/z; Figure S2. XPS total spectra of aerogels, (a) pure CS aerogel and (b) PCS aerogel; Figure S3. XRD pattern of CS and PCS; Figure S4. TG and DTG curves of the photothermal layer; Figure S5. N$_2$ adsorption–desorption isothermal curves of pure chitosan aerogel and PCS; Figure S6. Water contact angle of pure chitosan aerogels; Figure S7. Before and after comparison after ultrasound simulation; Figure S8. Photothermal evaporation experimental setup diagram; Figure S9. Infrared thermogram of dry sample under a strong sunlight, (a) photothermal layer with added tannin-Zn^{2+}, (b) PCS, and (c) photothermal layer with added HA; Table S1. PCH solar water evaporator material cost analysis; Video S1 showing PCS–water contact angle (mp4); Video S2 showing pure CS aerogel–water contact angle (mp4); Video S3 showing PCS water-driven recovery (mp4); Video S4 showing pure CS aerogel immersed in water (mp4); Video S5 showing pure CS aerogel water-driven recovery (mp4); Video S6 showing PCS immersion in water (mp4); Video S7 showing simulating an outdoor environment with an ultrasonic cleaner (mp4).

Author Contributions: Conceptualization, Y.L. and M.M.; writing—original draft preparation, Y.L. and M.M.; validation, Y.S.; writing—review and editing, D.W., Y.L. and M.M.; investigation, Z.Z. and X.W.; methodology, J.W. and J.P.; supervision, D.W., C.W. and J.L.; project administration, D.W.; funding acquisition, D.W. All authors have read and agreed to the published version of the manuscript.

Funding: The research was supported by the financial supports from the National Natural Science Foundation of China (GN: 31770593) and the Fundamental Research Funds for the Central Universities (GN:2572023AW50).

Data Availability Statement: Data are contained within the article.

Acknowledgments: The authors would like to acknowledge the technical support from Analysis and Testing Center of Northeast Forestry University.

Conflicts of Interest: The authors declare no competing financial interest.

References

1. Wu, W.; Xu, Y.; Ma, X.; Tian, Z.; Zhang, C.; Han, J.; Han, X.; He, S.; Duan, G.; Li, Y. Cellulose-based interfacial solar evaporators: Structural regulation and performance manipulation. *Adv. Funct. Mater.* **2023**, *33*, 202302351. [CrossRef]
2. Kim, H.; Rao, S.R.; Kapustin, E.A.; Zhao, L.; Yang, S.; Yaghi, O.M.; Wang, E.N. Adsorption-based atmospheric water harvesting device for arid climates. *Nat. Commun.* **2018**, *9*, 1191. [CrossRef]
3. Elimelech, M.; Phillip, W.A. The future of seawater desalination: Energy, technology, and the environment. *Science* **2011**, *333*, 712–717. [CrossRef] [PubMed]
4. He, J.; Zhang, Z.; Xiao, C.; Liu, F.; Sun, H.; Zhu, Z.; Liang, W.; Li, A. High-performance salt-rejecting and cost-effective superhydrophilic porous monolithic polymer foam for solar steam generation. *ACS Appl. Mater. Interfaces* **2020**, *12*, 16308–16318. [CrossRef] [PubMed]
5. Yao, H.; Zhang, P.; Yang, C.; Liao, Q.; Hao, X.; Huang, Y.; Zhang, M.; Wang, X.; Lin, T.; Cheng, H.; et al. Janus-interface engineering boosting solar steam towards high-efficiency water collection. *Energy Environ. Sci.* **2021**, *14*, 5330–5338. [CrossRef]
6. Wu, L.; Dong, Z.C.; Cai, Z.R.; Ganapathy, T.; Fang, N.X.; Li, C.X.; Yu, C.L.; Zhang, Y.; Song, Y.L. Highly efficient three-dimensional solar evaporator for high salinity desalination by localized crystallization. *Nat. Commun.* **2020**, *11*, 521. [CrossRef]
7. Wang, Y.; Wu, X.; Wu, P.; Zhao, J.; Yang, X.; Owens, G.; Xu, H. Enhancing solar steam generation using a highly thermally conductive evaporator support. *Sci. Bull.* **2021**, *66*, 2479–2488. [CrossRef] [PubMed]
8. Xu, Z.; Ran, X.; Zhang, Z.; Zhong, M.; Wang, D.; Li, P.; Fan, Z. Designing a solar interfacial evaporator based on tree structures for great coordination of water transport and salt rejection. *Mater. Horiz.* **2023**, *10*, 1737–1744. [CrossRef] [PubMed]
9. Kuang, Y.; Chen, C.; He, S.; Hitz, E.M.; Wang, Y.; Gan, W.; Mi, R.; Hu, L. A high-performance self-regenerating solar evaporator for continuous water desalination. *Adv. Mater.* **2019**, *31*, e1900498. [CrossRef]
10. Li, C.; Jiang, D.; Huo, B.; Ding, M.; Huang, C.; Jia, D.; Li, H.; Liu, C.-Y.; Liu, J. Scalable and robust bilayer polymer foams for highly efficient and stable solar desalination. *Nano Energy* **2019**, *60*, 841–849. [CrossRef]
11. He, F.; Han, M.; Zhang, J.; Wang, Z.; Wu, X.; Zhou, Y.; Jiang, L.; Peng, S.; Li, Y. A simple, mild and versatile method for preparation of photothermal woods toward highly efficient solar steam generation. *Nano Energy* **2020**, *71*, 104650. [CrossRef]
12. Liu, T.; Gou, S.; He, Y.; Fang, S.; Zhou, L.; Gou, G.; Liu, L. N-methylene phosphonic chitosan aerogels for efficient capture of Cu^{2+} and Pb^{2+} from aqueous environment. *Carbohydr. Polym.* **2021**, *269*, 118355. [CrossRef] [PubMed]
13. Zhang, H.; Li, Y.; Shi, R.; Chen, L.; Fan, M. A robust salt-tolerant superoleophobic chitosan/nanofibrillated cellulose aerogel for highly efficient oil/water separation. *Carbohydr. Polym.* **2018**, *200*, 611–615. [CrossRef] [PubMed]
14. Zhang, Y.; Liu, Y.; Guo, Z.; Li, F.; Zhang, H.; Bai, F.; Wang, L. Chitosan-based bifunctional composite aerogel combining absorption and phototherapy for bacteria elimination. *Carbohydr. Polym.* **2020**, *247*, 116739. [CrossRef] [PubMed]
15. Moghtader, F.; Solakoglu, S.; Piskin, E. Alginate- and chitosan-modified gelatin hydrogel microbeads for delivery of E. coli phages. *Gels* **2024**, *10*, 244. [CrossRef]
16. Venkatesan, R.; Vetcher, A.A.; Al-Asbahi, B.A.; Kim, S.-C. Chitosan-based films blended with tannic acid and moringa oleifera for application in food packaging: The preservation of strawberries (*Fragaria ananassa*). *Polymers* **2024**, *16*, 937. [CrossRef] [PubMed]
17. Jiang, X.; Zhang, J.; You, F.; Yao, C.; Yang, H.; Chen, R.; Yu, P. Chitosan/clay aerogel: Microstructural evolution, flame resistance and sound absorption. *Appl. Clay Sci.* **2022**, *228*, 106624. [CrossRef]
18. Lisuzzo, L.; Cavallaro, G.; Lazzara, G.; Milioto, S. Supramolecular systems based on chitosan and chemically functionalized nanocelluloses as protective and reinforcing fillers of paper structure. *Carbohydr. Polym. Technol. Appl.* **2023**, *6*, 100380. [CrossRef]
19. Ruiz-Hitzky, E.; Ruiz-García, C.; Fernandes, F.M.; Dico, G.L.; Lisuzzo, L.; Prevot, V.; Darder, M.; Aranda, P. Sepiolite-hydrogels: Synthesis by ultrasound irradiation and their use for the preparation of functional clay-based nanoarchitectured materials. *Front. Chem.* **2021**, *9*, 733105. [CrossRef]
20. Menshutina, N.; Majouga, A.; Uvarova, A.; Lovskaya, D.; Tsygankov, P.; Mochalova, M.; Abramova, O.; Ushakova, V.; Morozova, A.; Silantyev, A. Chitosan aerogel particles as nasal drug delivery systems. *Gels* **2022**, *8*, 796. [CrossRef]
21. Yudaev, P.; Semenova, A.; Chistyakov, E. Gel based on modified chitosan for oil spill cleanup. *J. Appl. Polym. Sci.* **2024**, *141*, e54838. [CrossRef]
22. Bidgoli, H.; Khodadadi, A.A.; Mortazavi, Y. A hydrophobic/oleophilic chitosan-based sorbent: Toward an effective oil spill remediation technology. *J. Environ. Chem. Eng.* **2019**, *7*, 103340. [CrossRef]

23. Li, H.; Yang, Y.; Mu, M.; Feng, C.; Chuan, D.; Ren, Y.; Wang, X.; Fan, R.; Yan, J.; Guo, G. MXene-based polysaccharide aerogel with multifunctional enduring antimicrobial effects for infected wound healing. *Int. J. Biol. Macromol.* **2024**, *261*, 129238. [CrossRef] [PubMed]
24. Chen, Y.; Xiang, Y.; Zhang, H.; Zhu, T.; Chen, S.; Li, J.; Du, J.; Yan, X. A multifunctional chitosan composite aerogel based on high density amidation for chronic wound healing. *Carbohydr. Polym.* **2023**, *321*, 121248. [CrossRef] [PubMed]
25. Wei, S.; Ching, Y.C.; Chuah, C.H. Synthesis of chitosan aerogels as promising carriers for drug delivery: A review. *Carbohydr. Polym.* **2020**, *231*, 115744. [CrossRef] [PubMed]
26. He, S.; Li, J.; Cao, X.; Xie, F.; Yang, H.; Wang, C.; Bittencourt, C.; Li, W. Regenerated cellulose/chitosan composite aerogel with highly efficient adsorption for anionic dyes. *Int. J. Biol. Macromol.* **2023**, *244*, 125067. [CrossRef]
27. Ko, E.; Kim, H. Preparation of chitosan aerogel crosslinked in chemical and ionical ways by non-acid condition for wound dressing. *Int. J. Biol. Macromol.* **2020**, *164*, 2177–2185. [CrossRef]
28. Sirviö, J.A.; Visanko, M.; Liimatainen, H. Synthesis of imidazolium-crosslinked chitosan aerogel and its prospect as a dye removing adsorbent. *RSC Adv.* **2016**, *6*, 56544–56548. [CrossRef]
29. Ozimek, J.; Pielichowski, K. Recent advances in polyurethane/poss hybrids for biomedical applications. *Molecules* **2022**, *27*, 40. [CrossRef]
30. Kuo, S.-W.; Chang, F.-C. POSS related polymer nanocomposites. *Prog. Polym. Sci.* **2011**, *36*, 1649–1696. [CrossRef]
31. Wang, B.T.; Zhang, Y.; Zhang, P.; Fang, Z.P. Functionalization of polyhedral oligomeric silsesquioxanes with bis(hydroxyethyl) ester and preparation of the corresponding degradable nanohybrids. *Chin. Chem. Lett.* **2012**, *23*, 1083–1086. [CrossRef]
32. Tang, L.; Qiu, Z. Effect of poly(ethylene glycol)-polyhedral oligomeric silsesquioxanes on the thermal and mechanical properties of biodegradable poly(l-lactide). *Compos. Commun.* **2017**, *3*, 11–13. [CrossRef]
33. Yu, B.; Yuen, A.C.Y.; Xu, X.; Zhang, Z.-C.; Yang, W.; Lu, H.; Fei, B.; Yeoh, G.H.; Song, P.; Wang, H. Engineering mxene surface with poss for reducing fire hazards of polystyrene with enhanced thermal stability. *J. Hazard. Mater.* **2021**, *401*, 123342. [CrossRef]
34. Teng, S.; Jiang, Z.; Qiu, Z. Effect of different POSS structures on the crystallization behavior and dynamic mechanical properties of biodegradable Poly(ethylene succinate). *Polymer* **2019**, *163*, 68–73. [CrossRef]
35. Hawashi, M.; Altway, A.; Widjaja, T.; Gunawan, S. Optimization of process conditions for tannin content reduction in cassava leaves during solid state fermentation using *Saccharomyces cerevisiae*. *Heliyon* **2019**, *5*, e02298. [CrossRef]
36. Yan, W.; Shi, M.; Dong, C.; Liu, L.; Gao, C. Applications of tannic acid in membrane technologies: A review. *Adv. Colloid Interface Sci.* **2020**, *284*, 102267. [CrossRef] [PubMed]
37. Lin, X.; Zhang, H.; Li, S.; Huang, L.; Zhang, R.; Zhang, L.; Yu, A.; Duan, B. Polyphenol-driving assembly for constructing chitin-polyphenol-metal hydrogel as wound dressing. *Carbohydr. Polym.* **2022**, *290*, 119444. [CrossRef] [PubMed]
38. Luo, X.; Ma, C.; Chen, Z.; Zhang, X.; Niu, N.; Li, J.; Liu, S.; Li, S. Biomass-derived solar-to-thermal materials: Promising energy absorbers to convert light to mechanical motion. *J. Mater. Chem. A* **2019**, *7*, 4002–4008. [CrossRef]
39. Karakochuk, C.D.; Murphy, H.M.; Whitfield, K.C.; Barr, S.I.; Vercauteren, S.M.; Talukder, A.; Porter, K.; Kroeun, H.; Eath, M.; McLean, J.; et al. Elevated levels of iron in groundwater in prey veng province in cambodia: A possible factor contributing to high iron stores in women. *J. Water Health* **2015**, *13*, 575–586. [CrossRef]
40. Zhu, Z.; Jiang, L.; Liu, J.; He, S.; Shao, W. Sustainable, highly efficient and superhydrophobic fluorinated silica functionalized chitosan aerogel for gravity-driven oil/water separation. *Gels* **2021**, *7*, 66. [CrossRef]
41. Sang, Y.; Miao, P.; Chen, T.; Zhao, Y.; Chen, L.; Tian, Y.; Han, X.; Gao, J. Fabrication and evaluation of graphene oxide/hydroxypropyl cellulose/chitosan hybrid aerogel for 5-fluorouracil release. *Gels* **2022**, *8*, 649. [CrossRef] [PubMed]
42. Teng, Z.; Wang, B.; Hu, Y.; Xu, D. Light-responsive nanocomposites combining graphene oxide with poss based on host-guest chemistry. *Chin. Chem. Lett.* **2019**, *30*, 717–720. [CrossRef]
43. Li, Y.; Zhang, Z.; Fu, Z.; Wang, D.; Wang, C.; Li, J. Fluorescence response mechanism of green synthetic carboxymethyl chitosan-Eu^{3+} aerogel to acidic gases. *Int. J. Biol. Macromol.* **2021**, *192*, 1185–1195. [CrossRef] [PubMed]
44. Lu, H.; Jiang, X.; Wang, J.; Hu, R. Degradable composite aerogel with excellent water-absorption for trace water removal in oil and oil-in-water emulsion filtration. *Front. Mater.* **2022**, *9*, 1093164. [CrossRef]
45. Ye, S.; Liu, Y.; Feng, J. Low-density, mechanical compressible, water-induced self-recoverable graphene aerogels for water treatment. *ACS Appl. Mater. Interfaces* **2017**, *9*, 22456–22464. [CrossRef] [PubMed]
46. Cao, Y.; Chen, X.; Li, Y.; Wang, Y.; Yu, H.; Li, Z.; Zhou, Y. Regulating and controlling the microstructure of nanocellulose aerogels by varying the intensity of hydrogen bonds. *ACS Sustain. Chem. Eng.* **2023**, *11*, 1581–1590. [CrossRef]
47. Guo, R.; Hou, X.; Zhao, D.; Wang, H.; Shi, C.; Zhou, Y. Mechanical stability and biological activity of Mg–Sr co-doped bioactive glass/chitosan composite scaffolds. *J. Non-Cryst. Solids* **2022**, *583*, 121481. [CrossRef]
48. Chen, L.; He, X.; Liu, H.; Qian, L.; Kim, S.H. Water adsorption on hydrophilic and hydrophobic surfaces of silicon. *J. Phys. Chem. C* **2018**, *122*, 11385–11391. [CrossRef]
49. Huang, Z.; Wu, J.; Zhao, Y.; Zhang, D.; Tong, L.; Gao, F.; Liu, C.; Chen, F. Starch-based shape memory sponge for rapid hemostasis in penetrating wounds. *J. Mater. Chem. B* **2023**, *11*, 852–864. [CrossRef]
50. Deng, Z.; Guo, Y.; Ma, P.X.; Guo, B. Rapid thermal responsive conductive hybrid cryogels with shape memory properties, photothermal properties and pressure dependent conductivity. *J. Colloid Interface Sci.* **2018**, *526*, 281–294. [CrossRef]
51. Xuan, H.; Du, Q.; Li, R.; Shen, X.; Zhou, J.; Li, B.; Jin, Y.; Yuan, H. Shape-memory-reduced graphene/chitosan cryogels for non-compressible wounds. *Int. J. Mol. Sci.* **2023**, *24*, 1389. [CrossRef] [PubMed]

52. Yang, X.; Cranston, E.D. Chemically cross-linked cellulose nanocrystal aerogels with shape recovery and superabsorbent properties. *Chem. Mater.* **2014**, *26*, 6016–6025. [CrossRef]
53. Peng, Z.; Yu, C.; Zhong, W. Facile preparation of a 3D porous aligned graphene-based wall network architecture by confined self-assembly with shape memory for artificial muscle, pressure sensor, and flexible supercapacitor. *ACS Appl. Mater. Interfaces* **2022**, *14*, 17739–17753. [CrossRef] [PubMed]
54. Li, Y.; Liu, Y.; Liu, Y.; Lai, W.; Huang, F.; Ou, A.; Qin, R.; Liu, X.; Wang, X. Ester crosslinking enhanced hydrophilic cellulose nanofibrils aerogel. *ACS Sustain. Chem. Eng.* **2018**, *6*, 11979–11988. [CrossRef]
55. Wang, C.; Wang, H.; Zou, F.; Chen, S.; Wang, Y. Development of polyhydroxyalkanoate-based polyurethane with water-thermal response shape-memory behavior as new 3D elastomers scaffolds. *Polymers* **2019**, *11*, 1030. [CrossRef] [PubMed]
56. Wu, Z.; Sun, H.; Xu, Z.; Chi, H.; Li, X.; Wang, S.; Zhang, T.; Zhao, Y. Underwater mechanically tough, elastic, superhydrophilic cellulose nanofiber-based aerogels for water-in-oil emulsion separation and solar steam generation. *ACS Appl. Nano Mater.* **2021**, *4*, 8979–8989. [CrossRef]
57. Wang, J.; Du, W.; Zhang, Z.; Gao, W.; Li, Z. Biomass/polyhedral oligomeric silsesquioxane nanocomposites: Advances in preparation strategies and performances. *J. Appl. Polym. Sci.* **2020**, *138*, 49641. [CrossRef]
58. Sun, X.; Tian, Q.; Xue, Z.; Zhang, Y.; Mu, T. The dissolution behaviour of chitosan in acetate-based ionic liquids and their interactions: From experimental evidence to density functional theory analysis. *RSC Adv.* **2014**, *4*, 30282–30291. [CrossRef]
59. Nguyen, H.G.; Nguyen, T.A.H.; Do, D.B.; Pham, X.N.; Nguyen, T.H.; Nghiem, H.L.T.; Nguyen, M.V.; Pham, T.T. Natural cellulose fiber-derived photothermal aerogel for efficient and sustainable solar desalination. *Langmuir* **2023**, *39*, 6780–6793. [CrossRef]
60. Li, M.; Zhu, Z.; Ni, L.; Qi, L.; Yang, Y.; Zhou, Y.; Qi, J.; Li, J. Nanofiber-constructed aerogel solar evaporator for efficient salt resistance and volatile removal. *ACS ES&T Eng.* **2023**, *3*, 2027–2037. [CrossRef]
61. Wu, Y.; Qiu, H.; Sun, J.; Wang, Y.; Gao, C.; Liu, Y. A silsesquioxane-based flexible polyimide aerogel with high hydrophobicity and good adsorption for liquid pollutants in wastewater. *J. Mater. Sci.* **2021**, *56*, 3576–3588. [CrossRef]
62. Gao, H.; Yang, M.; Dang, B.; Luo, X.; Liu, S.; Li, S.; Chen, Z.; Li, J. Natural phenolic compound–iron complexes: Sustainable solar absorbers for wood-based solar steam generation devices. *RSC Adv.* **2020**, *10*, 1152–1158. [CrossRef]
63. Zhang, P.; Piao, X.; Guo, H.; Xiong, Y.; Cao, Y.; Yan, Y.; Wang, Z.; Jin, C. A multi-function bamboo-based solar interface evaporator for efficient solar evaporation and sewage treatment. *Ind. Crop. Prod.* **2023**, *200*, 116823. [CrossRef]
64. He, M.; Alam, K.; Liu, H.; Zheng, M.; Zhao, J.; Wang, L.; Liu, L.; Qin, X.; Yu, J. Textile waste derived cellulose based composite aerogel for efficient solar steam generation. *Compos. Commun.* **2021**, *28*, 100936. [CrossRef]
65. Li, S.; He, Y.; Guan, Y.; Liu, X.; Liu, M.; Xie, M.; Zhou, L.; Wei, C.; Yu, C.; Chen, Y. Cellulose nanofibril-stabilized pickering emulsion and in situ polymerization lead to hybrid aerogel for high-efficiency solar steam generation. *ACS Appl. Polym. Mater.* **2020**, *2*, 4581–4591. [CrossRef]
66. Liu, J.; Liu, Q.; Ma, D.; Yuan, Y.; Yao, J.; Zhang, W.; Su, H.; Su, Y.; Gu, J.; Zhang, D. Simultaneously achieving thermal insulation and rapid water transport in sugarcane stems for efficient solar steam generation. *J. Mater. Chem. A* **2019**, *7*, 9034–9039. [CrossRef]
67. Liu, M.; He, X.; Gu, J.; Li, Z.; Liu, H.; Liu, W.; Zhang, Y.; Zheng, M.; Yu, J.; Wang, L.; et al. Cotton fiber-based composite aerogel derived from waste biomass for high-performance solar-driven interfacial evaporation. *Ind. Crop. Prod.* **2024**, *211*, 118220. [CrossRef]
68. Meng, R.; Lyu, J.; Zou, L.; Zhong, Q.; Liu, Z.; Zhu, B.; Chen, M.; Zhang, L.; Chen, Z. CNT-based gel-coated cotton fabrics for constructing symmetrical evaporator with up/down inversion property for efficient continuous solar desalination. *Desalination* **2023**, *554*, 116494. [CrossRef]
69. Qi, Q.; Wang, Y.; Wang, W.; Ding, X.; Yu, D. High-efficiency solar evaporator prepared by one-step carbon nanotubes loading on cotton fabric toward water purification. *Sci. Total Environ.* **2019**, *698*, 134136. [CrossRef]
70. Li, W.; Jian, H.; Wang, W.; Yu, D. Highly efficient solar vapour generation via self-floating three-dimensional Ti_2O_3-based aerogels. *Colloids Surf. A Physicochem. Eng. Asp.* **2021**, *634*, 128031. [CrossRef]
71. Zhang, X.; Pi, M.; Lu, H.; Li, M.; Wang, X.; Wang, Z.; Ran, R. A biomass hybrid hydrogel with hierarchical porous structure for efficient solar steam generation. *Sol. Energy Mater. Sol. Cells* **2022**, *242*, 111742. [CrossRef]
72. Chen, Y.; Cai, J.; Liu, D.; Liu, S.; Lei, D.; Zheng, L.; Wei, Q.; Gao, M. Zinc-based metal organic framework with antibacterial and anti-inflammatory properties for promoting wound healing. *Regen. Biomater.* **2022**, *9*, rbac019. [CrossRef] [PubMed]
73. Zhang, Y.; Deng, W.; Wu, M.; Liu, Z.; Yu, G.; Cui, Q.; Liu, C.; Fatehi, P.; Li, B. Robust, scalable, and cost-effective surface carbonized pulp foam for highly efficient solar steam generation. *ACS Appl. Mater. Interfaces* **2023**, *15*, 7414–7426. [CrossRef] [PubMed]

Disclaimer/Publisher's Note: The statements, opinions and data contained in all publications are solely those of the individual author(s) and contributor(s) and not of MDPI and/or the editor(s). MDPI and/or the editor(s) disclaim responsibility for any injury to people or property resulting from any ideas, methods, instructions or products referred to in the content.

Article

Preparation of a Novel Lignocellulose-Based Aerogel by Partially Dissolving Medulla Tetrapanacis via Ionic Liquid

Long Quan [1,†], Xueqian Shi [1,†], Jie Zhang [1], Zhuju Shu [1,*] and Liang Zhou [1,2,*]

[1] School of Materials and Chemistry, Anhui Agricultural University, Hefei 230036, China; longquan_ahau@163.com (L.Q.); 15357108782@163.com (X.S.); zhangjie@goldenchemical.com (J.Z.)
[2] Key Lab of State Forest and Grassland Administration on Wood Quality Improvement & Utilization, Hefei 230036, China
* Correspondence: shuzuju@ahau.edu.cn (Z.S.); mcyjs1@ahau.edu.cn (L.Z.)
† These authors contributed equally to this work.

Abstract: A novel lignocellulosic aerogel, MT-LCA, was successfully prepared from MT by undergoing partial dissolution in an ionic liquid, coagulation in water, freezing in liquid nitrogen, and subsequent freeze-drying. The MT-LCA preserves its original honeycomb-like porous structure, and the newly formed micropores contribute to increased porosity and specific surface area. FT-IR analysis reveals that MT, after dissolution and coagulation, experiences no chemical reactions. However, a change in the crystalline structure of cellulose is observed, transitioning from cellulose I to cellulose II. Both MT and MT-LCA demonstrate a quasi-second-order kinetic process during methylene blue adsorption, indicative of chemical adsorption. The Langmuir model proves to be more appropriate for characterizing the methylene blue adsorption process. Both adsorbents exhibit monolayer adsorption, and their effective adsorption sites are uniformly distributed. The higher porosity, nanoscale micropores, and larger pore size in MT-LCA enhance its capillary force, providing efficient directional transport performance. Consequently, the prepared MT-LCA displays exceptional compressive performance and efficient directional transport capabilities, making it well-suited for applications requiring high compressive performance and selective directional transport.

Keywords: Medulla tetrapanacis; lignocellulose-based aerogel; ionic liquid; absorption of dye

Citation: Quan, L.; Shi, X.; Zhang, J.; Shu, Z.; Zhou, L. Preparation of a Novel Lignocellulose-Based Aerogel by Partially Dissolving Medulla Tetrapanacis via Ionic Liquid. *Gels* 2024, 10, 138. https://doi.org/10.3390/gels10020138

Academic Editor: Pasquale Del Gaudio

Received: 23 January 2024
Revised: 3 February 2024
Accepted: 6 February 2024
Published: 9 February 2024

Copyright: © 2024 by the authors. Licensee MDPI, Basel, Switzerland. This article is an open access article distributed under the terms and conditions of the Creative Commons Attribution (CC BY) license (https://creativecommons.org/licenses/by/4.0/).

1. Introduction

Abundant storage stems of plants supply enormous quantities of lignocellulosic resources [1]. To deal with the problems arising from fossil-based energy and products, the lignocellulosic materials are attracting greater attention due to their biodegradable, low-cost, and versatile applications [1,2]. Since the living stems of plants play a vital role in transporting water from the earth to the crown or leaves and mechanical support, the structure of the stem is assembled in a porous and hierarchical manner built mainly of cellulose, hemi-cellulose, and lignin [3]. Therefore, various kinds of stem directly or indirectly, namely after certain adjustments or modifications, are used as porous functional materials [4]. *Tetrapanax papyriferus*, is a typical fast-growing bush with large leaves and its natural range is in southern China [5]. The stem pith of the plant, which is known as Medulla tetrapanacis (MT), is a traditional Chinese medicine called "da-tong-cao". It is considered helpful in promoting diuresis and alleviating edema and abdominal distention [6]. Moreover, MT also is a traditional resource for preparing painting paper, since it has a white color and reasonable luster. Recently, MT has been pyrolyzed into porous carbon materials for multiple applications as an absorbent or supercapacitor electrode [7]. It turns out that the original porous structure in itself is beneficial for the required properties of the output products [8].

Lignocellulose-based aerogels have been a subject of increasing interest in materials science and sustainable technologies, since they can replace or partially replace aerogels originated from synthetic polymers [9]. These aerogels are lightweight, highly porous materials made from lignocellulosic biomass sources such as wood, agricultural residues, and paper waste. They have a wide range of potential applications due to their biodegradability, low cost, and eco-friendliness [10,11]. Lignocellulose-based aerogels have demonstrated potential in the removal of dyes and pigments from aqueous solutions, making them useful in environmental remediation, wastewater treatment, and water purification [12,13]. Compared to the common synthetic polymer sorbents, the lignocellulosic aerogel is more environmentally friendly, biodegradable, and relatively low-cost [14,15]. In addition, heat or sound insulation, storage of energy, catalysis support and other versatile applications of lignocellulosic aerogels have also been reported recently. Lignocellulosic aerogels are generally prepared from wood, sourced through two independent strategies. The first one is dehydration from hydrogel, which is mostly harvested by coagulating the dissolving cellulose or lignocellulose solution, by freeze-drying or supercritical drying. The second strategy is to construct them with nanofilaments, which generally are composed of cellulose and lignocellulose. Moreover, Garemark proposed a different method to produce hierarchical and anisotropic lignocellulosic aerogel by keeping the main microstructural porous of wooden stem intact and constructing aerogels inside the lumen simultaneously by partially dissolving and coagulating cell walls [16]. It is found that such a structure is beneficial to fabricating anisotropic cellulose aerogels and can be considered as a universal substrate in multiple fields.

As mentioned earlier, Medulla tetrapanacis (MT) exhibits a characteristic porous microstructure, suggesting that its density should be lower than that of the wooden stem, as observed in Garemark's development of a lignocellulosic aerogel [16]. This observation served as an inspiration for us to adopt a similar approach in crafting anisotropic lignocellulosic aerogel from MT. To partially dissolve the MT efficiently, one kind of EmimOAc was selected, since it was reported that it had good dissolving ability not only for cellulose but also for lignocellulosic resources and it can maintain relatively low viscosity at room temperature, which benefits permeation [14]. Subsequently, we subjected the aerogel to testing to evaluate its capacity for adsorbing dyes or pigments from water. Additionally, we investigated the adsorption kinetics of the aerogel to comprehend the adsorption process. To elucidate the mechanism of adsorption, we conducted examinations of the chemical structure, microstructure, porosity, and thermal stability. We anticipate that the anisotropic lignocellulosic aerogel derived from MT through this method holds potential applications in both dye removal from water solutions and pigment adsorption in organic solvents. This development not only enhances the value of MT but also provides an economical alternative for anisotropic lignocellulosic aerogel in the field of environmental protection.

2. Results and Discussion

2.1. Morphological Features

To investigate the morphological changes of MT and MT-LCA before and after EmimOAc treatment, SEM observations were conducted. Figure 1a shows the cross-sectional SEM image of MT, which inherently possesses a honeycomb-like porous structure with an average pore size of approximately 150 μm and an average pore wall thickness of around 1 μm. Figure 1c depicts the longitudinal section of MT, revealing a well-organized orientation of pore channels. Compared to the transverse direction, the longitudinal pores in MT are larger, presenting a more open and uniform porous structure. The abundant porous structure of MT contributes to its high porosity (40%) and low density (0.05 g/cm^3). This characteristic not only facilitates rapid infiltration of EmimOAc into the pores during dissolution but also shortens the dissolution time while ensuring a more uniform dissolution process.

Figure 1b and Figure 1d, respectively, depict the cross-section and tangential-section SEM images of MT-LCA. MT-LCA not only retains the original transverse honeycomb-like porous structure and longitudinally oriented pore channels of MT but also exhibits an increased number of pores in the original cell walls. Simultaneously, a fibrous network structure appears within the cell lumens, creating numerous tiny pores. Under the effects of EmimOAc dissolution and swelling, the original cell wall structure of MT undergoes disruption, leading to the dissolution of some natural polymers in the EmimOAc solution. This dissolution results in the etching of the cell walls, forming distinct new pores. The natural polymers dissolved in the EmimOAc solution enter the cell lumens and, after coagulation with deionized water, form a hydrogel. Rapid freezing with liquid nitrogen forms small-sized ice crystals within the hydrogel. Following freeze-drying, both the pores created by EmimOAc etching and those formed by ice crystals are preserved.

Therefore, MT-LCA has been successfully prepared from MT through being partially dissolved in an ionic liquid, coagulated in water, frozen in liquid nitrogen and freeze-dried. It retains the original transverse honeycomb-like porous structure and longitudinally oriented pore channels while acquiring smaller micropores. The preservation of anisotropic structures imparts excellent compressive performance and efficient adsorption rates to MT-LCA. The newly generated micropores will contribute to an increased porosity in MT-LCA, enhancing its capability to adsorb dye molecules and other solvents [17].

Figure 1. SEM pictures of MT and MT-LCA: (**a**) Cross section of MT, (**b**) Cross section of MT-LCA, (**c**) Tangential section of MT, (**d**) Tangential section of MT-LCA.

2.2. FTIR Analysis

The infrared spectra of MT and MT-LCA are shown in Figure 2. The infrared spectra of MT and MT-LCA are quite similar, with no significant changes observed in the positions of the main absorption peaks. This suggests that during the process of dissolution and regeneration, there were no chemical reactions occurring in the lignocellulosic fibers of MT, and the molecular vibrational modes remained unchanged.

As shown in Figure 2a, the absorption peak around 3358 cm^{-1} corresponds to the -OH stretching vibration of cellulose in both MT and MT-LCA. In MT-LCA, the absorption peak at 3358 cm^{-1} shifts to a higher wavenumber, indicating that the hydrogen bonding within and between the original natural polymers (cellulose, hemicellulose, and lignin) in Medulla tetrapanacis (MT) has been somewhat disrupted, leading to a weakening of the hydrogen-bonding interactions after partial dissolution in EmimOAc [18]. The peaks at 2919 cm^{-1} and 2850 cm^{-1} correspond to the C-H stretching vibrations of cellulose. The absorption peak at 1601 cm^{-1} corresponds to the aromatic ring skeletal vibration of lignin, while the peak at 1100 cm^{-1} is associated with the syringyl propane structural unit in lignin [19]. Both peaks in MT-LCA are weakened, indicating a lower lignin content in MT-LCA compared to MT, possibly due to partial degradation of lignin during the dissolution and regeneration process. The infrared absorption peak at 1420 cm^{-1} is related to the crystalline spectrum of cellulose I (Figure 2b). The reduction in the characteristic absorption peak at 1420 cm^{-1} suggests a decrease in the cellulose I content in MT-LCA, possibly due to the dissolution of natural cellulose in MT [20]. The peaks at 1256 cm^{-1} and 897 cm^{-1} correspond to the C-O stretching vibration of cellulose and the characteristic absorption peak of the β-O-4 glycosidic bond in hemicellulose, respectively. These peaks show no significant changes after dissolution and regeneration. Through infrared spectroscopy, a comparison of the chemical structures of Medulla tetrapanacis (MT) before and after dissolution reveals a weakening of hydrogen bonds in the original natural polymers, partial degradation of lignin, and a transition of cellulose from a crystalline I structure to II structure in MT-LCA [21].

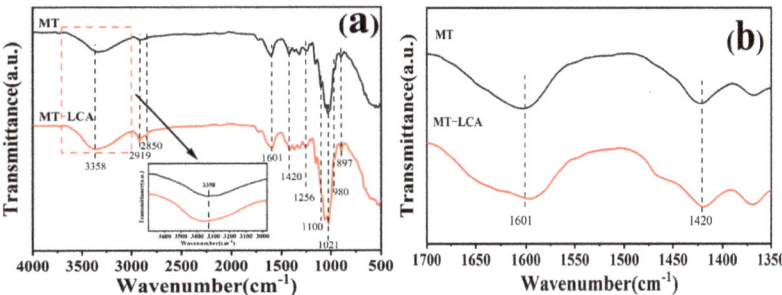

Figure 2. (a) FTIR spectra of MT and MT-LCA, (b) Magnified image from a to show the area between 1350 to 1700 cm^{-1}.

2.3. XRD Analysis

To investigate the crystalline structure of MT and MT-LCA, X-ray diffraction (XRD) was employed for sample characterization. The XRD test results for MT and MT-LCA are shown in Figure 3, where MT exhibits four diffraction peaks distributed at 14.7°, 15.1°, 21.7°, and 24.2°.

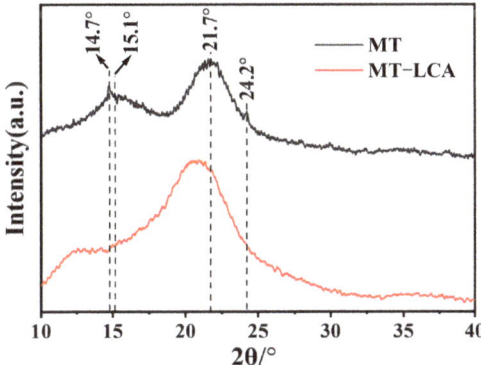

Figure 3. XRD patterns of MT and MT-LCA.

The diffraction peaks at 15.1° and 21.7° correspond to characteristic cellulose I-type diffraction peaks, representing the 101 and 200 crystallographic planes of cellulose [22], indicating that MT possesses a typical cellulose I crystalline structure. The diffraction peaks at 14.7° and 24.2° are characteristic diffraction peaks of calcium salts, suggesting the presence of a small amount of calcium salts in MT. In comparison to MT, the positions of the diffraction peaks in MT-LCA, prepared through dissolution and regeneration, have changed [4]. As shown in Figure 3, the diffraction peak at 21.7° in MT shifts to 20.6° after dissolution and regeneration. This shift indicates a transformation of the cellulose crystalline structure from the original cellulose I to cellulose II type [23]. For the same reason, the diffraction peak at 15.1° in MT also shifts to 12.5°. Comparing the XRD curves of MT and MT-LCA, the characteristic peaks related to calcium salts at 14.7° and 24.2° disappear after dissolution and regeneration, suggesting that calcium salts are dissolved and removed from the material during dissolution and washing processes [4].

2.4. Thermal Stability

To compare the thermal stability of MT and MT-LCA, thermogravimetric (TG) testing was conducted [24], and the results along with a differentiated thermogravimetric (DTG) curve are shown in Figure 4 and Table 1. The thermal weight loss curves of MT and MT-LCA are generally similar, indicating that their thermal decomposition patterns are roughly the same. The first stage, ranging from 30 °C to 100 °C, primarily involves the desorption weight loss of physically adsorbed water in wood cellulose and the removal of water molecules bound by hydrogen bonding in cellulose [25]. The second stage, occurring from 230 °C to 400 °C, involves the sequential thermal decomposition of hemicellulose and cellulose. As the temperature gradually rises, the rate of thermal weight loss accelerates, with the β-O-4 ether bonds of some glucose units in the cellulose main chain breaking [26], followed by the cleavage of the pyranose rings. The third stage, occurring after 400 °C, involves the decarboxylation of cellulose, leading to the formation of carbon black and the release of volatile gases [27]. During this stage, lignin also begins to undergo thermal degradation [28].

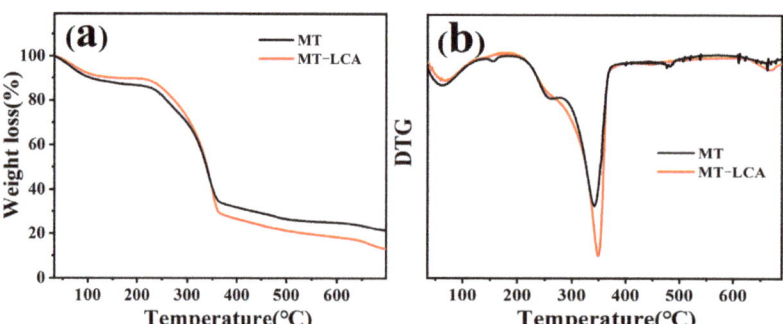

Figure 4. Thermogravimetric curves of MT and MT-LCA: (**a**) TGA curves, (**b**) DTG curves.

Table 1. Thermal performance data of MT and MT-LCA.

Samples	$T_{d5\%}$/°C	T_{Max}/°C
MT	64	166/364
MT-LCA	93	135/364

Upon careful comparison of the thermogravimetric curves of MT and MT-LCA, it is observed that the initial thermal decomposition temperature of cellulose in MT-LCA is lower, and its decomposition rate is faster compared to MT. This suggests that the thermal stability of MT-LCA is lower. The reason for this phenomenon may be attributed to the disruption of the crystalline structure of cellulose in MT-LCA after dissolution and regeneration [29]. As the temperature increases, the amorphous region of cellulose in MT-LCA begins to decompose. In comparison to the natural cellulose in MT, the regenerated cellulose in MT-LCA exhibits lower crystallinity, with a larger proportion of the amorphous region. The thermal stability of cellulose decreases with a reduction in crystallinity [30].

Through a comparison of the thermogravimetric curves, it is evident that MT-LCA has lower thermal stability relative to MT.

2.5. Compressive Tests

To investigate the compression performance of MT and MT-LCA, mechanical compression experiments were conducted. Figure 5 depicts the compression stress–strain curves for MT and MT-LCA.

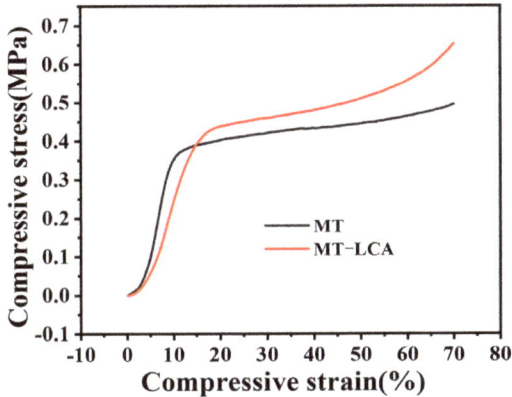

Figure 5. Compression stress–strain curves of MT and MT-LCA.

MT and MT-LCA undergo three stages during compression: the first stage involves a linear increase in compression stress with increasing compression strain in the elastic phase when the compression strain is relatively small. When the strain exceeds the yield strength, the regular pore structure within MT and MT-LCA starts to collapse. The internal network fibers come into contact and squeeze against each other, leading to plastic deformation in MT and MT-LCA. During this stage, as strain increases, stress increases more gradually and approaches a plateau, representing the yielding platform in the second stage. The third stage is the densification stage, where with continuous compression deformation of MT and MT-LCA, stress sharply increases, and the regular honeycomb-like pore structure is severely disrupted, indicating a trend toward densification of the internal three-dimensional network structure [31].

Compared to MT, MT-LCA has some of the natural polymers in the cell wall dissolved by EmimOAc, causing a certain degree of damage to its pore structure, resulting in reduced compression strength in the initial stage. As shown in Table 2, MT-LCA has a higher density than MT (densities of MT-LCA and MT are 0.084 g/cm^3 and 0.050 g/cm^3, respectively). With increasing compression strain, the stronger intermolecular forces in the higher-density MT-LCA contribute to its superior compressive performance compared to MT. Although MT-LCA exhibits slightly lower mechanical strength than MT in the initial stage, it outperforms cellulose aerogels prepared from bottom to top. MT-LCA retains the original honeycomb-like pore structure and regularly oriented pore channel structure of MT, while other cellulose aerogels, lacking honeycomb-like pore structures and regularly oriented pore channels, easily experience pore structure collapse and show inferior compression strength under pressure [19].

Table 2. Density and compressive mechanical property data of MT and MT-LCA.

Samples	Density (g/cm^3)	Compressive Modulus (MPa)	Compressive Strength (MPa)
MT	0.050 ± 0.006	5.223 ± 0.121	0.411 ± 0.011
MT-LCA	0.084 ± 0.008	2.484 ± 0.081	0.472 ± 0.051

2.6. Porosity Analysis

To investigate the changes in specific surface area before and after the partial dissolution of MT, nitrogen adsorption–desorption tests were conducted on MT and MT-LCA. The nitrogen adsorption–desorption curves for MT and MT-LCA are shown in Figure 6. The IUPAC classifies physical adsorption isotherms into six types, and the nitrogen adsorption–desorption curves of MT and MT-LCA fall into Type II, characteristic of mesoporous adsorbent materials with interactions between adsorbent and adsorbate. The presence of an H3 hysteresis loop indicates the existence of narrow mesoporous structures (2–50 nm) [4,5,7].

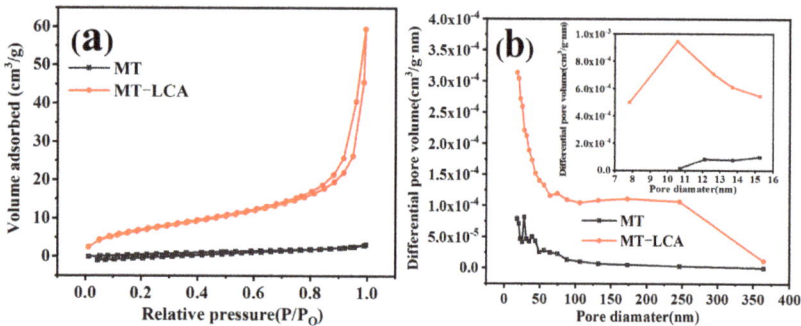

Figure 6. Nitrogen adsorption–desorption curves and pore size distribution diagram of MT and MT-LCA: (**a**) Nitrogen adsorption–desorption curves, (**b**) Pore distribution diagram.

At low pressures, the adsorption behavior of MT-LCA is mainly monolayer adsorption, transitioning to multilayer adsorption as the pressure increases. Comparing the nitrogen adsorption–desorption curves of MT and MT-LCA, at low relative pressures (<0.01), MT-LCA shows a distinct upward trend, indicating the presence of numerous micropores. In contrast, the curve for MT remains relatively unchanged, suggesting the absence of micropores in untreated MT. When the relative pressure exceeds 0.8 and the adsorbed N_2 transforms into a liquid state, the nitrogen adsorption–desorption curve for MT-LCA sharply rises, indicating the presence of some large pores where nitrogen condenses, causing the curve to increase [32].

The nitrogen adsorption–desorption curves were used to calculate their porosity, total pore volume, average pore diameter, and specific surface area, with the results shown in Table 3. The porosity of MT-LCA increases to 51%, and the average pore diameters of MT and MT-LCA are 36.71 nm and 22.17 nm, respectively. The total pore volume and specific surface area of MT are only 0.001 cm^3/g and 0.669 m^2/g, significantly lower than those of MT-LCA.

Table 3. Porosities and pore-related physical properties of MT and MT-LCA.

Samples	Porosity (%)	Total Pore Volume (cm^3/g)	Average Pore Size (nm)	Specific Surface Area (m^2/g)
MT	38	0.001	36.71	0.669
MT-LCA	51	0.041	22.17	36.576

To further compare the pore structures of MT and MT-LCA, pore size distribution calculations were performed using the BJH and HK models [4,8]. The pore size distribution results are shown in Figure 6b. In MT, the mesopore diameters are primarily concentrated in the range of 10 to 16 nm, and the macropore diameters are mainly in the range of 50 to 100 nm. However, the quantities of micropores, mesopores, and macropores in MT are relatively low. In contrast, MT-LCA exhibits a higher abundance of micropores, mesopores, and macropores. Additionally, the distribution ranges of mesopore and macropore diameters in MT-LCA are broader compared to MT. The mesopore diameters in MT-LCA are mainly distributed in the range of 8 to 16 nm, and the number of macropores with diameters in the range of 50 to 250 nm is also higher than in MT.

The changes in the pore structures described above are consistent with the observations from SEM. After dissolution and swelling with EmimOAc, the original cell wall structure of MT is disrupted. Some cellulose and hemicellulose dissolve in the EmimOAc solution, resulting in etching of the cell wall and the formation of new pores. The cellulose and hemicellulose dissolved in the EmimOAc solution enter the cell cavity and, after deionized water coagulation, form nano-fibers that aggregate into a hydrogel [15]. Rapid freezing with liquid nitrogen at this stage forms smaller ice crystals inside the hydrogel. After freeze-drying, the pores formed by the EmimOAc solution etching and the pores formed by the ice crystals in the hydrogel are retained. This is likely the main factor contributing to the improvement of the pore structure-related data for MT-LCA.

2.7. Absorption of Dye

Methylene blue, as a model dye, was used to compare the adsorption capacity of MT and MT-LCA for dyes, and the results are shown in Figure 7. To investigate the influence of solution pH on the adsorption performance of the adsorbents, the adsorption amounts of MT and MT-LCA for methylene blue were tested at different pH values. The relationship curve between solution pH values and adsorption amounts is depicted in Figure 6a. As the pH values gradually increase, the adsorption trend of both MT and MT-LCA for methylene blue solution is generally similar, showing a gradual increase. This is mainly because MT and MT-LCA contain carboxyl and hydroxyl groups, and the acidity and alkalinity of the solution affect the charge of these two functional groups in the aqueous solution [33]. When the solution is acidic, the adsorbent carries a positive charge, while methylene blue exists in the solution in cationic form. With a large number of H^+ ions in the solution occupying the surface of MT and MT-LCA, electrostatic repulsion occurs with the cationic form of methylene blue in the solution, resulting in a lower adsorption amount at lower pH values. However, as the pH value rises, and the solution becomes alkaline, the adsorbent's charge transforms into a negative charge. This leads to electrostatic attraction between the adsorbent and methylene blue dye, gradually increasing the adsorption amount [34].

To investigate the effect of adsorption time on the adsorption performance of MT and MT-LCA for methylene blue, the adsorption amounts of MT and MT-LCA for methylene blue were tested at different time intervals. The relationship curve between adsorption time and adsorption amount is shown in Figure 7b. Before reaching adsorption equilibrium, the adsorption amounts of methylene blue by both MT and MT-LCA increase with the extension of adsorption time. When the adsorption time is in the range of 0 to 400 min, the adsorption amount of the adsorbent increases rapidly. However, beyond 400 min, the growth of the adsorption amount becomes more gradual. Eventually, MT reaches adsorption equilibrium around 1100 min, while MT-LCA achieves adsorption equilibrium around 850 min. This is because, in the initial stage of adsorption, the surfaces of MT and MT-LCA have numerous adsorption sites, and there is a significant difference in methylene blue concentration between their interiors and exteriors, facilitating the rapid diffusion and attachment of methylene blue molecules to the adsorbent. As time progresses, the adsorption sites on the surface of the adsorbent are gradually occupied by methylene blue molecules, leading to a gradual reduction in the adsorption rate [35]. Ultimately, the effective adsorption sites of the adsorbent are fully occupied, and the adsorption capacity of MT and MT-LCA for dye molecules reaches saturation, achieving dynamic equilibrium in the adsorption and desorption of dye molecules by MT and MT-LCA [32].

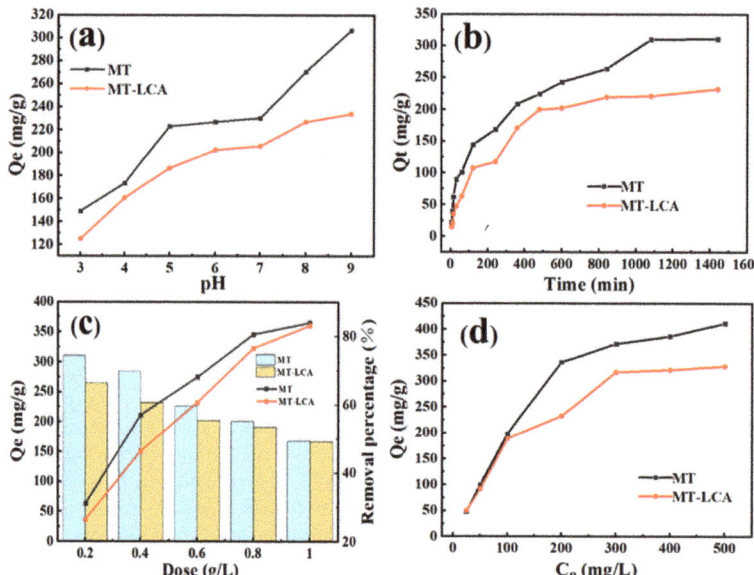

Figure 7. The effect of different conditions on the adsorption performance of adsorbents: (**a**) Solution pH value, (**b**) Adsorption time, (**c**) Amount of adsorbent added, (**d**) Initial concentration of solution.

To investigate the impact of adsorbent dosage on the adsorption performance of MT and MT-LCA for methylene blue, tests were conducted with different dosages of MT and MT-LCA to measure the adsorption amount and removal efficiency of methylene blue. The relationship between adsorbent dosage, removal efficiency, and adsorption amount is shown in Figure 7c. As the dosage of MT and MT-LCA increases, more adsorbent provides additional adsorption sites, facilitating the adsorption of more methylene blue molecules. Therefore, the removal efficiency of methylene blue by the adsorbent increases with the rising dosage of MT and MT-LCA. However, as the concentration of methylene blue in the solution decreases and adsorption approaches saturation, the adsorption sites of the adsorbent cannot be fully utilized [35]. Consequently, when the dosage increases to a certain extent, the increase in removal efficiency becomes more gradual. The adsorption amount and removal efficiency exhibit an inverse trend. With an increasing dosage of the adsorbent, the adsorption amount of MT and MT-LCA for methylene blue continuously decreases. This is due to an excessive number of adsorption sites in the solution, leading to increased competition among more adsorbents for the adsorption of dye molecules and resulting in a reduction in the adsorption amount per unit mass of adsorbent for dye molecules [34].

To investigate the impact of initial solution concentration on the adsorption performance of MT and MT-LCA for methylene blue, tests were conducted with MT and MT-LCA at different initial concentrations of methylene blue, and the relationship between the initial concentration of the solution and the adsorption amount is shown in Figure 7d. With an increase in the initial concentration of methylene blue solution, the adsorption amount of both MT and MT-LCA also increases, but the trend gradually levels off. The main reason for this is that as the concentration of methylene blue in the solution increases, the concentration difference between the solution and the adsorbent becomes larger, driving more methylene blue molecules into the interior of the adsorbent, where they combine with the adsorption sites of MT and MT-LCA. However, the effective adsorption sites on the adsorbent are limited. Therefore, as the concentration of methylene blue solution continues to increase, the effective adsorption sites of the adsorbent are almost entirely occupied by methylene blue molecules, leading to a tendency of the adsorption amount by the adsorbent for methylene blue solution to approach equilibrium [34,35].

2.8. Absorption Mechanism of Dye

In order to investigate the adsorption mechanisms of MT and MT on the aqueous solution of methylene blue, adsorption dynamic models were employed to predict the relationship between adsorption amount and adsorption time. This approach was utilized to explore the adsorption mechanisms involved in the process of MT and MT adsorbing methylene blue.

The dynamic simulation results obtained from Equations (3) and (4), as well as the adsorption fitting curves of the dynamic models, are shown in Figure 8 and Table 4. The correlation coefficients obtained from the quasi-second-order kinetic model for both MT and MT-LCA are higher than those obtained from the quasi-first-order kinetic model. Particularly, the correlation coefficient for the quasi-second-order kinetic model of MT-LCA reaches up to 0.99, providing a more accurate description of the adsorption process of methylene blue molecules by MT-LCA. The adsorption equilibrium amount obtained from the quasi-second-order kinetic model for MT-LCA also closely matches the experimentally measured adsorption amount. Consequently, the adsorption process of both MT and MT-LCA on methylene blue is better described by the quasi-second-order kinetic model. This indicates that the adsorption of methylene blue by MT and MT-LCA is primarily a chemical adsorption process, and the adsorption capacity of MT and MT-LCA is positively correlated with their adsorption sites [36]. The adsorption process involves both surface adsorption and internal diffusion [37].

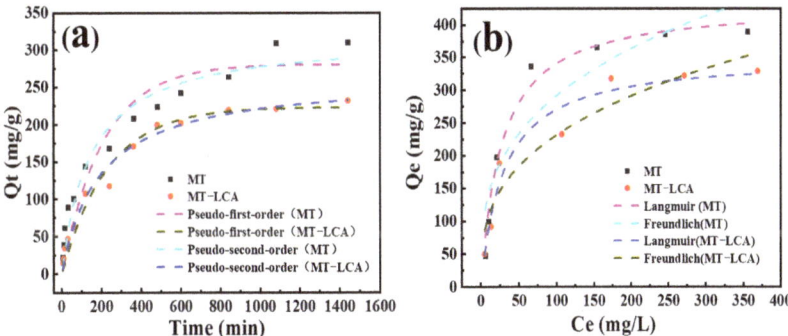

Figure 8. Kinetic adsorption fitting curves (**a**) and adsorption isotherms (**b**) of MT and MT-LCA.

Table 4. Parameters related to dynamic simulation and isothermal model of MT and MT-LCA.

	Models	Parameters	MT	MT-LCA
Kinetic adsorption fitting curves	Quasi-first-order kinetic models	Q_e K_1 R^2	280.68 0.0048 0.90	223.56 0.0044 0.96
	Quasi-second-order kinetic models	Q_e K_2 R^2	318.35 0.000024 0.95	263.78 0.000019 0.99
Adsorption isotherms	Langmuir	Q_{max} K_L R^2	425.86 0.040 0.98	408.15 0.038 0.95
	Freundlich	K_F n R^2	67.56 3.15 0.84	63.89 3.10 0.90

To investigate the interaction between MT, MT-LCA, and methylene blue, the Langmuir isotherm equation (Equation (5)) and Freundlich isotherm equation (Equation (6)) were employed to fit the data of methylene blue adsorption by MT and MT-LCA at different initial concentrations. The obtained adsorption isotherms and isotherm parameters are presented in Figure 8 and Table 4. In comparison to the correlation coefficients of the Freundlich model adsorption isotherms (0.84 and 0.90), the correlation coefficients of the Langmuir model adsorption isotherms for MT and MT-LCA (0.98 and 0.95) are higher. This suggests that the Langmuir model is more suitable for describing the adsorption of methylene blue by MT and MT-LCA [38]. Therefore, both adsorbents exhibit monolayer adsorption of methylene blue, and their effective adsorption sites are uniformly distributed on their surfaces.

2.9. Absorption of Oil

To further analyze the adsorption performance of MT and MT-LCA, their saturation adsorption capacities were tested for six different oils and organic solvents [30,39], as shown in Figure 9. For viscous oils such as soybean oil, engine oil, and paraffin oil, MT-LCA performs slightly worse than MT, but the difference is not significant. However, MT-LCA exhibits a significantly higher saturation adsorption capacity for organic solvents compared to MT. Due to its higher density, MT-LCA has a smaller volume than MT under the same conditions. Therefore, when immersed in viscous oils, the contact area of MT-LCA with the oil is smaller than that of MT. Meanwhile, since viscous oils provide high saturation adsorption capacity through surface adhesion, MT achieves a higher saturation adsorption capacity [40]. When MT and MT-LCA are immersed in organic solvents with lower viscosity, the volume advantage of MT is less pronounced. Additionally, organic solvents can migrate to the interior through surface tension, utilizing the material's porous structure. Given that the porous structure of MT-LCA is superior to MT, its saturation adsorption capacity for organic solvents is significantly higher than that of MT.

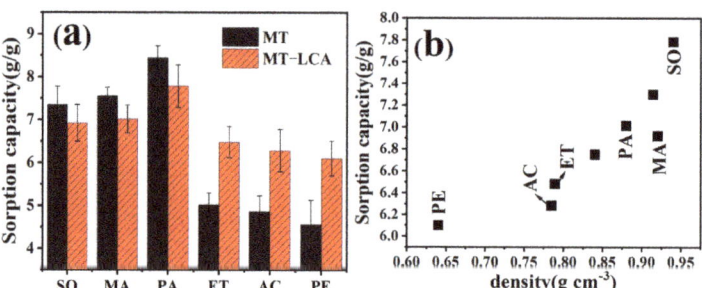

Figure 9. Saturated adsorption capacities of MT and MT-LCA: (**a**) Saturated adsorption capacities of MT and MT-LCA for different oils, (**b**) Relationship between saturated adsorption capacities of MT-LCA and oil density.

To further explore the adsorption mechanism of MT-LCA for oils, we analyzed the relationship between the saturation adsorption capacity of the reed cellulose aerogel and the density of oils and organic solvents. As shown in Figure 9b, the saturation adsorption capacity of MT-LCA for oils and organic solvents is positively correlated with the density of oils and organic solvents. The higher the density of oils and organic solvents, the greater the saturation adsorption capacity of MT-LCA. This is primarily because, with a constant volume of the adsorbent, the number of effective adsorption sites is fixed, and the volume change range for adsorbing different types of oils and organic solvents is relatively small [30]. Therefore, the higher the density of oils and organic solvents, the greater the mass of oils and organic solvents that the adsorbent can adsorb [30,39].

To explore the directional transport performance of MT and MT-LCA for solvents, the transport distance of soybean oil (Sudan III staining) in MT and MT-LCA was tested at different time intervals. Figures 10 and 11 illustrate the transport capabilities of MT and MT-LCA for soybean oil, and experimental results indicate a faster transport rate of soybean oil in MT-LCA. As shown in Figure 11, when the time reaches 15 min, not only does MT-LCA adsorb soybean oil more rapidly, but it also achieves a higher adsorption capacity. This phenomenon is primarily attributed to two factors. On the one hand, MT is formed by relatively closed cell stacking, without transport channels between cells. After partial dissolution and swelling in an ionic liquid, the original cell walls of MT are eroded, forming transport channels that facilitate the directional transport of the adsorbent. On the other hand, during the preparation of MT-LCA, a large number of macropores, mesopores, and micropores are generated, leading to increased porosity, surface area, and pore size. The higher porosity, nanoscale micropores, and larger pore size in MT-LCA enhance its capillary force, providing efficient directional transport performance [30,39]. Therefore, through partial dissolution and regeneration of MT to transform it into an adsorbent material with a hierarchical pore structure, excellent directional transport functionality is achieved. MT-LCA exhibits potential for selective directional transport.

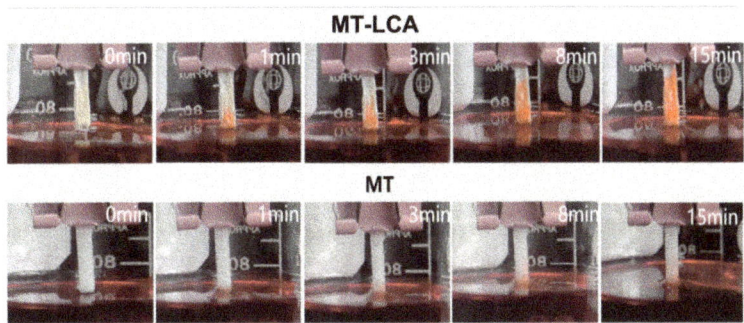

Figure 10. Absorbing behavior of the MT and MT-LCA in soybean oil (Sudan III dyed).

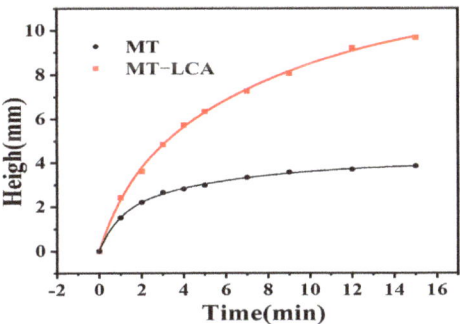

Figure 11. Relationship between height and time of adsorption of soybean oil by MT and MT-LCA.

3. Conclusions

A novel lignocellulosic aerogel, MT-LCA, has been successfully prepared from MT through partial dissolution in an ionic liquid, coagulation in water, freezing in liquid nitrogen and freeze-drying. The MT-LC preserves the original honeycomb-like porous structure in the lateral direction and the regularly oriented pore channel structure in the longitudinal direction. Additionally, smaller micropores were also found in cell walls and in the lumen. The retention of anisotropic structure imparts excellent compressive performance to MT-LCA, while the newly formed micropores enhance its porosity and specific surface area. FT-IR results indicate that MT, after dissolution and coagulation, undergoes no chemical reaction. However, there is a change in the crystalline structure of cellulose, transitioning from cellulose I to cellulose II. Both MT and MT-LCA exhibit a quasi-second-order kinetic process in the adsorption of methylene blue, belonging to chemical adsorption. The Langmuir model is more suitable for describing the adsorption of methylene blue. Both adsorbents show monolayer adsorption, and their effective adsorption sites are uniformly distributed. Adsorption experiments with oil and organic solvents indicate that the saturated adsorption capacity is positively correlated with the density of oil and organic solvents. The higher porosity, nanoscale micropores, and larger pore size enhance the capillary force of MT-LCA, providing efficient directional transport performance. The prepared MT-LCA exhibits outstanding compressive performance and efficient directional transport capabilities, making it suitable for applications in areas that demand high compressive performance and selective directional transport.

4. Materials and Methods

4.1. Materials

The MT was harvest at Guangdong, China. Ethanol, acetone, petroleum ether and paraffin were obtained from Xilong Chemical Co., Ltd. (Guangzhou, China). Methylene blue and Sudan III were provided by Kermel Chemical Reagent Co., Ltd. Tianjin, China. 3-ethyl-1-methyl- imidazolium acetate ([Emim]OAc, 98%) was purchased from Qingdao Ionike New Material Technology Co., Ltd. (Qingdao, China). Methyl silicone oil and engine oil were supplied by Shanghai Aladdin Biochemical Technology Co., Ltd. (Shanghai, China). Soybean oil was purchased from a local market.

4.2. Preparation of Lignocellulosic Aerogel

MT specimens, which were characterized by a density of 0.05 g/mm^3 and a diameter of 20.0 ± 0.1 mm, were prepared for fabrication of lignocellulosic aerogels. The samples were sliced into a thickness of 21 ± 1.5 mm and subjected to water-bath heating at 60 °C for 20 min. Afterward, the samples underwent rapid freezing in liquid nitrogen, followed by freeze-drying at −81 °C for 48 h to eliminate water content. Subsequently, the freeze-dried samples were immersed in an ionic liquid (EmimOAc) and heated in a vacuum drying oven at 90 °C for 4 h. To eliminate the ionic liquid (IL) from the partially dissolving samples, it was coagulated in distilled water for a duration of 7 days. Following this, the lignocellulosic hydrogels underwent another freezing in liquid nitrogen and were subsequently freeze-dried to yield the Medulla tetrapanacis-derived lignocellulosic aerogels (MT-LCA).

4.3. Characterizations

Morphological features of both MT and MT-LCA were observed using a scanning electron microscope (SEM) (Hitachi S-4800, Tokyo, Japan). Chemical structures of MT and MT-LCA were analyzed through Fourier-transform infrared spectroscopy (FTIR) (NEXUS-870, Nicolet, MN, USA). Thermal gravimetric analysis was conducted in a temperature range of 35 to 700 °C under a nitrogen atmosphere using a thermal analyzer (TA Q-600, New Castle, PA, USA). Mechanical compression properties were evaluated using a mechanical testing machine (Jindou AG-X plus, Tokyo, Japan). Specific surface area, pore size distribution, and nitrogen adsorption–desorption isotherms were determined with an automatic specific surface area and pore analyzer (Quantachrome AUTOSORB IQ, Boynton Beach, FL, USA). Crystal structures of MT and MT-LCA were examined using X-ray diffraction with an X-ray diffraction analyzer (Puxi D8 DISCOVER, Shanghai, China).

4.4. Density and Porosity

The following Equation was set to measure the density (ρ) of MT or MT-LCA.

$$\rho = M/V \tag{1}$$

where M and V are, separately, the mass and the volume of MT or MT-LCA. The porosity (P) of t MT or MT-LCA was measured according to the following equation.

$$P = (V_0 - V_1)/V_0 \times 100\% \tag{2}$$

where V_0 is the volume of MT or MT-LCA in natural state, and the V_1 is the absolute dense volume of MT or MT-LCA.

4.5. Adsorption of Dye

In order to compare the adsorption capacity of MT and MT-LCA for dyes and to investigate the adsorption mechanism, methylene blue was selected as a model dye for dye adsorption experiments [8]. In the experiment assessing methylene blue adsorption, various factors influencing adsorption performance, including pH value, adsorption time, adsorbent dosage, and initial methylene blue solution concentration, were systematically investigated through individual experiments [36,41]. The experimental procedures are detailed as follows:

Firstly, regarding the influence of pH value, a 100 mg/L methylene blue solution was prepared, and its pH was adjusted using hydrochloric acid and sodium hydroxide. The adjusted pH values ranged from 3 to 9. Samples weighing 5 mg were introduced into 50 mL methylene blue solutions at different pH levels, subjected to 24 h of oscillation in a constant-temperature water bath at 30 °C, and the upper clear liquid was extracted for absorbance measurement using a UV-visible spectrophotometer.

Next, to explore the effect of adsorption time, 5 mg samples were added to 100 mg/L methylene blue solutions, and the mixtures were oscillated in a constant-temperature water bath at 30 °C with a shaking speed of 120 rpm/min. At specified intervals, portions of the methylene blue solution were extracted, and their absorbance was measured using a UV-visible spectrophotometer.

For the investigation into the impact of adsorbent dosage, samples weighing 2 mg, 4 mg, 6 mg, 8 mg, and 10 mg were added to 100 mg/L methylene blue solutions, followed by 24 h of oscillation at 30 °C with a shaking speed of 120 rpm/min. Subsequently, portions of the methylene blue solution from the upper clear liquid were collected for absorbance measurement using a UV-visible spectrophotometer.

Lastly, the effect of the initial concentration of the methylene blue solution was studied by preparing solutions with concentrations of 50, 100, 200, 300, 400, and 500 mg/L. Samples weighing 5 mg were placed in these solutions, and after 24 h of oscillation at 30 °C, the absorbance of the upper clear liquid was measured. The objective was to comprehensively understand the interplay of these factors on the adsorption behavior.

4.6. Adsorption Mechanism and Kinetics

The quasi-first-order and quasi-second-order kinetic models were used to investigate the dye adsorption kinetics of MT and MT-LCA [4,8]. The following two groups of formulas represented the quasi-first-order dynamic model and the quasi-second-order dynamic model, respectively:

$$\ln(Q_e - Q_t) = \ln Q_e - K_1 t / 2.303 \tag{3}$$

$$t/Q_t = 1/K_2 \, Q_e^2 + 1/Q_e \tag{4}$$

where K_1 is the rate constant of a quasi-first-order dynamic simulation; K_2 is the rate constant of a quasi-second-order dynamics simulation; Q_e is the adsorption capacity of a sample after adsorption equilibrium; Q_t is the adsorption amount of a sample at time of t; t is the reaction time.

In order to study the adsorption mechanism of methylene blue on samples, we fitted the experimental data with two isothermal models: Langmuir and Freundlich [4,8]. The equations of the two adsorption isotherms are as follows:

$$\log Q_e = \log K_F - \log C_e / n \tag{5}$$

$$C_e/Q_e = 1/Q_m K_L + C_e/Q_m \tag{6}$$

where C_e is the dye concentration at adsorption equilibrium; Q_e is the amount of MB dye adsorbed by the sample at adsorption equilibrium; K_F and K_L are, separately, the adsorption constants of Freundlich and Langmuir; n is the empirical constant of adsorption capacity and adsorption strength; Q_m is the maximum adsorption capacity of a single molecular layer.

4.7. Oil Sorption Capacity

To further assess the adsorption capabilities of MT and MT-LCA, the saturation adsorption capacities for six common oils and organic solvents were determined. Specifically, soybean oil, engine oil, paraffin oil, ethanol, acetone, and petroleum ether were denoted as SO, MA, PA, ET, AC, and PE, respectively [42]. Prior to sorption, MT and MT-LCA were weighed and immersed in these liquids. To eliminate excess liquid, the oil-loaded MT and MT-LCA were removed and allowed to stand for an additional 1 min, ensuring that any oil on the material surface dripped off due to gravity. The oil sorption capacity (*Kc*) was calculated using the following equation [42].

$$K_C = (m_1 - m_0)/m_0 \tag{7}$$

where m_0 is the sample mass before adsorption, and m_1 is the mass of the sample when oil sorption equilibrium is reached.

Author Contributions: Conceptualization, L.Z. and Z.S.; methodology, L.Q. and X.S.; software, J.Z.; validation, L.Q. and X.S.; investigation, L.Q. and X.S.; resources, X.S.; data curation, X.S.; writing—original draft preparation, L.Q.; writing—review and editing, L.Z. All authors have read and agreed to the published version of the manuscript.

Funding: This research was co-funded by Natural Science Foundation of Anhui Province (No. 2008085MC84) and National Natural Science Foundation of China (No. 31770596).

Institutional Review Board Statement: Not applicable.

Informed Consent Statement: Not applicable.

Data Availability Statement: The original contributions presented in the study are included in the article, further inquiries can be directed to the corresponding authors.

Conflicts of Interest: The authors declare no conflicts of interest.

References

1. Zhang, J.; Fu, H.; Liu, Y.; Dang, H.; Ye, L.; Conejio, A.N.; Xu, R. Review on biomass metallurgy: Pretreatment technology, metallurgical mechanism and process design. *Int. J. Miner. Metall. Mater.* **2022**, *29*, 1133–1149. [CrossRef]
2. Deng, C.; Ma, F.; Xu, X.; Zhu, B.; Tao, J.; Li, Q. Allocation Patterns and Temporal Dynamics of Chinese Fir Biomass in Hunan Province, China. *Forests* **2023**, *14*, 286. [CrossRef]
3. Zhang, X.; Li, L.; Xu, F. Chemical Characteristics of Wood Cell Wall with an Emphasis on Ultrastructure: A Mini-Review. *Forests* **2022**, *13*, 439. [CrossRef]
4. Zhang, J.; Ji, H.; Liu, Z.; Zhang, L.; Wang, Z.; Guan, Y.; Gao, H. 3D Porous Structure-Inspired Lignocellulosic Biosorbent of Medulla tetrapanacis for Efficient Adsorption of Cationic Dyes. *Molecules* **2022**, *27*, 6228. [CrossRef]
5. Zhang, L.; Li, W.; Cao, H.; Hu, D.; Chen, X.; Guan, Y.; Tang, J.; Gao, H. Ultra-efficient sorption of Cu^{2+} and Pb^{2+} ions by light biochar derived from *Medulla tetrapanacis*. *Bioresour. Technol.* **2019**, *291*, 121818. [CrossRef] [PubMed]
6. Kwok, C.T.-K.; Chow, F.W.-N.; Cheung, K.Y.-C.; Zhang, X.-Y.; Mok, D.K.-W.; Kwan, Y.-W.; Chan, G.H.-H.; Leung, G.P.-H.; Cheung, K.-W.; Lee, S.M.-Y.; et al. *Medulla Tetrapanacis* water extract alleviates inflammation and infection by regulating macrophage polarization through MAPK signaling pathway. *Inflammopharmacology* **2023**. [CrossRef]
7. Liu, B.; Yang, M.; Yang, D.; Chen, H.; Li, H. Medulla tetrapanacis-derived O/N co-doped porous carbon materials for efficient oxygen reduction electrocatalysts and high-rate supercapacitors. *Electrochim. Acta* **2018**, *272*, 88–96. [CrossRef]
8. Cai, X.; Shi, T.; Yu, C.; Liao, R.-P.; Ren, J. Sorption Characteristics of Methylene Blue on Medulla Tetrapanacis Biochar and its Activation Technology. *Water Air Soil Pollut.* **2023**, *234*, 223. [CrossRef]
9. Melelli, A.; Jamme, F.; Beaugrand, J.; Bourmaud, A. Evolution of the ultrastructure and polysaccharide composition of flax fibres over time: When history meets science. *Carbohydr. Polym.* **2022**, *291*, 119584. [CrossRef]

10. Zhang, H.; Wang, J.; Xu, G.; Xu, Y.; Wang, F.; Shen, H. Ultralight, hydrophobic, sustainable, cost-effective and floating kapok/microfibrillated cellulose aerogels as speedy and recyclable oil superabsorbents. *J. Hazard. Mater.* **2021**, *406*, 124758. [CrossRef]
11. Song, J.; Chen, C.; Yang, Z.; Kuang, Y.; Li, T.; Li, Y.; Huang, H.; Kierzewski, I.; Liu, B.; He, S.; et al. Highly Compressible, Anisotropic Aerogel with Aligned Cellulose Nanofibers. *ACS Nano* **2018**, *12*, 140–147. [CrossRef]
12. Gu, P.; Liu, W.; Hou, Q.; Ni, Y. Lignocellulose-derived hydrogel/aerogel-based flexible quasi-solid-state supercapacitors with high-performance: A review. *J. Mater. Chem. A* **2021**, *9*, 14233–14264. [CrossRef]
13. Choi, H.-J. Assessment of sulfonation in lignocellulosic derived material for adsorption of methylene blue. *Environ. Eng. Res.* **2022**, *27*, 169–178. [CrossRef]
14. Khakalo, A.; Tanaka, A.; Korpela, A.; Orelma, H. Delignification and Ionic Liquid Treatment of Wood toward Multifunctional High-Performance Structural Materials. *ACS Appl. Mater. Interfaces* **2020**, *12*, 23532–23542. [CrossRef] [PubMed]
15. Aiello, A.; Cosby, T.; McFarland, J.; Durkin, D.P.; Trulove, P.C. Mesoporous xerogel cellulose composites from biorenewable natural cotton fibers. *Carbohydr. Polym.* **2022**, *282*, 119040. [CrossRef]
16. Garemark, J.; Yang, X.; Sheng, X.; Cheung, O.; Sun, L.; Berglund, L.A.; Li, Y. Top-Down Approach Making Anisotropic Cellulose Aerogels as Universal Substrates for Multifunctionalization. *ACS Nano* **2020**, *14*, 7111–7120. [CrossRef]
17. Sun, J.; Zhou, Y.; Zhou, J.; Yang, H. Filtration Capacity and Radiation Cooling of Cellulose Aerogel Derived from Natural Regenerated Cellulose Fibers. *J. Nat. Fibers* **2023**, *20*, 2181276. [CrossRef]
18. Xing, H.; Fei, Y.; Cheng, J.; Wang, C.; Zhang, J.; Niu, C.; Fu, Q.; Cheng, J.; Lu, L. Green Preparation of Durian Rind-Based Cellulose Nanofiber and Its Application in Aerogel. *Molecules* **2022**, *27*, 6507. [CrossRef]
19. Lei, E.; Gan, W.; Sun, J.; Wu, Z.; Ma, C.; Li, W.; Liu, S. High-Performance Supercapacitor Device with Ultrathick Electrodes Fabricated from All-Cellulose-Based Carbon Aerogel. *Energy Fuels* **2021**, *35*, 8295–8302. [CrossRef]
20. Al-Qahtani, S.D.; Snari, R.M.; Al-Ahmed, Z.A.; Hossan, A.; Munshi, A.M.; Alfi, A.A.; El-Metwaly, N.M. Novel halochromic hydrazonal chromophore immobilized into rice-straw based cellulose aerogel for vapochromic detection of ammonia. *J. Mol. Liq.* **2022**, *350*, 118539. [CrossRef]
21. Song, Y.; Wu, T.; Bao, J.; Xu, M.; Yang, Q.; Zhu, L.; Shi, Z.; Hu, G.-H.; Xiong, C. Porous cellulose composite aerogel films with super piezoelectric properties for energy harvesting. *Carbohydr. Polym.* **2022**, *288*, 119407. [CrossRef]
22. Park, S.; Baker, J.O.; Himmel, M.E.; Parilla, P.A.; Johnson, D.K. Cellulose crystallinity index: Measurement techniques and their impact on interpreting cellulase performance. *Biotechnol. Biofuels* **2010**, *3*, 10. [CrossRef]
23. Meng, X.; Ragauskas, A.J. Recent advances in understanding the role of cellulose accessibility in enzymatic hydrolysis of lignocellulosic substrates. *Curr. Opin. Biotechnol.* **2014**, *27*, 150–158. [CrossRef]
24. Baysal, G.; Aydin, H.; Köytepe, S.; Seçkin, T. Comparison dielectric and thermal properties of polyurethane/organoclay nanocomposites. *Thermochim. Acta* **2013**, *566*, 305–313. [CrossRef]
25. Wang, X.; Yang, X.; Wu, Z.; Liu, X.; Li, Q.; Zhu, W.; Jiang, Y.; Hu, L. Enhanced Mechanical Stability and Hydrophobicity of Cellulose Aerogels via Quantitative Doping of Nano-Lignin. *Polymers* **2023**, *15*, 1316. [CrossRef] [PubMed]
26. Shi, W.; Ching, Y.C.; Chuah, C.H. Preparation of aerogel beads and microspheres based on chitosan and cellulose for drug delivery: A review. *Int. J. Biol. Macromol.* **2021**, *170*, 751–767. [CrossRef] [PubMed]
27. Setyawan, H.; Fauziyah, M.a.; Tomo, H.S.S.; Widiyastuti, W.; Nurtono, T. Fabrication of Hydrophobic Cellulose Aerogels from Renewable Biomass Coir Fibers for Oil Spillage Clean-Up. *J. Polym. Environ.* **2022**, *30*, 5228–5238. [CrossRef]
28. Luo, M.; Wang, M.; Pang, H.; Zhang, R.; Huang, J.; Liang, K.; Chen, P.; Sun, P.; Kong, B. Super-assembled highly compressible and flexible cellulose aerogels for methylene blue removal from water. *Chin. Chem. Lett.* **2021**, *32*, 2091–2096. [CrossRef]
29. Zhang, J.; Koubaa, A.; Xing, D.; Liu, W.; Wang, Q.; Wang, X.; Wang, H. Improving lignocellulose thermal stability by chemical modification with boric acid for incorporating into polyamide. *Mater. Des.* **2020**, *191*, 108589. [CrossRef]
30. Li, Y.; He, X.; Liu, P. Hydrophobic Aerogel from Cotton Pulp: Reusable Adsorbents for Oil/Organic Solvent-Water Separation. *J. Polym. Environ.* **2023**, *31*, 2380–2387. [CrossRef]
31. Rostamitabar, M.; Seide, G.; Jockenhoevel, S.; Ghazanfari, S. Cellulose Aerogel Fibers for Wound Dressing Applications. *Tissue Eng. Part A* **2022**, *28*, S373.
32. Zhuang, J.; Pan, M.; Zhang, Y.; Liu, F.; Xu, Z. Rapid adsorption of directional cellulose nanofibers/3-glycidoxypropyltrimethoxy silane/polyethyleneimine aerogels on microplastics in water. *Int. J. Biol. Macromol.* **2023**, *235*, 123884. [CrossRef] [PubMed]
33. Wang, Z.; Song, L.; Wang, Y.; Zhang, X.-F.; Yao, J. Construction of a hybrid graphene oxide/nanofibrillated cellulose aerogel used for the efficient removal of methylene blue and tetracycline. *J. Phys. Chem. Solids* **2021**, *150*, 109839. [CrossRef]
34. Zhou, Y.; Lu, J.; Zhou, Y.; Liu, Y. Recent advances for dyes removal using novel adsorbents: A review. *Environ. Pollut.* **2019**, *252*, 352–365. [CrossRef]
35. Rápó, E.; Tonk, S. Factors Affecting Synthetic Dye Adsorption; Desorption Studies: A Review of Results from the Last Five Years (2017–2021). *Molecules* **2021**, *26*, 5419. [CrossRef] [PubMed]
36. Amaly, N.; El-Moghazy, A.Y.; Nitin, N.; Sun, G.; Pandey, P.K. Synergistic adsorption-photocatalytic degradation of tetracycline by microcrystalline cellulose composite aerogel dopped with montmorillonite hosted methylene blue. *Chem. Eng. J.* **2022**, *430*, 133077. [CrossRef]
37. Liu, Q.; Yu, H.; Zeng, F.; Li, X.; Sun, J.; Li, C.; Lin, H.; Su, Z. HKUST-1 modified ultrastability cellulose/chitosan composite aerogel for highly efficient removal of methylene blue. *Carbohydr. Polym.* **2021**, *255*, 117402. [CrossRef]

38. Ruan, C.; Ma, Y.; Shi, G.; He, C.; Du, C.; Jin, X.; Liu, X.; He, S.; Huang, Y. Self-assembly cellulose nanocrystals/SiO$_2$ composite aerogel under freeze-drying: Adsorption towards dye contaminant. *Appl. Surf. Sci.* **2022**, *592*, 153280. [CrossRef]
39. Zhang, Y.; Sam, E.K.; Liu, J.; Lv, X. Biomass-Based/Derived Value-Added Porous Absorbents for Oil/Water Separation. *Waste Biomass Valorization* **2023**, *14*, 3147–3168. [CrossRef]
40. Zheng, X.; Sun, W.; Li, A.; Wang, B.; Jiang, R.; Song, Z.; Zhang, Y.; Li, Z. Graphene oxide and polyethyleneimine cooperative construct ionic imprinted cellulose nanocrystal aerogel for selective adsorption of Dy(III). *Cellulose* **2022**, *29*, 469–481. [CrossRef]
41. Padmavathy, K.S.; Madhu, G.; Haseena, P.V. A study on Effects of pH, Adsorbent Dosage, Time, Initial Concentration and Adsorption Isotherm Study for the Removal of Hexavalent Chromium (Cr (VI)) from Wastewater by Magnetite Nanoparticles. *Procedia Technol.* **2016**, *24*, 585–594. [CrossRef]
42. Imran, M.; Islam, A.; Farooq, M.U.; Ye, J.; Zhang, P. Characterization and adsorption capacity of modified 3D porous aerogel from grapefruit peels for removal of oils and organic solvents. *Environ. Sci. Pollut. Res.* **2020**, *27*, 43493–43504. [CrossRef] [PubMed]

Disclaimer/Publisher's Note: The statements, opinions and data contained in all publications are solely those of the individual author(s) and contributor(s) and not of MDPI and/or the editor(s). MDPI and/or the editor(s) disclaim responsibility for any injury to people or property resulting from any ideas, methods, instructions or products referred to in the content.

Article

In Vitro Release of Glycyrrhiza Glabra Extract by a Gel-Based Microneedle Patch for Psoriasis Treatment

Ayeh Khorshidian [1,2], Niloufar Sharifi [2,3], Fatemeh Choupani Kheirabadi [2,4], Farnoushsadat Rezaei [5], Seyed Alireza Sheikholeslami [2,6], Ayda Ariyannejad [2,7], Javad Esmaeili [2,6,8,*], Hojat Basati [2,9] and Aboulfazl Barati [10,*]

1. Department of Biomedical Engineering, TISSUEHUB Co., Tehran 1956854977, Iran; ayehkhorshidian@gmail.com
2. Department of Tissue Engineering, TISSUEHUB Co., Tehran 1956854977, Iran; shrf.niloufar@gmail.com (N.S.); fateme.choopani@yahoo.com (F.C.K.); seyed_sheykholeslami69@yahoo.com (S.A.S.); reaalaydaa@gmail.com (A.A.); 3basati@gmail.com (H.B.)
3. School of Pharmaceutical Sciences, Zhengzhou University, Zhengzhou 450066, China
4. Department of Biomedical Engineering, Faculty of Engineering, Islamic Azad University, Tabriz 54911, Iran
5. Department of Chemical and Biomedical Engineering, University of Missouri, Columbia, MO 65211, USA; frkb2@umsystem.edu
6. Department of Chemical Engineering, Faculty of Engineering, Arak University, Arak 3848177584, Iran
7. Department of Marine Biology, Faculty of Life Science and Biotechnology, Shahid Beheshti University, Tehran 1983969411, Iran
8. Tissue Engineering Hub (TEHUB), Universal Scientific Education and Research Network (USERN), Tehran 1956854977, Iran
9. Department of Chemical Engineering, Faculty of Engineering, Tehran University, Tehran 3584014179, Iran
10. Center for Materials and Manufacturing Sciences, Department of Chemistry and Physics, Troy University, Troy, AL 36082, USA
* Correspondence: ja_esmaeili@yahoo.com (J.E.); abarati@troy.edu (A.B.)

Citation: Khorshidian, A.; Sharifi, N.; Choupani Kheirabadi, F.; Rezaei, F.; Sheikholeslami, S.A.; Ariyannejad, A.; Esmaeili, J.; Basati, H.; Barati, A. In Vitro Release of Glycyrrhiza Glabra Extract by a Gel-Based Microneedle Patch for Psoriasis Treatment. *Gels* 2024, *10*, 87. https://doi.org/10.3390/gels10020087

Academic Editors: Esmaiel Jabbari and David Díaz Díaz

Received: 30 June 2023
Revised: 1 October 2023
Accepted: 17 January 2024
Published: 23 January 2024

Copyright: © 2024 by the authors. Licensee MDPI, Basel, Switzerland. This article is an open access article distributed under the terms and conditions of the Creative Commons Attribution (CC BY) license (https://creativecommons.org/licenses/by/4.0/).

Abstract: Microneedle patches are attractive drug delivery systems that give hope for treating skin disorders. In this study, to first fabricate a chitosan-based low-cost microneedle patch (MNP) using a CO_2 laser cutter for in vitro purposes was tried and then the delivery and impact of Glycyrrhiza glabra extract (GgE) on the cell population by this microneedle was evaluated. Microscopic analysis, swelling, penetration, degradation, biocompatibility, and drug delivery were carried out to assess the patch's performance. DAPI staining and acridine orange (AO) staining were performed to evaluate cell numbers. Based on the results, the MNs were conical and sharp enough (diameter: 400–500 μm, height: 700–900 μm). They showed notable swelling (2 folds) during 5 min and good degradability during 30 min, which can be considered a burst release. The MNP showed no cytotoxicity against fibroblast cell line L929. It also demonstrated good potential for GgE delivery. The results from AO and DAPI staining approved the reduction in the cell population after GgE delivery. To sum up, the fabricated MNP can be a useful recommendation for lab-scale studies. In addition, a GgE-loaded MNP can be a good remedy for skin disorders in which cell proliferation needs to be controlled.

Keywords: microneedle patch; skin disorder; drug delivery; burst release; psoriasis

1. Introduction

The ongoing uncontrolled multiplication of cells is the primary aberration that leads to the emergence of illnesses. The symptoms and severity of skin conditions vary widely. They may be fleeting or long-lasting, painful or not. Some have environmental causes, while others could have genetic causes [1]. While most skin disorders are minor, others can indicate a more serious problem. There are diseases that are caused by changes in the cell population in the skin. Eczema, psoriasis, acne, rosacea, ichthyosis, vitiligo, hives, seborrheic, and dermatitis are the most common skin disorders [2].

Skin architecture and function depend on diverse populations of epidermal cells and dermal fibroblasts. Multiple epithelial stem cell (SC) populations have been shown to contribute to skin homeostasis [3]. Physical parameters such as tension, pressure, or temperature contribute to the cellular microenvironment, are sensed by distinct cell populations, and influence cellular fate [4]. "Proliferative skin disease" is a phrase that we define to include all skin diseases characterized by an abnormal proliferation of epidermal keratinocytes [5]. Although these proliferative diseases (e.g., psoriasis, cancer, eczema, and ichthyosis) have multiple causes, the universality of common physiological mechanisms means that the molecular basis of cell proliferation has mechanisms common to all cells [6].

The primary causes of psoriasis are understood to be excessive keratinocyte proliferation and aberrant keratinocyte differentiation, as well as infiltration of many inflammatory cells [7]. However, in most cases, the first prescribed treatment will be a topical treatment, such as vitamin D analogs or topical corticosteroids, creams, and ointments applied to the skin. If these are not effective, phototherapy may be prescribed [8]. By the way, controlling the abnormal proliferation of cells seems to be a useful mechanism in controlling or healing skin disorders.

Herbal remedies have been used in traditional medicine practices for centuries for a wide range of diseases and are a promising alternative, offering a substantial improvement in patient conditions and significantly decreasing disease symptoms [9]. Investigating the proliferation, differentiation, and cytotoxic effects of different herbal extracts on stem cells may provide in-depth insights into their disease-curing mechanisms. However, herbal extracts have shown much promise for cell proliferation and differentiation in many different studies [10,11].

Herbal extracts contain a plethora of phytochemicals such as polyphenols, flavonoids, and other plant-derived chemicals that synergistically aid in treating diseases in traditional medicine methods [10]. As these extracts are composed of naturally occurring medicinal herbs, which may be regularly consumed by local communities, these may cause the least side effects and have lower toxicity than current stimulants [12].

Medical devices with a micron scale called microneedle patches are used to administer drugs. To give vaccinations, medications, and other therapeutic substances, they are made up of small needles that pierce the skin. The usage of microneedles has expanded beyond transdermal drug administration to include intraocular, vaginal, transungual, cardiac, vascular, gastrointestinal, and intracochlear delivery [13]. Small molecular medications, macromolecular pharmaceuticals (proteins, mRNA, peptides, vaccines), and hydrophilic or hydrophobic drugs can all be delivered via microneedles.

There are several types of microneedle patches used for drug delivery and other applications including solid microneedles, coated microneedles, dissolving microneedles, hollow microneedles, and hydrogel microneedles. It is vital to remember that depending on the individual application and preferred drug delivery mechanism, several sorts of microneedles may have distinct benefits and disadvantages. To increase the effectiveness of medication administration and the patient experience, researchers are still investigating and creating new kinds of microneedle patches.

The main aim of this study was to evaluate the potential of the obtained herbal extract from the root of *Glycyrrhiza glabra* in controlling the cell population using a new low-cost fabricated microneedle patch (MNP). This study was conducted as an in vitro study to figure out if it is possible to slow down the cell proliferation rate, which can be a good candidate to treat appropriate skin disorders.

2. Results and Discussion

2.1. HPLC Analysis

Figure 1A shows the results of the HPLC analysis for the Gg extract. It is clear that some peaks existed in the extract that can be attributed to the different components in the extract. The most important one is related to glycyrrhizic acid (known as Enoxolone (ENx)). It has been discovered that Enoxolone is a pentacyclic triterpenic organic acid [14]. Licorice root also contains other phytocompounds such as glycyrrhizin, 18β-glycyrrhetinic acid, glabrin A and B, and isoflavones [15]. However, it was also reported that Enoxolone can be extracted from biological materials including rat blood, human urine, and rat bile [14]. It can be concluded that this agent exists in the human body and no side effect can be hypothesized for its release under transdermal delivery approaches. The primary active component of licorice root, which has been used for centuries as a sweetener and traditional herbal remedy, is glycyrrhizic acid. Asthma and arthritis are inflammatory disorders, and glycyrrhizic acid has been demonstrated to have anti-inflammatory properties. It has been discovered to have direct antiviral action against SARS-CoV-2 and the hepatitis virus. Glycyrrhizic acid may be used to treat cancer due to reports that it possesses anticancer properties. Given that glycyrrhizic acid has been demonstrated to have neuroprotective properties, it may be used to treat neurological conditions including Parkinson's and Alzheimer's. Asthma and hay fever are two allergic illnesses for which glycyrrhizic acid has been reported to have antiallergic benefits. The results from the HPLC confirm the presence of this ingredient. The study by Trupti W. Charpe et al. proves the presence of glycyrrhizic acid as the main ingredient [16].

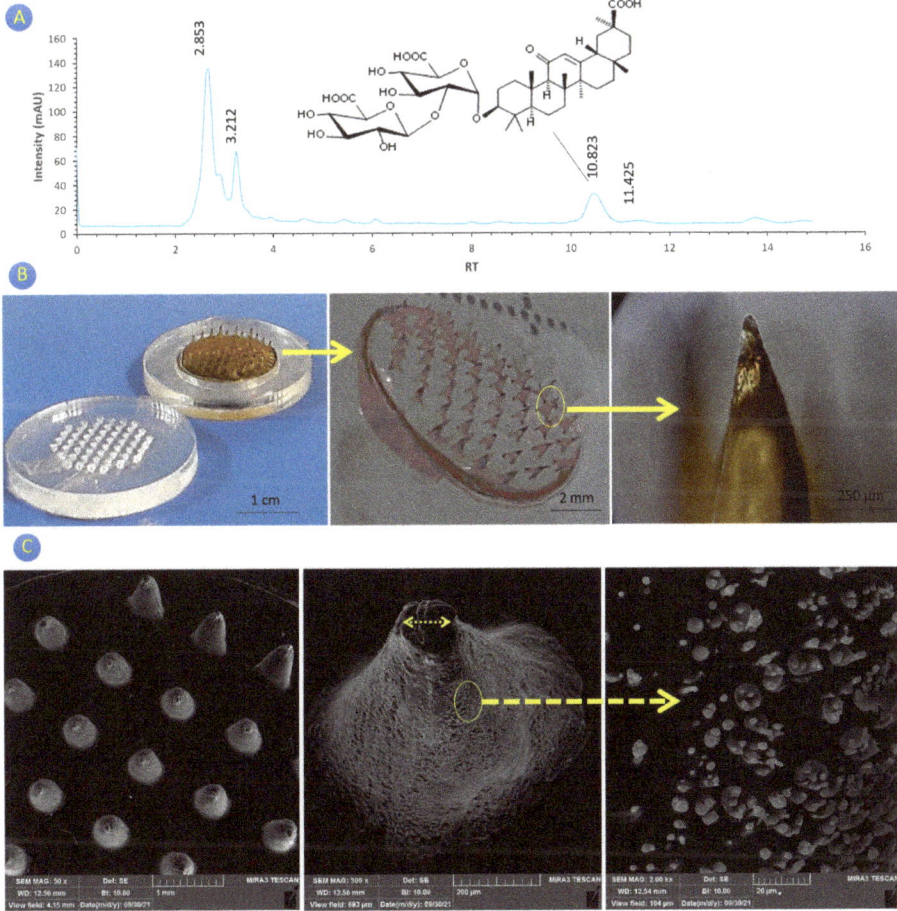

Figure 1. (**A**) HPLC chromatogram of GgE, (**B**) GgE-loaded MNP showing the sharpness level of the created microneedles, (**C**) scanning electron microscopic images of the GgE-loaded MNP.

2.2. Fabricated Patch

In this study, a simple, low-cost, and easily fabricated MNP was introduced. The fabricated microneedles were similar and nearly all had a diameter of 400–500 μm and a height of 700–900 μm. This variance in the dimensions can be attributed to the power of the CO_2 laser cutter during working and enhancing the temperature. Figure 1B demonstrates the GgE-loaded MNP and, as can be seen, all have a conical shape with an acceptable sharpness. However, previous studies have reported several types of microneedles with a distinct sharpness [17–20]. The light brown color of the microneedles confirms the presence of the GgE. The tip of all the microneedles looks dark, which can be justified by the accumulation of the GgE. This issue is hypothesized to be useful for a better and deeper release of the GgE. Compared to the previous studies, in this study, acrylic sheets were employed, while they used Poly dimethyl silicone, which resulted in more polymer diffusion into the cavities [21].

2.3. Structure of Microneedles

Figure 1C shows the microscopic images of the fabricated GgE-loaded MNP. Based on the SEM images, it can be concluded that the microneedles were conical with minimum roughness and a lack of porosity. This can prove the dense structure of the microneedles with no trapped air. The sharpness of the created microneedles seems to be the same. It also shows that the surface of the microneedles is smooth with tiny particles, which can be attributed to the deposition of the GgE.

2.4. Microneedle Penetration

For the evaluation of the microneedle penetration efficacy, compared to the previous studies, the GgE-loaded MNP could penetrate into the parafilm and also create holes (Figure 2A). The created holes were circular (200–250 μm in diameter), which confirms the microneedles' good mechanical strength to tear the parafilm. The penetration was also tested on real human skin (two soft and hard parts) (Figure 2B), and based on our visual evaluation, the created microneedles could rip the skin and leave holes (blood was observed in some points). It is also important to highlight the MNP failed in some areas and the reason can be attributed to the nonuniform pressure loading or low mechanical strength of needles in that zone. The question is how much penetration the MNP made through the skin, the answers to which can be found by more in vivo and pathological studies (outside of the main aim of our study).

2.5. Rhodamine Delivery

Because MNPs are intended to be a substitute for conventional needles, the drug release efficacy is a crucial concern. There are numerous publications on the potential of chitosan-based MNPs for the release of various biomolecules (nicely reviewed by Prausnitz et al. [18]). Regardless of the drug, a fundamental challenge in the fabrication of microneedles is to ensure that there is a significant drug load in the tiny needles. Therefore, ensuring drug release is necessary for microneedle designing. Figure 3A demonstrates the rhodamine release from the rhodamine-loaded MNP in a sodium alginate layer (considered as a semi-skin layer) under a continuous water flow for 30 min. It is necessary to mention that this layer just plays a role as a reservoir to test the drug release from the MNP. Certainly, to make a layer mimicking real skin, it needs more experiments and study to evaluate it from distinct aspects including the tensile strength, water content, degradation, porosity, and thickness. After 15 and 30 min, the MNP showed a notable rhodamine release into the SLD matrix. It depicts that the chitosan MNP had the potential for a burst release of rhodamine after 15 and 30 min. However, the size of the loaded biomolecule is critical in release efficacy [22]. Water uptake and swelling are the main features that can affect drug release because due to the variance in the drug concentration (in the MNP and tissue), water is replaced with the drug and drugs diffuse out from the MNP [23]. To make this issue more clear, the rhodamine-loaded MNP is dry, while the SLD is saturated with water, and after inserting the MNP in the SLD, the microneedles start water adsorption and swelling. The penetrated water dissolves rhodamine and due to the difference in rhodamine concentration inside and outside of the microneedles, rhodamine diffuses out of the microneedles (inside the SLD). In the case of insertion in human skin, a similar mechanism happens due to the skin interstitial fluid.

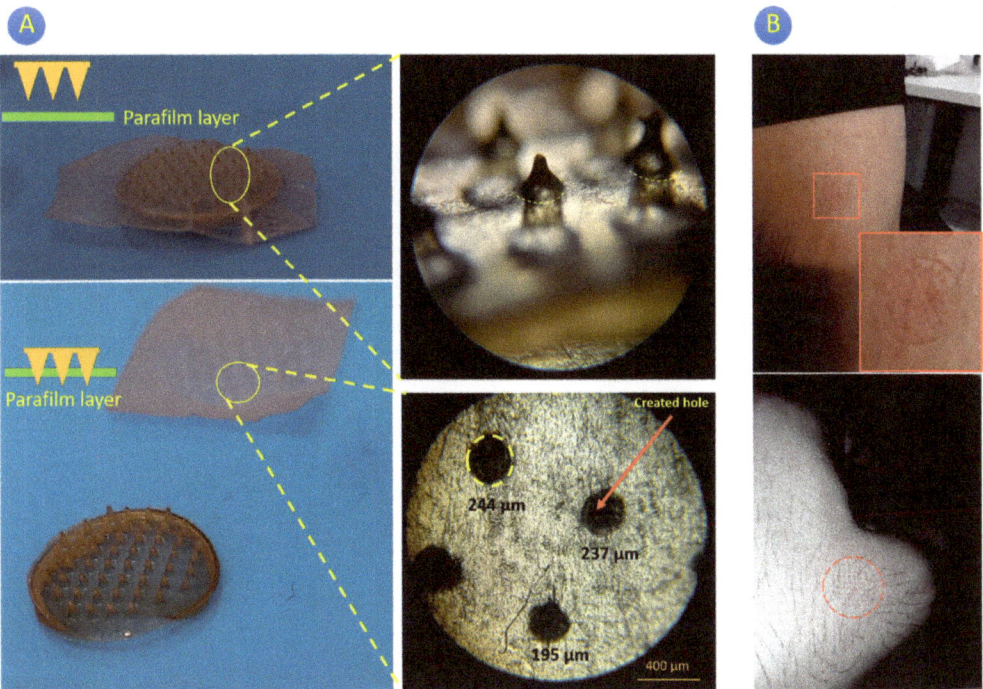

Figure 2. Penetration test using (**A**) parafilm and (**B**) human skin.

2.6. Swelling and Degradation

Figure 3B shows the swelling behavior of the GgE-loaded MNP every minute for 5 min, and then every 5 min for 20 min. It confirms the swelling behavior of the microneedles and based on Figure 3C, an increasing trend in swelling can be seen till 15 min later. As can be seen, 1 min after inserting the MNP, the microneedles are swollen (0.56%), which means after 1 min, GgE is expected to be released. After 5 min, the swelling ratio enhanced to 1.6%. During this time, it is obvious that the microneedles are swollen and due to this phenomenon, they are losing their mechanical strength, so it is recommended to never remove the patch from the skin until the specified time. Finally, after 10 min, it reached around 2%. After 15 min, no significant difference was observed compared to t = 10 min. After 15 min, no swelling was observed, so no data were reported at these intervals (Figure 3C). An interesting point is that the substrate did not swell during the first 5 min, but after that (t = 10 min), its deformation due to the swelling was obvious.

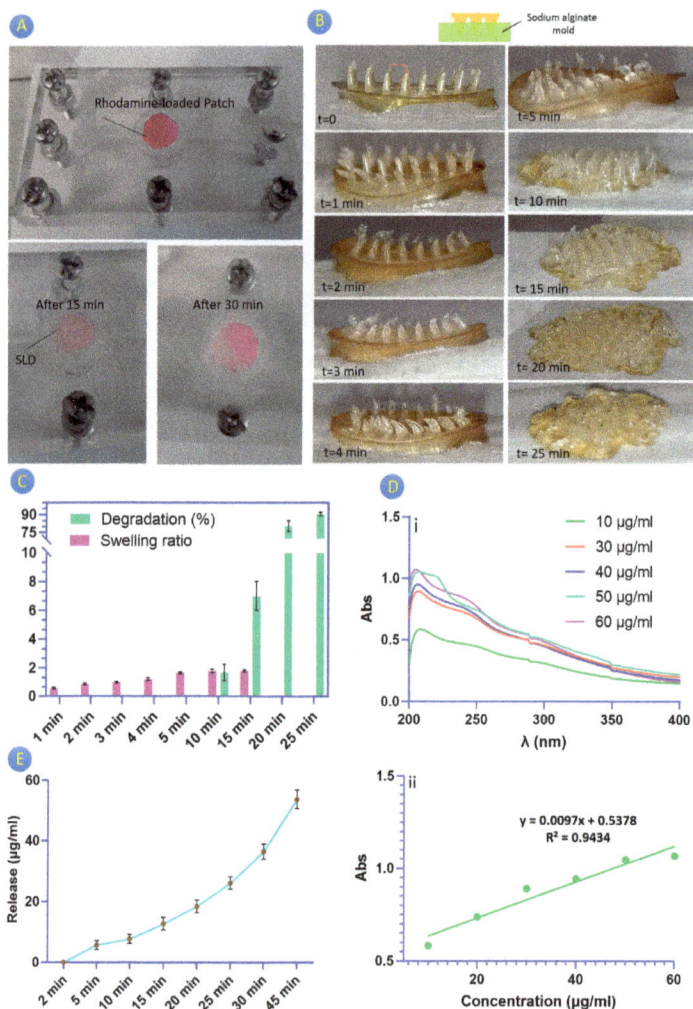

Figure 3. (**A**) Rhodamine release in SLD matrix evaluation using a microfluidic system, (**B**) visual approach to monitor swelling of microneedles, (**C**) swelling and degradation behavior of the GgE-loaded MNP, (**D**) release standard curve, (**E**) release profile of GgE from GgE-loaded MNP.

The effect of swelling on drug release can be significant and should be carefully considered in pharmaceutical formulations. When a drug delivery system comes into contact with a liquid, it may undergo swelling due to the absorption of the liquid by the polymer matrix. This swelling can affect the release of the drug from the delivery system. One important aspect to consider is the swelling kinetics. Different polymers have different swelling characteristics, and their ability to take up liquid can vary greatly. It is crucial to understand the kinetics of swelling for the specific polymer used in the drug delivery system. This can be determined experimentally by measuring the change in the size or weight of the polymer as it absorbs the liquid [24]. When the microneedles penetrate the skin, they induce a local immune response, leading to swelling. This swelling can potentially affect the release of drugs from the microneedles. Moreover, microneedle swelling can also modify the microenvironment around the microneedles. The increased swelling can lead to changes in the local pH, temperature, or moisture content. These changes can influence the solubility and stability of the drug molecules, further impacting their release [25]. It is important to note that the effect of microneedle swelling on drug release can vary depending on several factors, including the composition of the microneedles, the properties of the drug, and the specific application [26].

According to the observation, microneedles begin to degrade after 15 min, which causes them to become shorter. After 25 min, it is possible to assert that the microneedles have completely degraded, at which point it can be assumed that a 100% release has been accomplished. The main justification for the fast swelling and quick degradation of microneedles can be the lack of a crosslinking step in the MNP fabrication process. This design is suitable for drugs that must have a burst release. In contrast, this design can be modified for biomolecules that must be injected gradually [19,27]. (Note: Based on Figure 3B, it is observed that the patch substrate is swollen. This is due to the contact between the substrate and the SLD, which is saturated with water. As for the skin, the substrate is dry when it is in contact with the skin and does not swell, and only the microneedles penetrate the skin(. Because the size of the swollen microneedles is greater than the diameter of the hole that has been made in the skin, or the likelihood of the microneedles breaking during removal of the patch is increased, this swelling behavior can be useful in holding the patch on the skin without any external pressure.

Understanding the many elements that may cause medication degradation in microneedle patches is crucial. These variables include, among others, the temperature, pH, and humidity. In terms of drug release, degradation can significantly impact the release profile of the drug from the patches [28]. As the polymeric matrix degrades, the polymeric networks may change, which can affect its solubility and release kinetics. Therefore, microneedles gradually release drugs during their slow degradation under the skin. It highlights the point that the performance of a microneedle patch can be affected by the degradation of the carrier matrix, which is directly linked to drug diffusion, dissolution, and degradation.

2.7. GgE Release and Kinetic Study

The quantitative release of GgE from MNPs has been one of the main concerns in this study. Although GgE exists in both the microneedles and substrate, generally, the release is going to start from the microneedles. However, GgE diffusion from the substrate into the microneedles and then into the SLD matrix cannot be neglected. By the way, by using a microfluidic device, monitoring the correct amount of drug release from the MNP with a diameter of 1 cm with 52 microneedles was tried. It was assumed that the amount of released GgE from the substrate was ignorable. Furthermore, it has been agreed that the release study should last for around 45 min based on the findings of the swelling analysis, which showed that the substrate virtually began swelling after 15 min.

Figure 3D demonstrates the standard curve to determine the exact concentration of the released GgE. Figure 3E depicts the UV spectrum of the samples taken at different intervals. The graph has been plotted based on time (h) and accumulation release. Based on the results, no GgE was detected until 5 min. An amount of 6 µg/mL of GgE was detected at t = 5 min. The concentration of the released GgE at t = 10 min reached 8 µg/mL, and after 15 and 20 min, it reached 0.11 µg/mL and 19 µg/mL, respectively. By prolonging the process until 25 and 30 min, the concentrations of GgE were reported equal to 26 and 34 µg/mL, respectively. Finally, after 45 min, specific changes occurred in the GgE content (0.51 µg/mL), which can be due to the swelling of substrates. These results are in accordance with that swelling.

Figure 4 depicts the fitting between the experimental results and data from the models. Considering the zero-order model, there was a good fit between the experimental and model results with R^2 equal to 0.987 and the K value equal to 1.247.

$$F = M_t/M_0 = Kt = 1.247t \qquad (1)$$

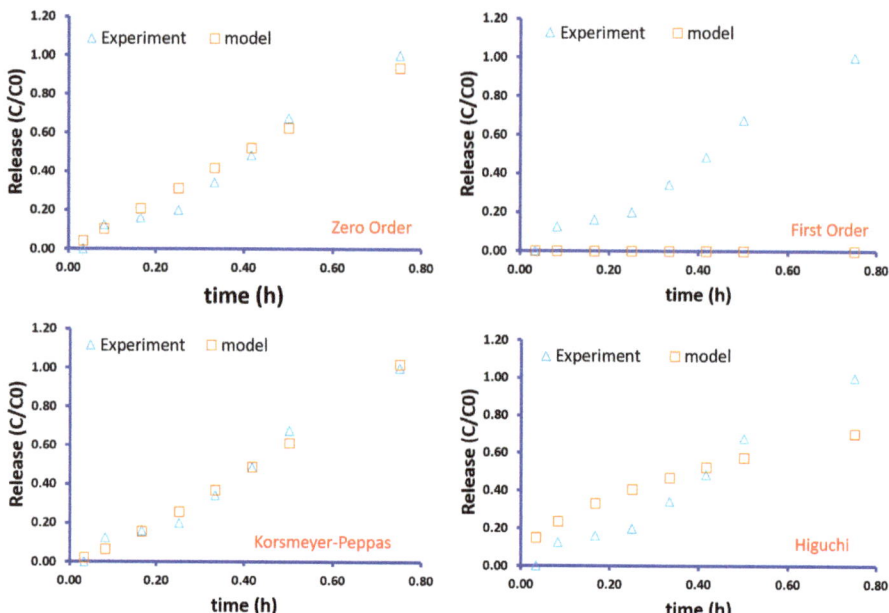

Figure 4. Comparing the experimental results with the obtained results from theoretical models.

In the case of the first-order model, there was no notable fitting and R^2 was not in an acceptable range. The Korsmeyer–Peppas model showed that it can be a good fit for the experimental results with R^2 equal to 0.992. This model has two main parameters, known as the constant K value and n as the power of time. These values were equal to 1.460 and 1.254, respectively.

$$F = M_t/M_0 = Kt^n = 1.46t^{1.254} \quad (2)$$

The Higuchi showed a close fitting compared to the Korsmeyer–Peppas model, but in general, it was not accepted as a good fitting model. R^2 was equal to 0.947. The Higuchi is similar to the Korsmeyer–Peppas model with $n = 0.5$. The K value for the Higuchi model was equal to 0.811.

$$F = M_t/M_0 = Kt^{0.5} = 0.811t^{0.5} \quad (3)$$

In general, it can be claimed that the release kinetic for the GgE-loaded patch follows the Korsmeyer–Peppas model. The Korsmeyer–Peppas model is a widely used mathematical model for describing drug release from hydrogels. It is based on the assumption that the drug release follows a diffusion mechanism. In this model, the release rate is related to time (t), the release exponent (n), and other parameters such as the initial drug loading and the geometry of the hydrogel matrix. In this model, $n < 0.43$ indicates Fickian diffusion, $0.43 < n < 0.85$ indicates relaxation-controlled, and $n > 0.85$ indicates non-Fickian (anomalous) transport [29]. In our study, $n = 1.461$, from which it can be claimed that the GgE release follows non-Fickian diffusion.

2.8. Cell Viability

Previous studies reported the high biocompatibility of chitosan as scaffolds [30], drug delivery systems [31], wound dressing [32], nanocomposites [33], and even MNPs [34]. Chitosan is biodegradable and biocompatible, and its immune-stimulating activity can increase both cellular and humoral responses. The fabricated MNPs in this study were evaluated after crosslinking the MNPs by TPP (0.25 mg/mL) and washing them using sterile PBS. Crosslinking was performed to prevent the fast degradation of chitosan MNPs during cell culture for 48 h. Figure 5A provides the biocompatibility of the pure MNP (without GgE) against the L929 cell line after 24 h, 48 h, and 72 h. As well as the previous studies, high cell viability was observed for the fabricated MNPs.

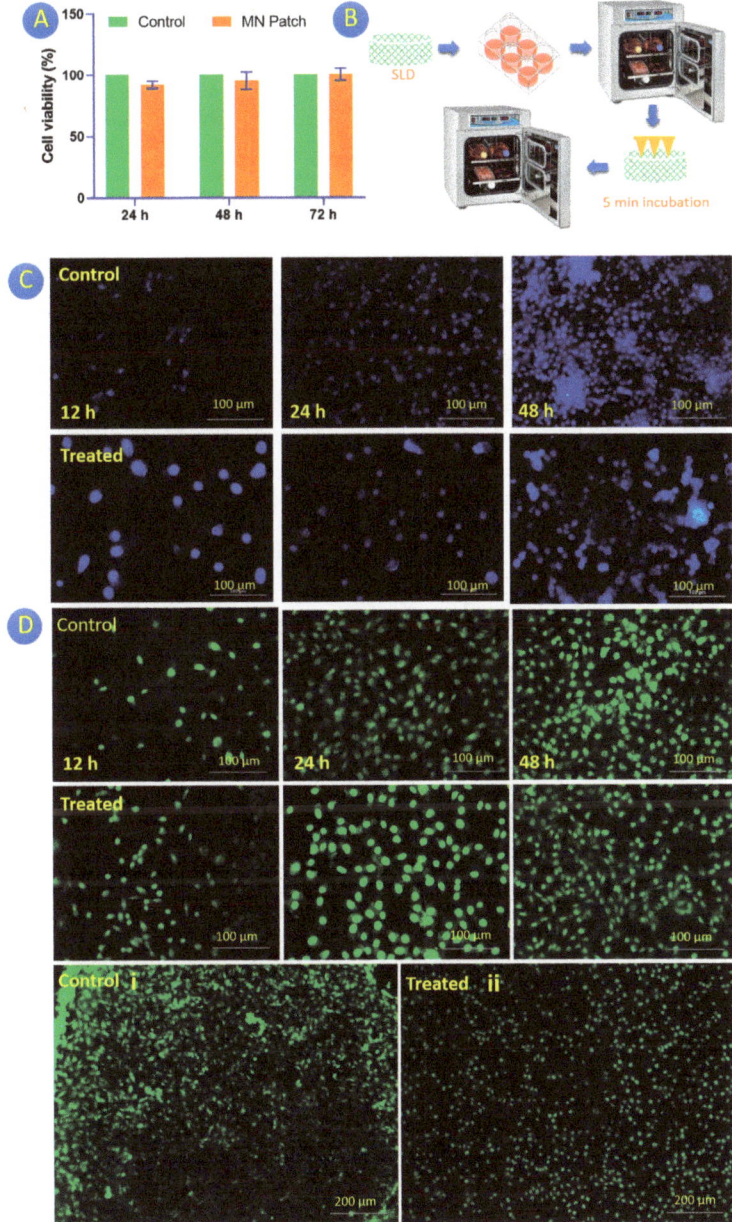

Figure 5. (**A**) Cytotoxicity results for the MNP without GgE, (**B**) schematic procedure for testing GgE-loaded MNP, (**C**) fluorescence micrographs of the DAPI-stained L929 cells, and (**D**) fluorescence micrographs of the acridine orange-stained L929 cells on SLD for control and treated SLDs.

The impact of GgE and its delivery in the matrix of the SLD using the GgE-loaded extract was carried out according to Figure 5B. In this technique, the control group received no GgE, while the treated group was exposed to GgE for 5 min after 12 and 24 h. Based on the release analysis, it can be predicted that cells receive nearly 6 μg of GgE per each MNP administration. Biological staining is a good candidate to check the number of living cells. Changes in the cell population have been monitored via DAPI staining (Figure 5C) and acridine orange (AO) staining (Figure 5D).

One of the purposes of DAPI staining is to study the cell cycle, determine the index of mitosis in an organism, or count cells [35]. The DAPI staining results revealed that both groups' cell populations expanded properly, with distinct cell nuclei and no discernible dead cells (Figure 5C). In the control group, the cells seemed to be attached to the SLD and with a higher concentration after 48 h in contrast to 24 h. Based on Figure 5C, the GgE delivery reduced cell proliferation compared to the control group. This finding indicates that in skin disorders like Psoriasis, in which cells have non-stop proliferation, a transdermal delivery system like the GgE microneedle patch can reduce the growth of cells and control this disease.

Comparing the control group with the treated group (Figure 5(Di) and Figure 5(Dii)), it can be seen that the rate of cell proliferation was slower when the GgE was released. Comparing the AO results at different intervals depicted that the presence of the GgE slowed down cell proliferation (Figure 5D).

There is a question about how GgE could lower the rate of cell proliferation. More biological testing is required to provide a mechanism to answer this question, and finding its answer was beyond the major focus of this study. According to our analysis of the literature, ENx, also known as glycyrrhetinic acid and derived from the herb licorice, is one of the key components of GgE. It has been reported that it has been found to induce G1-phase cell cycle arrest in human non-small-cell lung cancer cells through the endoplasmic reticulum stress pathway [36].

Based on earlier findings, the ENx protective activity was assumed by exerting anti-inflammatory, anti-catabolic, oxidative stress-decreasing effects, gastroprotective, antiviral, cardioprotective, anti-tumor, neuroprotective, and hepatoprotective activities in animal models [37]. ENx can be used as a flavoring agent in food and to mask the bitter taste of drugs such as aloe and quinine. ENx has also been found to modulate vascular injury and atherogenesis.

The ENx anti-inflammatory effects are based on an inhibitory effect on neutrophils, which are the main mediators of inflammation, producing superoxide radicals. It also provided anti-inflammatory effects by reducing the levels of inflammatory nitric oxide, prostaglandin E2, and intracellular reactive oxygen species and by suppressing the expression of pro-inflammatory traits by inhibiting NF-κB and phosphoinositide-3-kinase activity [38]. Furthermore, it has antibacterial and antifungal properties due to its steroid-like structure.

For its firming, moisturizing, whitening, and antiaging actions, ENx is utilized in skin cosmeceuticals to maintain the health and condition of the skin. Recent studies have revealed that ENx and its derivatives have some anticancer effect against a variety of cancer cells and can also cause cell death [39]. Notably, these compounds were found as potent inhibitors of transcription factor NF-κB. NF-κB is a transcript factor discovered by Sen and co-workers in 1986, which is usually overexpressed and constitutively activated in many types of malignancies [37]. The 1–4 NF-κB, as a family of related protein hetero- or homodimers, promotes the downstream protein expression of anti-apoptosis (XIAP, Bcl-2, and Bcl-xL), proliferation (c-Myc) and invasion (MMP-2 and MMP-9), and exhibits remarkable capabilities for regulating the transcription of hundreds of target genes [40].

Increasing evidence indicated that the aberrant activation of NF-κB pathways allows cancer cells to escape apoptosis, invasion, and metastasis, which contributes to cancer drug resistance. Because activation of NF-κB is an essential feature of the survival of cancer cells during treatment, which results in treatment resistance, considerable researchers have focused on targeting NF-κB for cancer therapy [41].

ENx inhibition of TNF-α-activated JNK/c-Jun and IκB/NF-κB signaling pathways, which regulate, respectively, AP-1- and NF-κB-mediated gene transcription, contributes to the suppression of ICAM-1 (playing a key role in the early stage of inflammatory response) [42]. NF-κB regulates TNF-α stimulated ICAM-1 expression at the transcriptional level in vascular endothelial cells. Its movement is interceded by homodimeric or heterodimeric combinations of NF-κB family proteins, such as p50, p65, and c-Rel. In its resting state, NF-κB is present in association with its cytoplasmic inhibitor, IκB [42]. Once activated by an inflammatory cytokine, such as TNF-α, IκB was rapidly phosphorylated and degraded, leading to the translocation of activated NF-κB from the cytoplasm to the nucleus24. ENx observably inhibits IκB degradation and NF-κB p65 translocation and subsequently reduces NF-κB DNA binding activity, recommending a decrease in ICAM-1 expression by ENx by blocking the classic inflammatory pathway of the IκB/NF-κB system. This reduced expression may, at least in part, account for the mechanism of ENx exerting its anti-inflammatory effects. Furthermore, ENx as well as NF-κB and JNK inhibitors suppress TNF-α-induced ICAM-1 protein expression, suggesting that NF-κB and AP-1 are important in regulating cytokine-induced ICAM-1 expression. In addition, ENx inhibition of ICAM-1 is mediated by its down-regulation of NF-κB and JNK signaling pathways [42,43].

3. Conclusions and Future Prospective

The main backbone of this study comprised addressing two issues: firstly, recommending a low-cost MNP fabrication for in vitro studies. Secondly, the evaluation of the GgE impact on slowing down the cell proliferation rate. Based on the results, the fabricated microneedles were conical and sharp enough, and showed notable penetration into parafilm, hydrogel, and skin. The release of GgE also showed good control over the cell population. However, more biological studies are recommended to prove the potential of GgE to heal skin disorders like psoriasis.

It may be a good idea to use GgE for in vivo investigations and other drug delivery methods in the future. There may be additional justifications for introducing GgE as a herbal treatment for skin disorders/diseases if cell cycle and cell signaling pathways are studied.

Understanding the main mechanism can result in new ideas to optimize the concentration and also develop new delivery systems.

Employing a new MNP with additional microneedles and various diameters in this situation may provide greater difficulties.

4. Materials and Method

4.1. Materials

Sodium alginate (medium viscosity, Sigma, Saint Louis, MO, USA), chitosan (low molecular weight, ≥75% deacetylation, Sigma), Thiazolyl Blue Tetrazolium Bromide (Sigma), and sodium triphosphate (TPP, Merck, Rahway, NJ, USA) were purchased from a local supplier (Alborz Shimi Co., Tehran, Iran). Phosphate-buffered saline, 4,6-diamidino-2-phenylindole dihydrochloride (Invitrogen, Waltham, MA, USA), acridine orange ethidium bromide (Invitrogen) were provided by KalaZist Co. (Tehran, Iran). All reagents were analytical grades.

4.2. Extraction

An amount of 10 g of the air-dried root of the *Glycyrrhiza glabra* plant was mixed with 90 mL of 90% hydroalcoholic solvent under stirring (500 rpm) at 60 °C for 24 h. The *Glycyrrhiza glabra* extract (GgE) was concentrated using a rotary evaporator under reduced pressure at 40 °C. Next, the obtained viscose solution was lyophilized using a freeze-dryer (DORSA TECH CO., Iran, Alborz). The final concentration was equal to 0.06 g/mL of extract. The final product was analyzed by HPLC (Agilent 1260 series, Santa Clara, CA, USA). The employed column was an Eclipse XBD-C18 reversed-phase column (250 mm × 4.6 mm, 5 m). The mobile phase consisted of a mixture of 30% acidified water (1% acetic acid) and 70% methanol. The flow rate was maintained at 1 mL/min.

4.3. Preparation of GgE-Loaded MNP

4.3.1. Microneedle Mold Fabrication

The preparation of the MNP mold was performed according to the instructions reported by Rezaei Nejad with a little modification (Figure 6A) [21]. Briefly, acrylic disks with a diameter of 15 mm were created using a CO_2 cutter (Vera) with a maximum power of 100 W. Then, using the CO_2 laser, the laser beam was adjusted on the surface of the acrylic disk. The laser power and pulse width were the main parameters to be optimized for desirable engraving and almost the same depth per run. The engraving was performed by a simple laser shot (17 W and 5 millisecond) on the selected points designed by CorelDraw. The depth of the created cone could be easily adjusted by changing the power and pulse width. Power is a crucial laser parameter since it determines the quantity of energy that is used throughout the procedure and maybe even how deep the cut will be. The time interval between a single laser pulse's beginning and ending is referred to as the pulse width. Higher cutting speeds and improved laser cutting quality are made possible by using the proper pulse widths. The ideal pulse width varies depending on the workpiece material. Increasing the power and the rate of engraving resulted in a too-depth cone (>1 mm). The fabrication process continued to complete the mold. The engraved acrylic mold was washed with isopropanol several times and then rewashed with deionized water to remove the dust and particles from the mold. The mold was vacuum-dried for 30 min. An acrylic cylinder with an inner diameter of 10 mm and an outer diameter of 15 mm was also fabricated using the CO_2 laser cutter. A plastic washer (for sealing) with the same diameter (according to the cylinder) was also fabricated. This plastic washer was placed on the main microneedle mold and then, the cylinder was located on the plastic washer (Figure 6B).

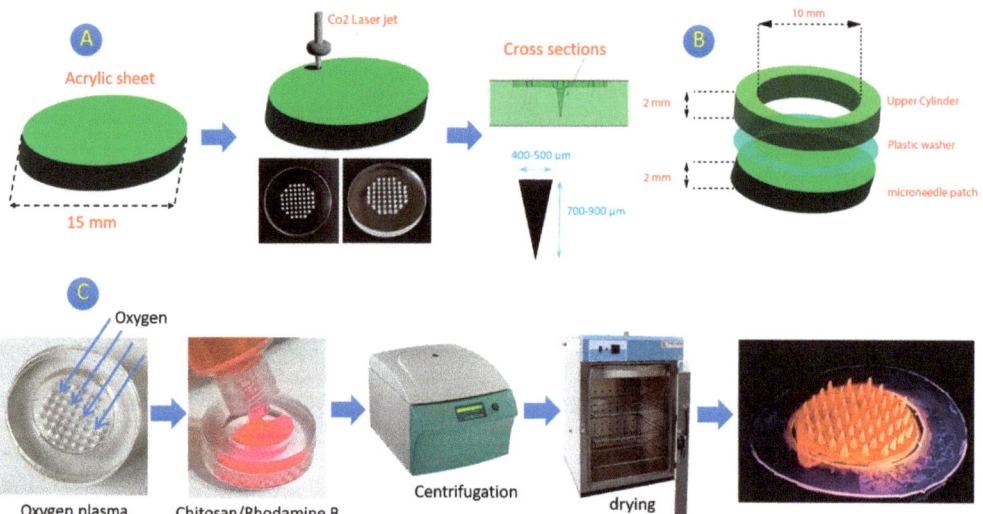

Figure 6. (**A**) The schematic protocol to create a microneedle mold, (**B**) components of the microneedle mold, and (**C**) the schematic process of creating MNP (preparing rhodamine-loaded MNP).

4.3.2. Preparation of Chitosan/Extract Solution

An amount of 1 g chitosan powder was gradually dissolved in 9 mL 1% *v/v* acetic acid at room temperature (200 rpm) for 6 h. Then, the unreacted acetic acid was removed using a dialysis bag (12 KD) in distilled water overnight. An amount of 100 mg of the dried GgE was dissolved in the final purred chitosan solution and mixed for 2 h at RT. The final chitosan/extract solution had a concentration of 10 mg/mL. The obtained solution was transferred into a 5 mL syringe and was left to remove the bubbles. Finally, the syringe was stored in a refrigerator at 10 °C.

4.3.3. Preparation of MNP

Fabrication of degradable microneedles was carried out using the chitosan/extract solution (10 mg/mL) according to Figure 6C. Initially, the acrylic mold was exposed to oxygen plasma for 5 min to create a more hydrophilic surface for better diffusion of the chitosan/GgE solution into the cavities. Frequently, 0.5 mL of the chitosan/GgE solution was dripped over the mold and allowed to cover the surface of the mold. Then, the mold was centrifuged (using specific holders) at 4000 rpm for 10 min. The centrifuge plays a vital role in the fabrication of the patch with sharp microneedles. Next, the mold was dried at 40 °C overnight. Finally, the patch was detached and stored at 4 °C.

4.4. SEM

To study the morphology and dimension of the created microneedles, SEM analysis (scanning electron microscopy) was performed using Philips XL30 (Philips, Amsterdam, The Netherlands, 25 kV). The patch was coated with a layer of gold using a spotting coater [44].

4.5. Semi-Skin Preparation

To prepare a skin-like disk (SLD), 1 g of sodium alginate was dissolved in 9 mL of water to prepare a 10% sodium alginate solution. Then, the solution was poured into cylindrical molds with an inner diameter of 10 mL and a depth of 2 mm. The molds were then frozen at -20 °C for 24 h. Next, the freeze-dried sodium alginate disks were immersed and crosslinked in a 10% w/v calcium chloride solution for 30 min at 10 °C. Finally, the crosslinked sodium alginate disks were washed and freeze-dried. The final SLDs were stored in the refrigerator.

4.6. Penetration Test

Following the recommended protocol by Rezaei Nejad et al., with a little modification, parafilm was nominated to test the penetration of the MNP. A piece of parafilm in 2.5×2.5 cm^2 was prepared. The patch was placed on the film and pushed by hand. This test was repeated three times and the number of created holes was counted [21].

4.7. Release Analysis

Color tracing: To study the efficacy of the patch in the delivery of pharmaceutical agents, a rhodamine-loaded patch was fabricated according to Figure 6C. A microfluidic device according to Figure S1 was employed [45]. The system comprised a mini pump, reservoir, chip, and cap. The chip was designed simply and made of a membrane, one straight microchannel, and a circular zone (diameter:12 mm, depth: 4 mm). The membrane was placed at the bottom of the circular zone and the microchannel passed under the membrane. The SLD was placed on the membrane. Finally, a rhodamine-loaded patch was placed on the SLD and pushed (to make sure that the needles were penetrated in the SLD matrix), and sealed by the cap. This system provides enough moisture in the SLD and due to the difference in the concentration of rhodamine in the patch and SLD, rhodamine is supposed to diffuse into the SLD. After 15 and 30 min, the SLD was checked for diffused rhodamine.

GgE tracing: Using the same approach (microfluidic system), GgE-loaded patches were located in the microfluidic device, and sampling was performed from the reservoir (1 mL) at different intervals (2, 5, 10, 15, 20, 25, 30, 45 min). Using a UV–vis spectrophotometer, the adsorption of the sample was measured in the range of 200–400 nm. After measurement, the sample was added to the reservoir.

The final release profile was studied to find out the most suitable model based on Korsmeyer–Peppas, Higuchi, first-order, and zero-order models.

4.8. Swelling

Swelling of the microneedles plays a vital role in drug delivery and drug diffusion. To aim for this, a GgE-loaded patch was pushed on a swollen SLD. After t = 1, 2, 3, 4, 5, 10, 15, 20, and 25 min, the patch was removed and analyzed visually. The width of the swollen microneedle was measured using ImageJ software (version 1.52v). The degree of swelling was calculated by Equation (4):

$$\% \, swelling = \frac{d_2 - d_1}{d_1} \times 100 \quad (4)$$

In which, d_1 and d_2 are the diameters of the microneedles at t_0 and t.

4.9. Biodegradability

The degree of degradation was measured at RT based on the disappearance of the microneedles by analyzing the height of microneedles using the ImageJ software (version 1.52v). After t = 1, 2, 3, 4, 5, 10, 15, 20, and 25 min, the patch was removed and analyzed visually. During degradation, the height of the microneedle was decreased. The degree of degradation was calculated by Equation (5). This approach seems to be more reliable than measuring the weight of the whole patch.

$$\% \, Degradation = \frac{h_1 - h_2}{h_1} \times 100 \quad (5)$$

In which, h_1 and h_2 are the diameters of the microneedles at t_0 and t.

4.10. Biocompatibility

The patches were sterilized by immersion in 70% ethanol for 24 h and then, by UV rays (254 nm) for 1 h. The patches were then washed several times with sterile PBS. Then, the patches were seeded with 1×10^4 fibroblast cell line L929 cells/well in a 96-well plate. Next, the patches were incubated for 48 h at 37 °C and 5% CO_2. After 48 h, 10 µL of the MTT reagent (0.5 mg/mL) was added to each well and incubated again for 4 h. Afterward, 100 µL of the solubilization solution was added to each well. The plate was then left for incubation overnight. The cells in the wells without patches were treated as the positive control. The purple formazan crystals were checked, and the absorbance was measured at 540 nm by a Wareness Technology Microplate-Reader. All experiments were repeated three times [46].

4.11. Efficacy of GgE Delivery on Cell Population

To evaluate if the release of GgE via an MNP can slow down cell proliferation, the following approach was performed according to ISO 10993-5. Firstly, the SLD, which was a hydrogel, was washed with sterile PBS. Then, it was seeded with 1×10^4 L929 cell/well in a 24-well plate and incubated for 12 h at 37 °C and 5% CO_2. After 12 h, a sterile GgE-loaded patch was pushed on the cell-seeded SLD and incubated for 5 min. During this time, it was predicted that GgE is released and diffused into the cell-seeded SLD. After 5 min, the patch was removed and the cell-seeded SLDs were incubated for 24 h. Again, another sterile GgE-loaded patch was pushed on the cell-seeded SLD and incubated for 5 min. After 5 min, the patch was removed and the cell-seeded SLDs were incubated for 24 h. The SLDs were analyzed by DAPI (4′,6-diamidino-2-phenylindole) staining and acridine orange staining after 12, 24, and 48 h.

4.11.1. DAPI Staining

To make sure about the cell proliferation, DAPI staining was performed. With 1% PBS, the SLDs were washed for 5 min (repeated two times). The SLDs were transferred into a 4% paraformaldehyde solution for 10 min and frequently washed with PBS. Then, 0.1% Triton solution was added to each sample. After 5 min, the SLDs were rewashed with PBS. Next, 150 λ of 4, 6-diamidino-2-phenylindole dihydrochloride (10 µg/mL, DAPI, Sigma, as a nuclear stain) was dripped over the SLDs for 5 min. Then, the SLDs were rinsed three times with PBS. The stained SLDs were kept in the dark and PBS until fluorescence microscopy.

4.11.2. Live–Dead Assay

Acridine orange staining was carried out for the SLDs after 12, 24, and 48 h. Dual fluorescence staining solution containing 100 µg/mL acridine orange ethidium bromide (Sigma Aldrich, St. Louis, MO, USA) was added to samples and washed with PBS after 5 min. The living L929 cells were imaged using a fluorescence microscope (LM, Leica 090-135002, Wetzlar, Germany).

4.12. Statistical Analysis

GraphPad Prism software (version 12) and ANOVA analysis were used to statistically analyze the results. The results were expressed as mean ± standard deviation, and $p < 0.05$ was considered a significant difference.

Supplementary Materials: The following supporting information can be downloaded at: https://www.mdpi.com/article/10.3390/gels10020087/s1, Figure S1: The employed release-test system based on microfluidic technology.

Author Contributions: Conceptualization, J.E. and A.K.; methodology, J.E.; software, A.K.; validation, A.B., N.S. and J.E.; formal analysis, F.C.K.; investigation, A.K.; resources, J.E.; data curation, F.C.K.; writing—original draft preparation, J.E., A.K., S.A.S. and F.C.K.; writing—review and editing, A.B. and J.E.; visualization, N.S. and F.R.; supervision, J.E. and A.B.; project administration, J.E., A.K, F.C.K., F.R., S.A.S., A.A. and H.B.; funding acquisition, J.E., A.K, F.C.K., F.R., N.S. and A.A. All authors have read and agreed to the published version of the manuscript.

Funding: This research received no external funding.

Institutional Review Board Statement: Not applicable.

Informed Consent Statement: Not applicable.

Data Availability Statement: The data presented in this study are available on request from the corresponding author. The data are not publicly available due to ethical.

Acknowledgments: We highly appreciate TISSUEHUB Co. for supporting us scientifically to carry out this project.

Conflicts of Interest: Authors Ayeh Khorshidian, Niloufar Sharifi, Fatemeh Choupani Kheirabadi, Seyed Alireza Sheikholeslami, Ayda Ariyannejad, Javad Esmaeili and Hojat Basati were employed by the company TISSUEHUB Co. The remaining authors declare that the research was conducted in the absence of any commercial or financial relationships that could be construed as a potential conflict of interest.

References

1. Sybert, V.P. *Genetic Skin Disorders*; Oxford University Press: Oxford, UK, 2017.
2. Halder, R.M.; Nootheti, P.K. Ethnic skin disorders overview. *J. Am. Acad. Dermatol.* **2003**, *48*, S143–S148. [CrossRef]
3. Blanpain, C.; Fuchs, E. Epidermal homeostasis: A balancing act of stem cells in the skin. *Nat. Rev. Mol. Cell Biol.* **2009**, *10*, 207–217. [CrossRef]
4. Rognoni, E.; Watt, F.M. Skin cell heterogeneity in development, wound healing, and cancer. *Trends Cell Biol.* **2018**, *28*, 709–722. [CrossRef]
5. Voorhees, J.J.; Chambers, D.A.; Duell, E.A.; Marcelo, C.L.; Krueger, G.G. Molecular mechanisms in proliferative skin disease. *J. Investig. Dermatol.* **1976**, *67*, 442–450. [CrossRef]
6. Kültz, D. Molecular and evolutionary basis of the cellular stress response. *Annu. Rev. Physiol.* **2005**, *67*, 225–257. [CrossRef]

7. Lowes, M.A.; Suarez-Farinas, M.; Krueger, J.G. Immunology of psoriasis. *Annu. Rev. Immunol.* **2014**, *32*, 227–255. [CrossRef]
8. Zhang, P.; Wu, M.X. A clinical review of phototherapy for psoriasis. *Lasers Med. Sci.* **2018**, *33*, 173–180. [CrossRef]
9. Steele, T.; Rogers, C.J.; Jacob, S.E. Herbal remedies for psoriasis: What are our patients taking? *Dermatol. Nurs.* **2007**, *19*, 448.
10. Udalamaththa, V.L.; Jayasinghe, C.D.; Udagama, P.V. Potential role of herbal remedies in stem cell therapy: Proliferation and differentiation of human mesenchymal stromal cells. *Stem Cell Res. Ther.* **2016**, *7*, 110. [CrossRef]
11. Kornicka, K.; Kocherova, I.; Marycz, K. The effects of chosen plant extracts and compounds on mesenchymal stem cells—A bridge between molecular nutrition and regenerative medicine-concise review. *Phytother. Res.* **2017**, *31*, 947–958. [CrossRef]
12. George, P. Concerns regarding the safety and toxicity of medicinal plants-An overview. *J. Appl. Pharm. Sci.* **2011**, *1*, 40–44.
13. Avcil, M.; Çelik, A. Microneedles in Drug Delivery: Progress and Challenges. *Micromachines* **2021**, *12*, 1321. [CrossRef]
14. Coleman, T.J.; Parke, D.V. A Spectrophotometric Method for the Determination of β-Glycyrrhetic Acid (Enoxolone) and its Esters in Biological Materials. *J. Pharm. Pharmacol.* **1963**, *15*, 841–845. [CrossRef]
15. Pastorino, G.; Cornara, L.; Soares, S.; Rodrigues, F.; Oliveira, M.B.P.P. Liquorice (*Glycyrrhiza glabra*): A phytochemical and pharmacological review. *Phytother. Res.* **2018**, *32*, 2323–2339. [CrossRef]
16. Charpe, T.W.; Rathod, V.K. Separation of glycyrrhizic acid from licorice root extract using macroporous resin. *Food Bioprod. Process.* **2015**, *93*, 51–57. [CrossRef]
17. Jamaledin, R.; Yiu, C.K.; Zare, E.N.; Niu, L.N.; Vecchione, R.; Chen, G.; Gu, Z.; Tay, F.R.; Makvandi, P. Advances in antimicrobial microneedle patches for combating infections. *Adv. Mater.* **2020**, *32*, 2002129. [CrossRef]
18. Prausnitz, M.R. Engineering microneedle patches for vaccination and drug delivery to skin. *Annu. Rev. Chem. Biomol. Eng.* **2017**, *8*, 177–200. [CrossRef]
19. Chen, M.-C.; Ling, M.-H.; Lai, K.-Y.; Pramudityo, E. Chitosan microneedle patches for sustained transdermal delivery of macromolecules. *Biomacromolecules* **2012**, *13*, 4022–4031. [CrossRef]
20. Castilla-Casadiego, D.A.; Carlton, H.; Gonzalez-Nino, D.; Miranda-Muñoz, K.A.; Daneshpour, R.; Huitink, D.; Prinz, G.; Powell, J.; Greenlee, L.; Almodovar, J. Design, characterization, and modeling of a chitosan microneedle patch for transdermal delivery of meloxicam as a pain management strategy for use in cattle. *Mater. Sci. Eng. C* **2021**, *118*, 111544. [CrossRef]
21. Nejad, H.R.; Sadeqi, A.; Kiaee, G.; Sonkusale, S. Low-cost and cleanroom-free fabrication of microneedles. *Microsyst. Nanoeng.* **2018**, *4*, 17073. [CrossRef]
22. Vora, L.K.; Courtenay, A.J.; Tekko, I.A.; Larrañeta, E.; Donnelly, R.F. Pullulan-based dissolving microneedle arrays for enhanced transdermal delivery of small and large biomolecules. *Int. J. Biol. Macromol.* **2020**, *146*, 290–298. [CrossRef]
23. Li, S.; Xia, D.; Prausnitz, M.R. Efficient Drug Delivery into Skin Using a Biphasic Dissolvable Microneedle Patch with Water-Insoluble Backing. *Adv. Funct. Mater.* **2021**, *31*, 2103359. [CrossRef]
24. Carbinatto, F.M.; de Castro, A.D.; Evangelista, R.C.; Cury, B.S.F. Insights into the swelling process and drug release mechanisms from cross-linked pectin/high amylose starch matrices. *Asian J. Pharm. Sci.* **2014**, *9*, 27–34. [CrossRef]
25. Qi, Z.; Yan, Z.; Tan, G.; Jia, T.; Geng, Y.; Shao, H.; Kundu, S.C.; Lu, S. Silk Fibroin Microneedles for Transdermal Drug Delivery: Where Do We Stand and How Far Can We Proceed? *Pharmaceutics* **2023**, *15*, 355. [CrossRef]
26. Sabbagh, F.; Kim, B.-S. Ex vivo transdermal delivery of nicotinamide mononucleotide using polyvinyl alcohol microneedles. *Polymers* **2023**, *15*, 2031. [CrossRef]
27. Zhu, Z.; Luo, H.; Lu, W.; Luan, H.; Wu, Y.; Luo, J.; Wang, Y.; Pi, J.; Lim, C.Y.; Wang, H. Rapidly dissolvable microneedle patches for transdermal delivery of exenatide. *Pharm. Res.* **2014**, *31*, 3348–3360. [CrossRef]
28. Nazary Abrbekoh, F.; Salimi, L.; Saghati, S.; Amini, H.; Fathi Karkan, S.; Moharamzadeh, K.; Sokullu, E.; Rahbarghazi, R. Application of microneedle patches for drug delivery; doorstep to novel therapies. *J. Tissue Eng.* **2022**, *13*, 20417314221085390. [CrossRef]
29. Wu, I.Y.; Bala, S.; Škalko-Basnet, N.; Di Cagno, M.P. Interpreting non-linear drug diffusion data: Utilizing Korsmeyer-Peppas model to study drug release from liposomes. *Eur. J. Pharm. Sci.* **2019**, *138*, 105026. [CrossRef]
30. VandeVord, P.J.; Matthew, H.W.; DeSilva, S.P.; Mayton, L.; Wu, B.; Wooley, P.H. Evaluation of the biocompatibility of a chitosan scaffold in mice. *J. Biomed. Mater. Res. Off. J. Soc. Biomater. Jpn. Soc. Biomater. Aust. Soc. Biomater. Korean Soc. Biomater.* **2002**, *59*, 585–590. [CrossRef] [PubMed]
31. Rodrigues, S.; Dionísio, M.; Remunan Lopez, C.; Grenha, A. Biocompatibility of chitosan carriers with application in drug delivery. *J. Funct. Biomater.* **2012**, *3*, 615–641. [CrossRef] [PubMed]
32. Ahmed, S.; Ikram, S. Chitosan based scaffolds and their applications in wound healing. *Achiev. Life Sci.* **2016**, *10*, 27–37. [CrossRef]
33. Hsu, S.-h.; Chang, Y.-B.; Tsai, C.-L.; Fu, K.-Y.; Wang, S.-H.; Tseng, H.-J. Characterization and biocompatibility of chitosan nanocomposites. *Colloids Surf. B Biointerfaces* **2011**, *85*, 198–206. [CrossRef]
34. Gorantla, S.; Dabholkar, N.; Sharma, S.; Rapalli, V.K.; Alexander, A.; Singhvi, G. Chitosan-based microneedles as a potential platform for drug delivery through the skin: Trends and regulatory aspects. *Int. J. Biol. Macromol.* **2021**, *184*, 438–453. [CrossRef]
35. Hamada, S.; Fujita, S. DAPI staining improved for quantitative cytofluorometry. *Histochemistry* **1983**, *79*, 219–226. [CrossRef]
36. Zhu, J.; Chen, M.; Chen, N.; Ma, A.; Zhu, C.; Zhao, R.; Jiang, M.; Zhou, J.; Ye, L.; Fu, H.; et al. Glycyrrhetinic acid induces G1-phase cell cycle arrest in human non-small cell lung cancer cells through endoplasmic reticulum stress pathway. *Int. J. Oncol.* **2015**, *46*, 981–988. [CrossRef]

37. Jin, L.; Zhang, B.; Hua, S.; Ji, M.; Huang, X.; Huang, R.; Wang, H. Glycyrrhetinic acid derivatives containing aminophosphonate ester species as multidrug resistance reversers that block the NF-κB pathway and cell proliferation. *Bioorganic Med. Chem. Lett.* **2018**, *28*, 3700–3707. [CrossRef]
38. Hong, G. Enoxolone suppresses apoptosis in chondrocytes and progression of osteoarthritis via modulating the ERK1/2 signaling pathway. *Arch. Med. Sci.* **2020**, *16*, 1–15. [CrossRef]
39. Sheng, L.-X.; Huang, J.-Y.; Liu, C.-M.; Zhang, J.-Z.; Cheng, K.-G. Synthesis of oleanolic acid/ursolic acid/glycyrrhetinic acid-hydrogen sulfide donor hybrids and their antitumor activity. *Med. Chem. Res.* **2019**, *28*, 1212–1222. [CrossRef]
40. Biswas, S.K.; Lewis, C.E. NF-κB as a central regulator of macrophage function in tumors. *J. Leukoc. Biol.* **2010**, *88*, 877–884. [CrossRef] [PubMed]
41. Suh, J.; Rabson, A.B. NF-κB activation in human prostate cancer: Important mediator or epiphenomenon? *J. Cell. Biochem.* **2004**, *91*, 100–117. [CrossRef] [PubMed]
42. Chang, Y.-L.; Chen, C.-L.; Kuo, C.-L.; Chen, B.-C.; You, J.-S. Glycyrrhetinic acid inhibits ICAM-1 expression via blocking JNK and NF-κB pathways in TNF-α-activated endothelial cells. *Acta Pharmacol. Sin.* **2010**, *31*, 546–553. [CrossRef]
43. Yan, S.; Zhang, X.; Zheng, H.; Hu, D.; Zhang, Y.; Guan, Q.; Liu, L.; Ding, Q.; Li, Y. Clematichinenoside inhibits VCAM-1 and ICAM-1 expression in TNF-α-treated endothelial cells via NADPH oxidase-dependent IκB kinase/NF-κB pathway. *Free. Radic. Biol. Med.* **2015**, *78*, 190–201. [CrossRef]
44. Bhatnagar, S.; Bankar, N.G.; Kulkarni, M.V.; Venuganti, V.V.K. Dissolvable microneedle patch containing doxorubicin and docetaxel is effective in 4T1 xenografted breast cancer mouse model. *Int. J. Pharm.* **2019**, *556*, 263–275. [CrossRef]
45. Esmaeili, J.; Barati, A.; Salehi, E.; Ai, J. Reliable Kinetics for Drug Delivery with a Microfluidic Device Integrated with the Dialysis Bag. *Mol. Pharm.* **2023**, *20*, 1129–1137. [CrossRef]
46. Esmaeili, J.; Jadbabaee, S.; Far, F.M.; Lukolayeh, M.E.; Kırboğa, K.K.; Rezaei, F.S.; Barati, A. Decellularized Alstroemeria flower stem modified with chitosan for tissue engineering purposes: A cellulose/chitosan scaffold. *Int. J. Biol. Macromol.* **2022**, *204*, 321–332. [CrossRef]

Disclaimer/Publisher's Note: The statements, opinions and data contained in all publications are solely those of the individual author(s) and contributor(s) and not of MDPI and/or the editor(s). MDPI and/or the editor(s) disclaim responsibility for any injury to people or property resulting from any ideas, methods, instructions or products referred to in the content.

Article

Non-Aqueous Poly(dimethylsiloxane) Organogel Sponges for Controlled Solvent Release: Synthesis, Characterization, and Application in the Cleaning of Artworks

Francesca Porpora [1], Luigi Dei [1], Teresa T. Duncan [2], Fedora Olivadese [1], Shae London [3], Barbara H. Berrie [4], Richard G. Weiss [3] and Emiliano Carretti [1,5,*]

[1] Department of Chemistry "Ugo Schiff" & CSGI Consortium, University of Florence, Via della Lastruccia, 3-13, 50019 Sesto Fiorentino, Italy; francesca.porpora@unifi.it (F.P.); luigi.dei@unifi.it (L.D.); fedora.olivadese@stud.unifi.it (F.O.)
[2] Scientific Analysis of Fine Art, LLC, Berwyn, PA 19312, USA
[3] Department of Chemistry and Institute for Soft Matter Synthesis and Metrology, Georgetown University, 37th and O Streets NW, Washington, DC 20057, USA; ssl63@georgetown.edu (S.L.); weissr@georgetown.edu (R.G.W.)
[4] Department of Scientific Research, National Gallery of Art, 2000 South Club Drive, Landover, MD 20785, USA; b-berrie@nga.gov
[5] National Research Council—National Institute of Optics (CNR-INO), Largo E. Fermi 6, 50125 Florence, Italy
* Correspondence: emiliano.carretti@unifi.it; Tel.: +39-0554-573-046

Citation: Porpora, F.; Dei, L.; Duncan, T.T.; Olivadese, F.; London, S.; Berrie, B.H.; Weiss, R.G.; Carretti, E. Non-Aqueous Poly(dimethylsiloxane) Organogel Sponges for Controlled Solvent Release: Synthesis, Characterization, and Application in the Cleaning of Artworks. *Gels* **2023**, *9*, 985. https://doi.org/10.3390/gels9120985

Academic Editor: Shiyang Li

Received: 10 November 2023
Revised: 7 December 2023
Accepted: 11 December 2023
Published: 15 December 2023

Copyright: © 2023 by the authors. Licensee MDPI, Basel, Switzerland. This article is an open access article distributed under the terms and conditions of the Creative Commons Attribution (CC BY) license (https://creativecommons.org/licenses/by/4.0/).

Abstract: Polydimethylsiloxane (PDMS) organogel sponges were prepared and studied in order to understand the role of pore size in an elastomeric network on the ability to uptake and release organic solvents. PDMS organogel sponges have been produced according to sugar leaching techniques by adding two sugar templates of different forms and grain sizes (a sugar cube template and a powdered sugar template), in order to obtain materials differing in porosity, pore size distribution, and solvent absorption and liquid retention capability. These materials were compared to PDMS organogel slabs that do not contain pores. The sponges were characterized by Fourier-transform infrared spectroscopy with attenuated total reflectance (FTIR-ATR) and compared with PDMS slabs that do not contain pores. Scanning electron microscopy (SEM) provided information about their morphology. X-ray micro-tomography (XMT) allowed us to ascertain how the form of the sugar templating agent influences the porosity of the systems: when templated with sugar cubes, the porosity was 77% and the mean size of the pores was ca. 300 µm; when templated with powdered sugar, the porosity decreased to ca. 10% and the mean pore size was reduced to ca. 75 µm. These materials, porous organic polymers (POPs), can absorb many solvents in different proportions as a function of their polarity. Absorption capacity, as measured by swelling with eight solvents covering a wide range of polarities, was investigated. Rheology data established that solvent absorption did not have an appreciable impact on the gel-like properties of the sponges, suggesting their potential for applications in cultural heritage conservation. Application tests were conducted on the surfaces of two different lab mock-ups that simulate real painted works of art. They demonstrated further that PDMS sponges are a potential innovative support for controlled and selective cleaning of works of art surfaces.

Keywords: Polydimethylsiloxane (PDMS) organogel sponges; template effects; viscoelastic properties; solvent swelling/de-swelling; cleaning works of art

1. Introduction

Recently, significant research has focused on the exploitation of the properties of some polymeric materials with porous structures (i.e., porous organic polymers, POPs) because they possess large surface areas and well-defined and tuneable pore size distributions [1,2]. The combination of these variable properties can provide synergic effects that are suitable for many technological applications [3]. POPs can be synthesised according to several routes with many functionalisations and are easily processed. Thus, the significant attention that POPs have received during the last decades can be attributed to their fundamental scientific characteristics as well as the extremely broad range of their realized and potential applications (these include greenhouse gas adsorption, proton conduction, energy storage and conversion, sensing, separation, catalysis, drug delivery and release, and tissue engineering) [3–10]. Polydimethylsiloxane (PDMS) organogel sponges [11–27] are important examples of POPs due to their elasticity, hydrophobicity, low surface tension, controllable porosity, low-toxicity, low-flammability, high thermal and electrical resistance, low bulk density, mechanical robustness, and high transmittance and low absorption of UV radiation [28–33].

A preliminary goal of this paper has been the assessment of a suitable strategy for the synthesis of PDMS-based systems using sugar as a templating agent [3,22,34] and exploiting the possibility of producing a large number of PDMS porous systems (i.e., sponges) [16,34]. Two different synthetic routes based on this strategy were followed. In one, sugar cubes with an edge of ca. 1 cm and comprised of grains with dimensions of a few hundred microns were chosen as templating agents; in another the templating agent was made from powdered sugar with grains having dimensions in the order of tens of microns. In both cases, the chemical and physicochemical properties of PDMS slabs (i.e., pure PDMS cast without templating agents) and of the two sponges (obtained using sugar cubes and fine powdered sugar as the templating agents) were investigated. Chemical analyses were performed using Fourier-transform infrared spectroscopy in the attenuated total reflectance mode (FTIR-ATR). Physicochemical characterizations were carried out using scanning electron microscopy (SEM) to obtain information about the morphology, and X-ray micro-tomography (XMT) was employed for the determination of the 3D structures and calculations of the porosity and of the pore size distributions. Analyses of these data have been useful for designing these materials for specific applications. Another aim of this work has been to obtain PDMS based systems with different porosity and pore size distributions that would present different and controllable properties in terms of solvent swelling, retention, and release. A possible application for such materials is as physically "gentle" cleaning agents for painted surfaces of historical and artistic interest.

By exploiting (1) the ability to modulate the porosity of the PDMS systems by varying the granularity of the templating agent [3], and (2) the well-known compatibility of PDMS with a wide range of mid-to-low-polarity solvents [35], materials were produced with varying interactions between the systems explored and the organic solvents chosen here (many of which have been used for the cleaning of painted surfaces). These interactions were examined by measuring the amounts of the solvents absorbed [36] as a function of both solvent polarity and of the porosity of the PDMS). The rheology of the PDMS systems was investigated before and after loading solvents to obtain information about their effect on the viscoelastic properties, with particular attention to the elasticity and the complex viscosity of the systems. In fact, the ability to control these mechanical features is critical in determining two fundamental properties that a cleaning system tailored for painted surfaces of works of art should possess: (1) the retention of the solvents within the slabs/sponges and (2) adequate contact with the surface of the work-of-art.

Despite the large differences among distinct classes of physical and chemical gels that have been tailored for the selective removal of coating materials on surfaces of painted works of art [37–42], few organogels and sponge-like systems have been reported for cleaning of works of art. Among these, a new class of polyhydroxybutyrate-based physical organogels have been prepared and successfully employed for the selective cleaning of bronze and water-sensitive painted surfaces of historical and artistic interest [43–46]. These systems can be easily prepared and loaded with organic solvents at different polarities (i.e., lactones and alkyl carbonates). The introduction of solvent-swelled PDMS gels adds to the range of solvents that can be applied via gel to works of art; moreover, the ability to tune the porosity implements a strategy that could potentially be applied to a range of gelled systems currently used in conservation. Thus, an aim of this research is a preliminary evaluation of the performance of PDMS-based water-free organogels. To investigate the use of these gels for conservation applications, cleaning tests were carried out on two lab mock-ups: a fresco and an easel painting that have naturally-aged, previously-applied surface coatings.

2. Results and Discussion

The chemical characterization of the PDMS slabs (PDMS systems obtained without the use of any templating agent, see Section 4.2) and PDMS_SC (PDMS systems obtained using a sugar cube as templating agent, see Section 4.2) and PDMS_PS sponges (PDMS systems obtained using powdered sugar as templating agent, see Section 4.2) using FTIR-ATR spectra are reported in Figure 1. All of the characteristic peaks of both the two components, the 'base' A and 'curing agent' B, and the cross-linked polymer are in agreement with the data reported in the literature [47,48]. Moreover, no meaningful differences in the profiles of the FTIR spectra of the three PDMS-based systems are observed, indicating that their chemical compositions are the same and that the templating agents have no influence on that. The absence of signals associated with sugar suggests that no detectable residues of the sugar templating agent are present. In addition, no glucose was detected by a Fehling [49] test carried out on the central portions of PDMS_SC and PDMS_PS fragments.

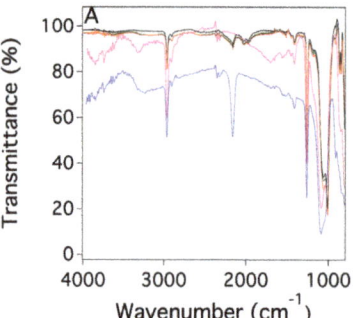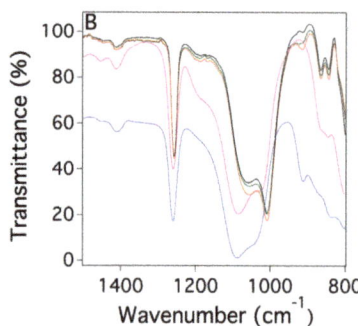

Figure 1. FTIR-ATR spectra of the A (magenta) and B (blue) components of the PDMS slab (ratio component A/component B = 20:1), the cross-linked polymer (black), and of the two different PDMS sponges (component ratio A/B = 10:1 for both systems); that with sugar cube as template (PDMS_SC red) and that with powdered sugar as template (PDMS_PS green). (**A**) reports the FTIR-ATR spectra in the range 4000–800 cm^{-1}; (**B**) reports the FTIR-ATR spectra in the range 1500–800 cm^{-1}.

The morphology of the two porous sponges was investigated by SEM. Figure 2 shows the micrographs of the two PDMS sponges, PDMS_SC (A and B) and PDMS_PS (C and D).

The images in Figure 2 indicate that the two templating agents result in very different morphologies, with a porosity much higher for sponges synthesized using sugar cubes than for systems obtained using the powdered sugar template.

To obtain more detailed information about the porosity of the two systems, XMT measurements were carried out, and the 3D images obtained through the procedure described in Section 4.2 are displayed in Figure 3. These images confirm that the use of the two different templating agents produces sponges with different porosities [16,34]. By employing the procedure described in Section 4.3, it was possible to deduce the pore size distribution (for pores larger than 5 μm) for both sponges. These values are reported in Figure 4A,B for PDMS_SC and PDMS_PS, respectively. The results confirm that the pore size of the PDMS_SC (ca. 303 μm) is larger than that of PDMS_PS (ca. 76 μm). Table 1 summarizes the results from the porosity analysis performed on the two sponges.

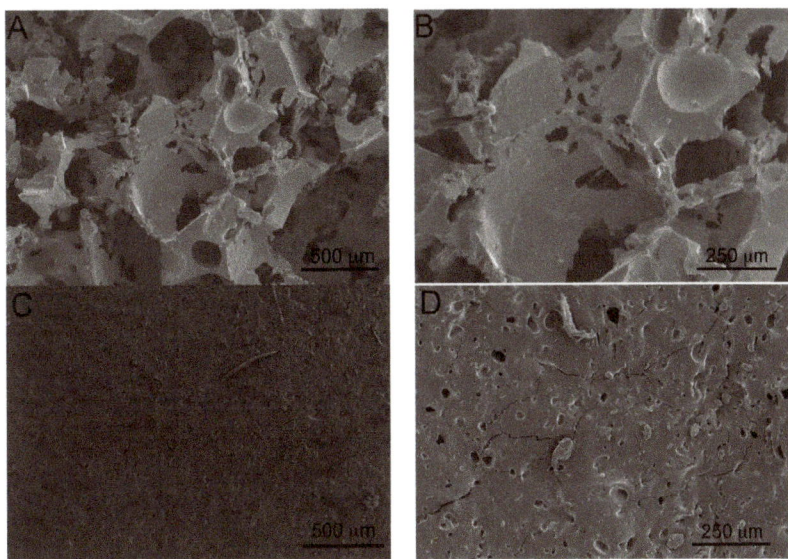

Figure 2. SEM micrographs of two different PDMS sponges: (**A**) PDMS_SC magnification 50×; (**B**) PDMS_SC magnification 100×; (**C**) PDMS_PS magnification 50×; (**D**) PDMS_PS magnification 100×. For both samples, the ratio of components A/B was 10:1.

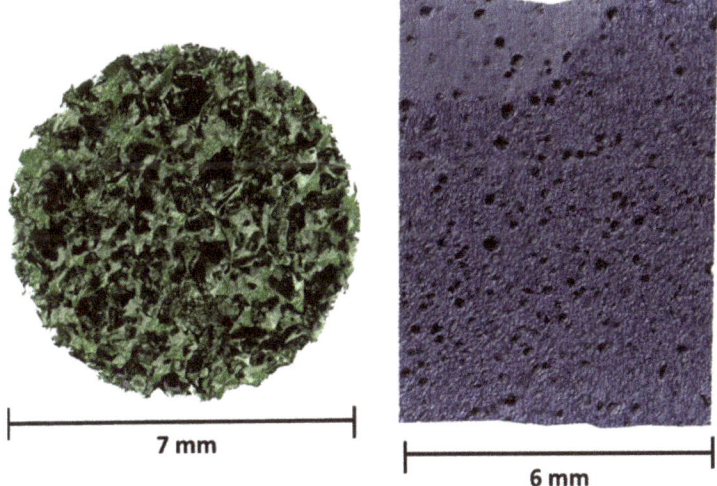

Figure 3. XMT 3D images for two different PDMS sponges in which the A/B component ratio was 10:1: (**left**) PDMS_SC; (**right**) PDMS_PS.

A fundamental aspect to investigate here, in view of a possible application of these systems as cleaning tools for painted surfaces, is a comparison of solvent absorption by the two sponges and that of the PDMS slab, focusing on both the swelling of the slab and on the capillary action of the porous PDMS sponges. This experiment is fundamental to understand the maximum amount of solvent that can be loaded into the three different systems in the view of a possible application as cleaning tools for painted surfaces; this is the reason why the absorption tests have been carried out over up to 25 h (Figure 5).

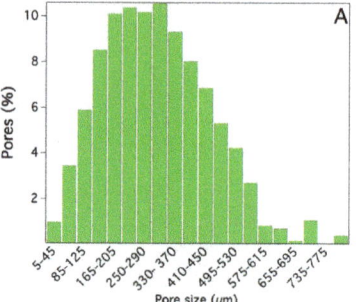
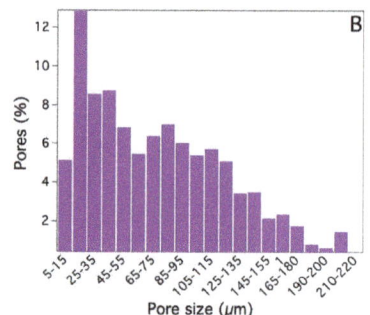

Figure 4. Histograms of the pore size distributions for two different PDMS sponges (10 × 5 × 5 mm) derived from XMT measurements: (**A**) PDMS_SC; (**B**) PDMS_PS. For both samples, the A/B component ratio was 10:1.

Table 1. Porosities and mean pore diameters (mm) of the pores of PDMS_SC and PDMS_PS sponges as calculated from the % of polymers.

Sample	Polymer (%)	Porosity (%)	Mean Pore Diameter (µm)
PDMS_SC	22.6 ±0.7	77.4 ± 5.2	303 ± 15
PDMS_PS	90.2 ±1.5	9.8 ± 0.38	76 ± 5

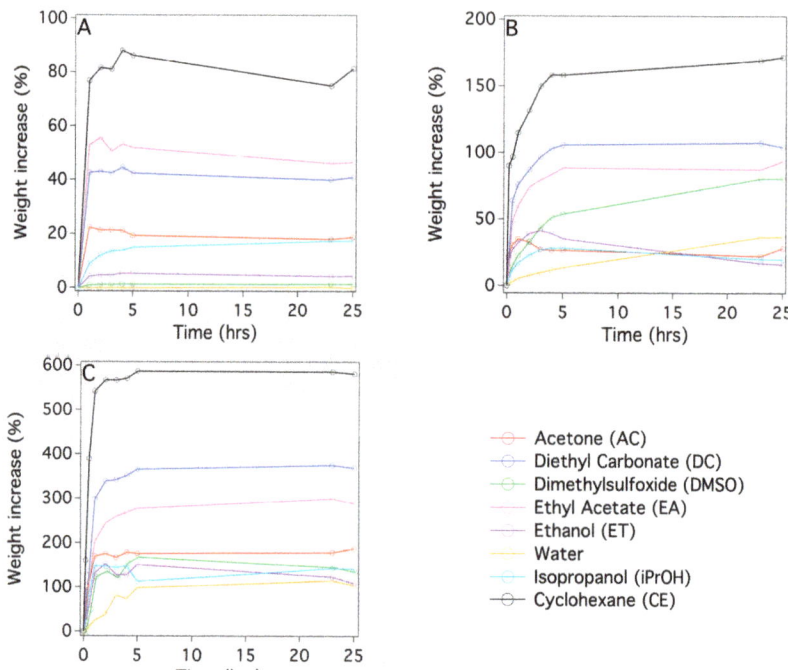

Figure 5. Kinetics curves for solvent absorption of the two different PDMS sponges and of the PDMS slab for various solvents (ratio component A/component B = 10:1): (**A**) PDMS slab, (**B**) PDMS_PS, (**C**) PDMS_SC.

In that regard, Figure 5 shows the kinetics of absorption by the three systems with various organic solvents, including some that are used for cleaning purposes in cultural heritage conservation. Figure 6A,B report the maximum amounts of the solvents absorbed by the PDMS-based systems and the initial absorption rates, respectively. From Figures 5 and 6A, the PDMS_SC, with the highest porosity and the larger pore size, absorbed up to four times the amount of solvent as absorbed by the PDMS_PS sponge with a higher rate (Figure 6B). This indicates that porosity is the key factor in determining the kinetics of solvent absorption. This is also confirmed by the lowest absorption amount is observed for the nonporous PDMS slabs (Figure 5A). Another result, that is in agreement with the literature data [35], is the strong dependence between the rates of the absorption and the magnitude of the dielectric constants ε of the solvents (Figure 6A,B). As expected, because PDMS is a low-polarity polymer, the maximum absorption amount and rate were observed for solvents with low ε values, such as cyclohexane; conversely, in polar solvents such as water, ethanol and DMSO, absorption was smaller and slower. This behavior is due to the large difference between the physico-chemical affinities of the solvents and the hydrophobic network of the PDMS matrix. Absorption tests have been carried out with cyclohexane and DMSO, solvents that should not be used in conservation due to health concerns, in these experiments to demonstrate that it is possible to load solvents having very different polarities into the PDMS based systems.

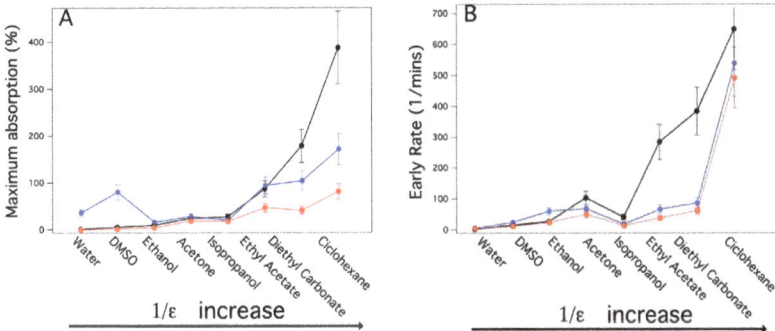

Figure 6. Maximum amount of organic solvent absorbed (**A**) and swelling rate (**B**) by PDMS slab (red dots), PDMS_PS, (blue dots), PDMS_SC (black dots). The solvents on X axis are presented as a function of $1/\varepsilon$, where ε is the dielectric constant of the solvents.

Rheological measurements were carried out to obtain information about the viscoelastic properties of the systems. These data indicate a solid-like rheological behavior for all the investigated samples ($G' \gg G''$ over the whole range of frequencies investigated; see Figure 7), and, according to the literature, [50] they can be mechanically classified as solid-like materials. Since one of the main goals of the study is the application of these systems as cleaning tools for works of art surfaces, it was crucial to ascertain how the viscoelastic properties could be affected by the presence of absorbed solvent in the swollen sponges. Therefore, frequency sweeps were carried out after the absorption of some organic solvents that were chosen among the ones used for cleaning purposes in cultural heritage conservation. Four solvents covering an extended range of polarities, were selected: ET, DMSO, DC, and EA (acetone was not used here due to its low boiling point). Figure 8, as an example, reports the trend of G' and G'' versus frequency sweep for the two sponges containing two solvents, among the four tested, at the extremes of the polarity scale, that are DC and DMSO. The other frequency sweep diagrams of PDMA_SC and PDMS_PS loaded with ET and EA are reported in Figure S1. These data clearly indicate that none of the four solvents tested alters the porosity or the gel-like nature of the two sponges independently (see Figures 7 and S1).

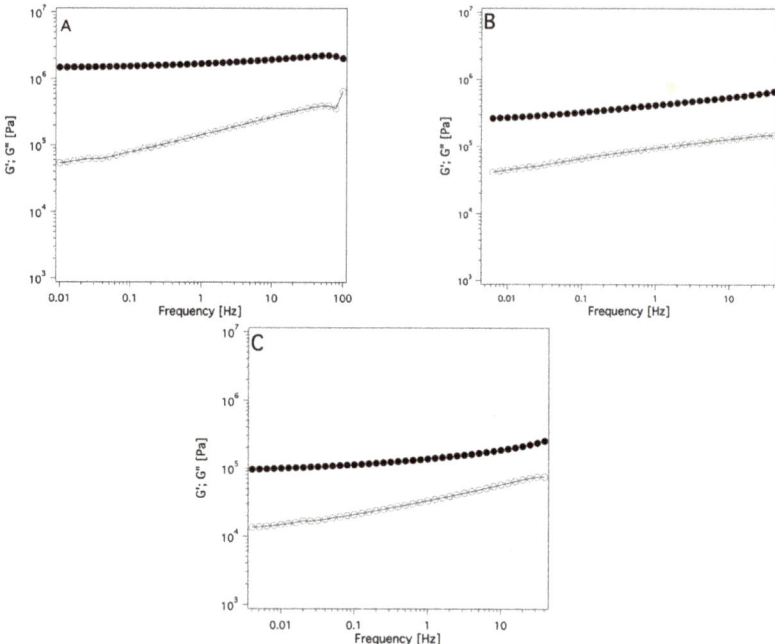

Figure 7. Frequency sweep diagrams of the (**A**) PDMS slab, (**B**) PDMS_PS, (**C**) PDMS_SC systems. Filled circles indicate G′; open circles indicate G″ (component ratio A/B = 10:1).

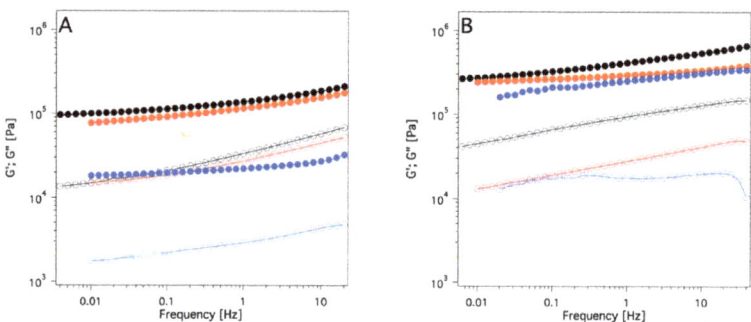

Figure 8. Frequency sweep diagram of the: (**A**) PDMS_SC sponge alone (black), with absorbed DC (blue), with absorbed DMSO (red); (**B**) PDMS_PS sponge alone (black), with absorbed DC (blue), and with absorbed DMSO (red). Absorption time: 24 h. Filled circles indicate G′; open circles indicate G″.

The main finding from the rheology measurements (Figures 8 and S1) is a decrease of the G′ value for both the investigated systems (PDMS_SC and PDMS_PS) after loading the four solvents. This effect is higher for DC than for DMSO and, with the same solvent this effect is more pronounced for the PDMS_SC. This is attributable to the higher porosity of the PDMS_SC system that allows the absorption of a larger amount of this solvent (Figure 5B,C). The trend of the complex viscosity η* (see Figure 9A,B) for the pure sponges and the swollen sponges by the four solvents after 24 h of absorption confirms the findings of frequency sweep measurements: a pronounced decrease of η* due to the presence of a purely viscous liquid phase within the pores is observed. Moreover, this effect seems to be inversely proportional to the polarity of the confined solvents as indicated by the lowest η* values of the DC and EA-based systems.

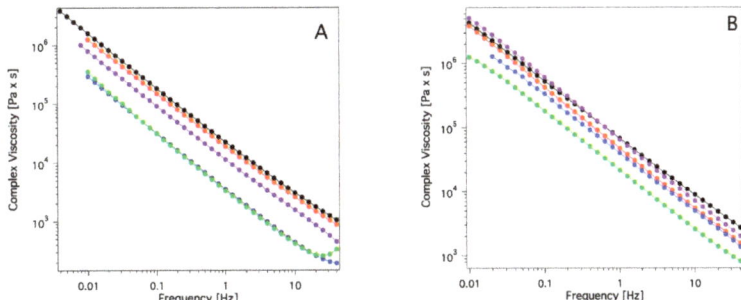

Figure 9. Complex viscosity of the PDMS_SC (**A**) and PDMS_PS (**B**) sponges alone (black) and with absorbed ethyl acetate (green), diethyl carbonate (blue), DMSO (red) and ethanol (purple).

From an application standpoint, one of the main advantages of using gels for cleaning surfaces of porous matrices of historical and artistic interest is linked to the high retention of the solvents confined within them. This allows control of the cleaning action which ideally remains limited to the contact area between the gel itself and the surface of the object to be cleaned. To obtain information for the PDMS-based systems investigated here, absorption tests of the solvents loaded inside them were carried out on sheets of Whatman® paper (which was chosen as a model porous system). The results are reported in Figure 10.

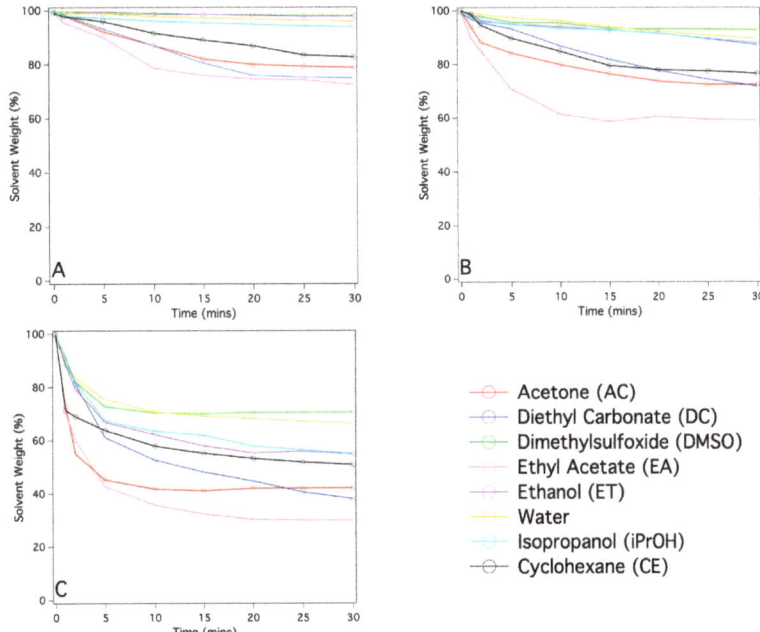

Figure 10. Kinetic curves describing the weight loss of the PDMS based systems soaked with organic solvents and placed in direct contact with a stack of Whatman® papers: (**A**) PDMS slab, (**B**) PDMS_PS, (**C**) PDMS_SC. The A/B component ratio is 10:1.

The data reported in Figure 10 clearly indicate that the extent of solvent release is proportional to the porosity of the PDMS-based networks and, therefore, to the maximum amount of solvent that can be loaded inside them. In fact, for the PDMS-SC system, release values ranging approximately between 30 and 65 wt% are observed, while for the PDMS slab and PDMS-PS, the values never exceeded 50 wt%.

The efficacy of the PDMS based systems as cleaning agents for painted surfaces was first tested on the surface of a fresco mock-up with a 20-year-old surface layer of poly(EMA/MA) 70:30. Figure 11A shows a region of the painting affected by this surface coating. Different tests were carried out by increasing the contact times up to 12 min. The grazing-light image on the right of Figure 11A, collected after the application of a PDMS_PS loaded with EA (ca. 10 wt%) for 12 min, shows the complete disappearance of the glossy effect typical of the copolymer film in the area of the cleaning test. This image indicates macroscopically that the cleaning system was effective for removing the aged coating. Figure 11B reports the FTIR spectra of the organic fraction extracted from mortar samples collected from areas of the fresco where the cleaning tests was carried out (for the detailed experimental procedure, see Section 4). In particular, the FTIR analysis of the cleaned surface (see Figure 11B) confirmed that the poly(EMA/MA) 70:30 layer was no longer detectable after a 12 min treatment with the PDMS_PS sponge, based on the absence of the peak at 1735 cm^{-1}, which is a marker of poly(EMA/MA) 70:30. Furthermore, Figure 11C shows the trend of the ratio between the intensity of the peaks at 1732 cm^{-1} and 2092 cm^{-1} (stretching of the C≡N bond of the Prussian blue used as internal standard) as a function of the application time. The decrease in the value of this parameter indicates the progressive cleaning action of the PDMS_PS sponge by increasing the contact time with the fresco surface.

The surface morphology was investigated by scanning electron microscopy (SEM); it is typical of wall paintings affected on the surface by polymeric coatings that make the surface "smooth" (i.e., with very low roughness and open porosity (Figure S2A)). After application of the cleaning system, the texture of the surface appeared much rougher with a higher porosity that is typical of an original mortar (Figure S2B).

Another potential application for these systems is the removal of aged organic coatings from the surface of easel paintings, as shown in the case study involving the removal of a 25 year naturally aged polymeric ketone resin varnish from the surface of a canvas mock-up. Even in this case, the cleaning test was carried out by means of a PDMS_PS loaded with EA (ca. 10 wt%) with a contact time of 5 min. In fact, from the Teas diagrams [51] it can be deduced that EA is a good solvent for the removal of this class of varnishes. Figure 12A shows the area of the painting treated. The picture on the right, collected after the application of the PDMS_PS system, indicates the visual absence of the varnish layer in the area involved in the cleaning test. ATR-FTIR spectra of the artwork surface before and after cleaning are consistent with this conclusion: the cleaning (Figure 12B, black line) shows strong absorbances centered at 1716 cm^{-1}, 1454 cm^{-1} and 1070 cm^{-1} attributable to the ketone resin [52]; in the spectrum collected after the cleaning test (Figure 12B, red line), the intensity of these diagnostic peaks is strongly decreased, indicating the removal of the surface coating. Also, SEM analysis (Figure S3) of the treated area confirms qualitatively that the cleaning test was successful. Note that the morphology of the paint surface after the cleaning test (Figure S3C) is similar to that of the area of the mock-up where the surface coating was absent (Figure S3A) and that it appears less homogeneous and rougher than the region treated by the surface varnish layer before the application of the PDMS_PS system (Figure S3B).

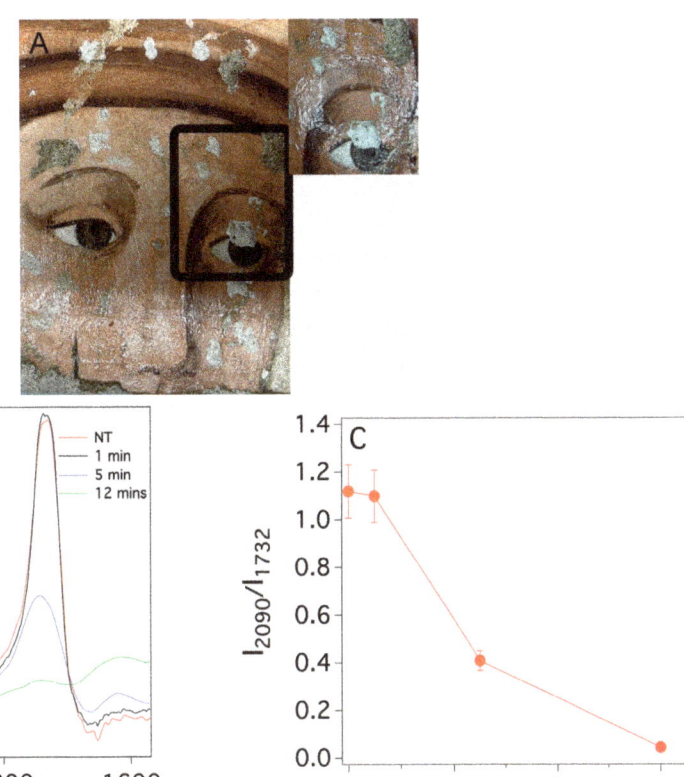

Figure 11. Results of the cleaning test carried out on a fresco mock-up coated with a surface layer of naturally aged poly(EMA/MA) 70:30 by means of a PDMS_PS sponge loaded with ca 10 wt% of EA: (**A**) image of the mock-up before (big picture) and after the cleaning (small picture showing a detail of the area where the 12 min test was carried out): (**B**) FTIR spectra of the resin extracted from mortar fragments collected in the area where the cleaning tests were carried out before (NT, red line) and after the PDMS_PS system application with different contact times: 1 min (black line), 5 min (blue line) and 12 min (green line). (**C**) ratio between the intensity of the peaks at 1732 cm^{-1} and at 2092 cm^{-1} as a function of the application time.

In both cases FTIR spectra did not indicate the presence of PDMS residues on the cleaned surfaces.

The first aspect to be discussed is the two synthetic routes selected to obtain the two POPs PDMS that have different total porosity and pore size distributions. Although the results are consistent with results already reported in the literature for using a sugar cube [34], the material prepared using powdered sugar was more compact with much lower porosity. The FTIR-ATR analysis reported in Figure 1 indicates that there is no meaningful difference in the composition of the PDMS sponges with respect to non-porous slabs (i.e., that no sugar residues were present).

Thus, for these three systems, the non-porous PDMS slab, porous PDMS_SC and PDMS_PS, we conclude that the chemical nature of the materials is the same and the templating agents did not leave any discernible residues.

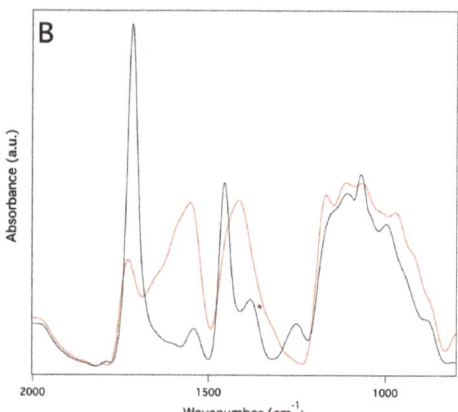

Figure 12. Cleaning of a canvas painting mock-up by means of a PDMS_PS system loaded with about 10 wt% of EA. (**A**) Image of the mock-up before and after the cleaning (small insert on the right) showing a detail of the area where the test was carried out; (**B**) ATR-FTIR collected in the cleaning test area before (black line) and after (red line) the application of the PDMS_PS system.

From a comparison of the morphology and porosity data reported in Table 1 and Figures 3 and 4, it is evident that the two sponges, although chemically the same, have different porosities: PDMS_SC has a much higher porosity than PDMS_PS derived from the large dimensions of the grains constituting the sugar cube (in the order of few hundreds of microns). Moreover, the mechanism of polymerization in the presence of the two templating agents (sugar cube or powdered sugar) differed. In the first case, the appropriately mixed A and B components of the polymerization were absorbed by the porous structure of the sugar cube, so that the polymerization occurred within the fixed macropores of the sugar cube. In the second case, the polymerization occurred at the random interstices among the fine sugar grains that were mixed with the A and B components. Both the different size of the crystal grains for cube and powdered sugar and the larger macropores present in the cube with respect to the smaller ones, consisting of interstices among fine powder grains, generated a large difference in the two PDMS sponges.

The swelling kinetics of the PDMS slab, PDMS_SC and PDMS_PS sponges was a clear indicator of the large difference in porosity among the three materials. Figure 5 shows that PDMS_SC has both the fastest kinetics of solvent absorption and the largest amount of swelling at the asymptotes. This was ascribed to its very high porosity with respect to PDMS_PS and to the non-porous PDMS. In this regard, we reiterate that solvent absorption by the polymer was due simply to swelling phenomena for the PDMS slab but, two synergistic effects were present (i.e., swelling of the polymer enhanced by capillary action mediated by the pores) for the POPs PDMS sponges. These results agree with observations from the literature [35] which show that the driving factor for solvent absorption by PDMS is the magnitude of the hydrophobic interaction which is largest for solvents with the lowest dielectric constants (Figure 5). This is a key point in view of a possible use of these systems as cleaning tools for surfaces of works of art. The swelling/absorption data indicate that these materials may be suitable as cleaning solvents which are able to be released in a controlled way which minimizes both their spreading into the work of art porous support, and any water contamination that is typical for most of the commonly used chemical hydrogels.

Another aspect investigated is their rheological properties (reported in Section 4 and in Supplementary File data). All the investigated systems are characterized by a solid-like behavior, so it is possible to classify them as chemical organogel sponges. As far as we know, these systems may be truly innovative tools for cleaning works of art, especially where even the presence of traces of water during the cleaning procedure represents a possibility of serious damage for the surface. A further interesting finding was that the amount of solvent absorbed by the PDMS-based systems did not meaningfully change the viscoelastic behavior of the PDMS sponges and slabs. As a result, the application tests could open new perspectives for cultural heritage conservation in the field of selective cleaning.

In that regard, the action of the PDMS_PS system loaded with the 10 wt% of acetone, and two different polymeric coatings present on the surface of two lab mock-ups—poly(EMA/MA) 70:30 and a ketone resin on the surface of a fresco and of a canvas painting respectively—indicates the selective extraction of these molecules within the cleaning agent. FTIR analysis carried out on the painting surface before and after the application of the cleaning systems clearly highlights the removal of both the coatings and the recovery of the original morphology typical of uncoated paint layers. Moreover, once the cleaning action is carried out, the PDMS_PS systems have been completely removed, without leaving any instrumentally detectable traces on the cleaned area. This avoids both the use of free-flowing organic solvents and any mechanical action on the paint surface. This feature represents a fundamental improvement over many other the traditional cleaning methods that employ physical gels [53].

3. Conclusions

Synthetic routes have been employed to obtain two types of PDMS sponges with large differences in both their total porosity and mean pore size. These sponges show strong promise as systems to be tried in cleaning tests for the surfaces of works of art. They allow the roles of porosity and pore size in determining solvent/release retention, its spreading within the works of art support, and the solubilization ability against coatings to be removed.

Chemical characterization by FTIR-ATR permitted a comparison of the properties of PDMS slabs with the above-mentioned sponges and to ascertain that the chemical composition of the cross-linked polymer was the same for all the investigated systems. Moreover, no traces of the sugar templating agents used for the preparation of the sponges were present after polymerization. Physicochemical characterizations, carried out by means of SEM, XTM, and porosity determinations, showed that while the sponges obtained using sugar cubes as a templating agent have a porosity around 77% with a mean pore size ca. 300 μm (with a distribution of pore diameters ranging from a few microns to several hundred microns), the porosity of the sponges synthesized using powdered sugar decreased up to 10% with a mean pore size of ca. 75 μm. In this case the distribution of pore diameters ranges from a few microns up to 200 microns.

The very low polarity of PDMS is a key factor in determining the amount of absorbed solvent through polymer swelling and capillary action. They decrease as the dielectric constant of the liquid is increased.

The rheology data indicate that all the systems tested are characterized by a gel-like behavior. G' was larger than G'' over the measured frequency range. In materials with higher porosity, the large amount of bulk and free solvent entrapped within the large pores caused a decrease of G' compared to the dry sponge. This effect is strictly related to the polarity of the solvent according to the greater absorption of low polarity solvents within the pores due to the high hydrophobicity of the PDMS. At lower porosities, the mean sizes of the pores were strongly reduced, and the bulk and free solvent absorbed by the porous polymers were strongly decreased. As a result, the effect on the elastic behavior of the sponges was limited too.

Application of the two porous PDMS based systems containing a small amount of a solvent commonly used for the removal of foreign patinas from painted surfaces of historical and artistic interest (i.e., acetone) allowed selective removal of aged varnishes from the surfaces of two mock-ups that simulate artworks with different substrates (i.e., a fresco and a canvas), without negatively altering the original substrates and without leaving any discernible traces of the PDMS_PS system. These preliminary results/proofs of concept, indicate that the PDMS systems loaded with different organic solvents presented in this paper should be explored further as selective cleaning tools for painted surfaces of historical and artistic interest, even for the cleaning of paintings having different compositions (i.e., modern and contemporary paintings).

The systems presented in this paper are complementary to other physical organogels presented in the literature [43–46]. In particular, due to their physicochemical properties, the PDMS-based organogels and sponges can be loaded with solvents (tunable by varying the contact time between the sponge and the solvent to be loaded, see Figure 5) having different dielectric constants (varying from 2 for cyclohexane to 47 for DMSO). Moreover, thanks to their strong cohesive internal forces (i.e., covalent bonds) they minimize residues left onto the painted surface as indicated by the FTIR data collected on the surfaces after cleaning.

Moreover, the PDMS-based organogels and sponges represent a potentially important alternative to the most used physical cleaning techniques, such as laser or micro-sandblasting, especially in those cases where the cleaning operation must have a high selectivity and a gentle mechanical action.

4. Materials and Methods

4.1. Materials

Polydimethylsiloxane, Sylgard 184® from Dow Corning Corporation (Wiesbaden, Germany), was composed of the base (component A) and a curing agent (component B). Commercial sugar cubes (1 × 1 × 0.7 cm, Dietor Vantaggio®) were pure saccharose grains having average dimensions of about 200 µm and the powdered sugar (dimensions of the grains on the order of tens of microns) were supplied by Sperlari S.r.l., Cremona, Italy. Solvents were from Sigma-Aldrich c/o Merck Life Science S.r.l., Milano, Italy (purity > 99%) and used without further purification. Dow Sylgard™ 184 was supplied by DOW Italy, Milan, Italy. Water was purified by a Millipore MilliQ Direct-Q® & Direct-Q UV water purification system (Water Resistivity: 18.2 MΩ at 25 °C) purchased from Merck-Millipore, Milano, Italy. Polyethyl methacrylate/methyl methacrylate 70/30 (ParaloidB72®) was purchased from Zecchi, Firenze, Italy. The ketone resin was purchased from Phase s.r.l., Firenze, Italy. Melinex®was purchased from Phase s.r.l., Firenze, Italy. Watman®paper was purchased from Cytiva, Little Chalfont, Kent, UK.

4.2. Synthesis Methods

The best ratio to obtain a sufficiently elastic material for applicative purposes was found to be 10:1 A/B by weight. The synthesis was performed as follows:

1. a total amount of 2 g of A and B were vigorously stirred in a Petri dish for approximately 2 min;
2. after homogenization, the Petri dish was placed in a dryer under vacuum at room temperature (ca. 20 °C) to eliminate air bubbles;
3. the polymerisation occurred at room temperature (ca. 20 °C) and soft filter paper was placed onto the Petri dish to avoid dust deposition;
4. after three days, the resulting material, transparent to the eye and elastic, was formed (see Figure S4).

The analogous procedure was performed inside an oven at 60–80 °C for 2 h instead of at room temperature for 3 days. The scheme of the reaction is reported in the literature [54–56].

The same procedure was used also for the synthesis of the porous sponges by using the sugar template method [34] with a further step (also for both the PDMS-based sponges, PDMS_SC and PDMS_PS, the A:B ratio was chosen as 10:1). For the sugar cube sponges (PDMS_SC), the synthesis procedure is reported below:

1. after air bubbles were eliminated, a sugar cube (1.4 g and 1 × 1.5 × 1 cm composed of grains having dimensions ranging in the order of 150–300 µm) was put onto the A/B mixture (A:B ratio was 10:1) at the center of the Petri dish;

2. the Petri dish containing the A/B mixture and the sugar cube was put into a dryer under vacuum for 30 min at room temperature (ca. 20 °C) and then kept under static vacuum for 2 h to allow complete infiltration of the A/B mixture within the porous structure of the sugar cube;
3. then, the Petri dish was put into an oven at 60 °C for 2 h. The curing time was determined by monitoring the trend of the value of the elastic modulus G' at 60 °C as a function of the time for 3 h at constant frequency (1 Hz) and amplitude strain (1%). Figure S5 shows that even after 1.5 h G' reached an asymptotic value indicating that the polymerization reaction was complete.
4. After this time, it was possible to put the sugar cube imbibed with PDMS into a vial filled in with MilliQ water at 40 °C to entirely solubilize the sugar and obtain only the porous PDMS_SC sponge (see Figure S6).

For sponges made using a powdered sugar as template (PDMS_PS), the synthesis procedure is reported below:

1. 2.8 g of powdered sugar (composed of grains having dimensions of a few tens of microns) was incorporated within the A/B mixture until it became a homogeneous system and was placed in a silicon mold ($1 \times 1.2 \times 1.5$ cm);
2. the mold was put inside a Petri dish and dried under vacuum for 30 min to eliminate any air bubbles;
3. thereafter, it was put into an oven at 60 °C for 2 h to complete the polymerization.
4. it was possible to remove the powdered sugar by keeping the sample (PDMS_PS cube) for 15 h at 90 °C in a beaker filled with MilliQ water to entirely solubilize the powdered sugar; this procedure left only the porous PDMS_PS sponge (see Figure S7).

After the preparations, all the systems were washed 3 times for 8 h in a mixture hexane:acetone 1:1 to completely remove any PDMS that was not polymerized. After the third wash, the FTIR spectra collected from the residues of the hexane/acetone mixtures indicated the absence of detectable PDMS residues.

In order to verify the absence of sugar residues the Fehling test was carried out onto 1 g of PDMS sponges fragments following the procedure reported in literature [49].

4.3. Physicochemical Characterization

FTIR spectra of the PDMS slabs and sponges were collected using a Shimadzu, Milan, Italy, IRAffinity-1S Fourier transform infrared spectrometer in transmittance mode (KBr pellets) for the liquid A and B components (Dow Sylgard™ 184) and using the MIRacle single reflection horizontal ATR accessory equipped with a diamond/ZnSe performance flat tip crystal plate. All spectra were then normalized for transmittance (%) intensity as a function of wavenumbers (cm^{-1}). The resolution was 4 cm^{-1} and the number of scans was 64 for both the spectra collected in transmittance (KBr) and ATR mode.

The SEM images were collected on a FEG-SEM ΣIGMA (Carl Zeiss, Jena, Germany) using an acceleration potential of 5 kV at a working distance of 5 mm. The metallization used gold vapor under vacuum; the magnifications were $50\times$ and $100\times$ for all the samples and there was no evidence of decomposition induced by the electron beam.

XMT analysis was carried out with a Skyscan 1172 high-resolution MicroCT system at CRIST Centre, University of Florence (Italy) on a sample of ~$10 \times 5 \times 5$ mm equipped with an X-ray tube having a focal spot size of 5 μm. The X-rays tube was equipped with a tungsten anode operating at 100 kV and 100 μA. 2D X-ray images (acquisition time is 3 s for each image) were captured over 180-degree on a rotating sample with a slice-to-slice rotation angle of 0.3 by placing the sample between the X-ray source and the CCD detector. The spatial resolution was ca. 5 μm. The 3D images were reconstructed from the tomographic projections using Micro Photonics Nrecon® software version 1.6.10.4. Analysis of the obtained 3D images by means of the CTAn software v.1.18 allowed the pore size distribution of each sample subjected to XMT analysis to be obtained. Two different batches for each sample were considered and the error associated both to porosity and to the mean pores diameter was calculated by averaging these two parameters over the volume of the two samples.

The kinetics of solvent absorption (swelling and capillary action) of PDMS slab, PDMS_SC and PDMS_PS sponges were monitored by the weight increase of the slab and sponges as a function of time. PDMS slabs and PDMS_SC or PDMS_PS sponges were cut appropriately to obtain samples of 0.35 g (the exact weights $W_{dry(time=0)}$ were determined on an analytical balance). The samples were kept immersed in 3 cm^3 of each solvent in a capped vial and removed at appropriate time intervals. Then, the samples were softly and rapidly dabbed on a Whatman filter paper and immediately weighed ($W_{wet(time=ti)}$). ATR-FTIR spectra were collected on the paper after its contact with the PDMS-based systems to determine whether part of the PDMS was removed during this operation. However, no evidence of the presence of PDMS residues was observed. The value of Weight Increase in percent $(WI\%)_{time=ti}$ was calculated according to the formula

$$(WI\%)_{time=ti} = \frac{W_{wet\,(time=ti)} - W_{dry\,(time=0)}}{W_{dry\,(time=0)}} \times 100$$

This WI% was plotted as a function of time to obtain the kinetics curve of solvent absorption. The solvents tested were acetone (AC), ethanol (ET), isopropanol (PR), dimethylsulfoxide (DMSO), propylene carbonate (PC), water (W), diethyl carbonate (DC), cyclohexane (CH), ethyl acetate (EA). From these curves, the initial absorption rates were extracted as the slopes obtained by the linear fitting of the early-time absorption values, and the maximum amounts of the absorbed solvents were calculated from the asymptotic values of the curves reported.

The weight loss of the three classes of PDMS-based systems saturated with the solvents indicated above (the initial weight was $W_{wet(time=0)}$) and placed in contact with sheets of Whatman® paper, was monitored as a function of time. For these measurements, to minimize the evaporation of the solvents, the PDMS systems were covered with a polyester sheet (Melinex®). Evaporation of the solvent was monitored over time; even in this case, the PDMS systems were covered with a polyester sheet (Melinex®) and the evaporated solvent at time t_i was measured too.

Then, the amount of solvent absorbed by the Whatman® paper was calculated as follows:

$$\Delta W_{TOT\,(time=ti)} = \Delta W_{ABS\,(time=ti)} - \Delta W_{EVAP\,(time=ti)}$$

where:

$$\Delta W_{TOT\,(time=ti)} = 100 - \left(\frac{W_{wet\,(time=0)} - W'_{wet\,(time=ti)}}{W_{wet\,(time=0)}}\right) \times 100$$

$$\Delta W_{ABS\,(time=ti)} = 100 - \left(\frac{W_{wet\,(time=0)} - W''_{wet\,(time=ti)}}{W_{wet\,(time=0)}}\right) \times 100$$

$$\Delta W_{EVAP\,(time=ti)} = 100 - \left(\frac{W_{wet\,(time=0)} - W'''_{wet\,(time=ti)}}{W_{wet\,(time=0)}}\right) \times 100$$

Here $W'_{wet\,(time=ti)}$ is the total weight loss, $W''_{wet\,(time=ti)}$ is the weight loss due to the solvent absorption by the Whatman® paper and $W'''_{wet\,(time=ti)}$ is the weight loss due to solvent evaporation.

The rheology measurements were carried out using a TA Discovery HR-3 hybrid rheometer according to the following procedure. Frequency sweep measurements were performed in the linear viscoelastic range, to monitor the behavior of the two parameters G′ (elastic modulus) and G″ (viscous modulus) as a function of the oscillation frequency at constant oscillation amplitude. The check of this range was determined through amplitude sweep measurements of G′ and G″ at constant frequency sweep (1 Hz) as a function of the oscillation amplitude (Figure S8). The normal force was set equal to 0.5 N for all measurements. The evaporation of the solvent was minimized by using a solvent trap system provided by TA Instruments (New Castle, DE, USA). All the measurements were repeated at least three times in order to verify the reproducibility.

4.4. Application Tests

In order to determine the efficacy of this methodology, two different cleaning tests were carried out using a PDMS_PS sponge soaked with acetone onto the surface of a fresco mock-up coated with a 20 years old layer of poly(EMA/MA) 70:30 and a canvas painting mock-up coated with a ketone resin. Two different cleaning procedures were followed: for the fresco, three applications on three different areas were carried out on the fresco surface with the following contact times: 1 min, 5 min and 12 min (a small sample of the cleaned area was collected of the uncleaned area and after each application), while a single application was carried out on the canvas sample with a contact time of 12 min.

FTIR spectra of samples from the fresco mock-up were collected in transmittance mode using the following procedure: four small (few mg) fragments of fresco, collected before the cleaning and after each application of the PDMS_PS sponge soaked with acetone, were immersed in chloroform for 24 h in order to selectively extract the poly(EMA/MA) 70:30 film residues. Then, 25 drops of the resulting solutions were placed into an agate mortar and the solvent allowed to evaporate. The spectrum of the obtained solid residue was recorded using a pellet made from 200 mg of KBr containing 0.0125 wt% of the residue and Prussian blue as internal standard. The Prussian blue was added in order to normalize the intensity of the peak at 1732 cm^{-1} (that is a marker of the poly(EMA/MA) 70:30) with the Prussian blue signal at 2092 cm^{-1}. This allowed a semiquantitative comparison among the different samples [57]. Spectra are the average of 64 scans recorded in the absorbance mode using a BioRad (Milan, Italy) model FTS-40 spectrometer with a resolution of 4 cm^{-1}.

The FTIR spectra of the area of interest on canvas in the cleaning tests with the PDMS_PS sponge soaked with acetone were collected in ATR mode through the procedure described above.

Supplementary Materials: The following supporting information can be downloaded at: https://www.mdpi.com/article/10.3390/gels9120985/s1. Figure S1: Frequency sweep diagram of the PDMS sponges; Figure S2: Scanning Electron Microscopy (SEM) images collected before and after the application of PDMS_PS sponge loaded with ca. 10 wt% of EA onto the surface of a fresco mock-up; Figure S3: Scanning Electron Microscopy (SEM) images collected in a region of the canvas painting mock-up where the coating was absent and in a region affected by the surface coating before and after the application of PDMS_PS sponge loaded with ca. 10 wt% of EA; Figure S4: Photographs of the obtained PDMS slab; Figure S5: Trend of the elastic modulus G' for a mixture composed by the base and by the curing agent of the Sylgard 184® kit as a function of time; Figure S6: Photograph of the obtained PDMS_SC sponge; Figure S7: Photograph of the obtained PDMS_PS sponge; Figure S8: Amplitude sweep diagram of the PDMS based systems.

Author Contributions: Conceptualization, E.C., R.G.W., L.D., T.T.D. and F.P.; methodology, E.C., R.G.W., L.D., T.T.D., B.H.B., F.O. and F.P.; validation, F.P., F.O., S.L., T.T.D. and E.C.; resources, E.C., R.G.W. and B.H.B.; data curation, E.C., R.G.W., F.P. and L.D.; writing—original draft preparation, L.D., E.C., F.P. and R.G.W.; writing—review and editing, L.D., E.C., F.P. and R.G.W.; visualization, L.D., E.C., F.P. and R.G.W.; supervision, E.C., L.D., F.P., T.T.D., B.H.B. and R.G.W.; project administration, E.C., L.D., F.P., T.T.D. and R.G.W.; funding acquisition, E.C., L.D. and R.G.W. All authors have read and agreed to the published version of the manuscript.

Funding: This research was funded by University of Florence, the CSGI Consortium and by Italian Ministry of University and Research (MUR) PRIN-2022 2022XX8BRT "Reversible adsorbent smart materials for molecular archaeology to disclose palaeolithic stone tools as bio-archives-SMarT4BioArCH", CUP Master & CUP B53DZ3014020006.

Informed Consent Statement: Informed consent was obtained from all subjects involved in the study.

Data Availability Statement: The data presented in this study are openly available in article.

Acknowledgments: Emma Dini is acknowledged for her help in the preparation of the PDMS based systems.

Conflicts of Interest: The authors declare no conflict of interest. Author Teresa T. Duncan was employed by the company Scientific Analysis of Fine Art, LLC. The remaining authors declare that the research was conducted in the absence of any commercial or financial relationships that could be construed as a potential conflict of interest.

References

1. Yang, X.-Y.; Chen, L.-H.; Li, Y.; Rooke, J.C.; Sanchez, C.; Su, B.-L. Hierarchically Porous Materials: Synthesis Strategies and Structure Design. *Chem. Soc. Rev.* **2017**, *46*, 481–558. [CrossRef]
2. Das, S.; Heasman, P.; Ben, T.; Qiu, S. Porous Organic Materials: Strategic Design and Structure–Function Correlation. *Chem. Rev.* **2017**, *117*, 1515–1563. [CrossRef] [PubMed]
3. Zhu, D.; Handschuh-Wang, S.; Zhou, X. Recent Progress in Fabrication and Application of Polydimethylsiloxane Sponges. *J. Mater. Chem. A* **2017**, *5*, 16467–16497. [CrossRef]
4. Ding, S.-Y.; Wang, W. Covalent Organic Frameworks (COFs): From Design to Applications. *Chem. Soc. Rev.* **2013**, *42*, 548–568. [CrossRef] [PubMed]
5. Kaur, P.; Hupp, J.T.; Nguyen, S.T. Porous Organic Polymers in Catalysis: Opportunities and Challenges. *ACS Catal.* **2011**, *1*, 819–835. [CrossRef]
6. Kim, S.; Lee, Y.M. Rigid and Microporous Polymers for Gas Separation Membranes. *Prog. Polym. Sci.* **2015**, *43*, 1–32. [CrossRef]
7. Sun, J.-K.; Xu, Q. Functional Materials Derived from Open Framework Templates/Precursors: Synthesis and Applications. *Energy Environ. Sci.* **2014**, *7*, 2071. [CrossRef]
8. Sun, L.-B.; Liu, X.-Q.; Zhou, H.-C. Design and Fabrication of Mesoporous Heterogeneous Basic Catalysts. *Chem. Soc. Rev.* **2015**, *44*, 5092–5147. [CrossRef]
9. Wang, W.; Zhou, M.; Yuan, D. Carbon Dioxide Capture in Amorphous Porous Organic Polymers. *J. Mater. Chem. A* **2017**, *5*, 1334–1347. [CrossRef]
10. Xiang, Z.; Cao, D. Porous Covalent–Organic Materials: Synthesis, Clean Energy Application and Design. *J. Mater. Chem. A* **2013**, *1*, 2691–2718. [CrossRef]
11. Choi, S.-J.; Kwon, T.-H.; Im, H.; Moon, D.-I.; Baek, D.J.; Seol, M.-L.; Duarte, J.P.; Choi, Y.-K. A Polydimethylsiloxane (PDMS) Sponge for the Selective Absorption of Oil from Water. *ACS Appl. Mater. Interfaces* **2011**, *3*, 4552–4556. [CrossRef] [PubMed]
12. McCall, W.R.; Kim, K.; Heath, C.; La Pierre, G.; Sirbuly, D.J. Piezoelectric Nanoparticle–Polymer Composite Foams. *ACS Appl. Mater. Interfaces* **2014**, *6*, 19504–19509. [CrossRef] [PubMed]
13. Yuen, P.K.; Su, H.; Goral, V.N.; Fink, K.A. Three-Dimensional Interconnected Microporous Poly(Dimethylsiloxane) Microfluidic Devices. *Lab. Chip.* **2011**, *11*, 1541. [CrossRef] [PubMed]
14. Zheng, Q.; Zhang, H.; Mi, H.; Cai, Z.; Ma, Z.; Gong, S. High-Performance Flexible Piezoelectric Nanogenerators Consisting of Porous Cellulose Nanofibril (CNF)/Poly(Dimethylsiloxane) (PDMS) Aerogel Films. *Nano Energy* **2016**, *26*, 504–512. [CrossRef]
15. Park, J.; Wang, S.; Li, M.; Ahn, C.; Hyun, J.K.; Kim, D.S.; Kim, D.K.; Rogers, J.A.; Huang, Y.; Jeon, S. Three-Dimensional Nanonetworks for Giant Stretchability in Dielectrics and Conductors. *Nat. Commun.* **2012**, *3*, 916. [CrossRef] [PubMed]
16. Wang, J.; Guo, J.; Si, P.; Cai, W.; Wang, Y.; Wu, G. Polydopamine-Based Synthesis of an In(OH) 3 –PDMS Sponge for Ammonia Detection by Switching Surface Wettability. *RSC Adv.* **2016**, *6*, 4329–4334. [CrossRef]
17. Kim, D.H.; Jung, M.C.; Cho, S.-H.; Kim, S.H.; Kim, H.-Y.; Lee, H.J.; Oh, K.H.; Moon, M.-W. UV-Responsive Nano-Sponge for Oil Absorption and Desorption. *Sci. Rep.* **2015**, *5*, 12908. [CrossRef]
18. Yu, C.; Yu, C.; Cui, L.; Song, Z.; Zhao, X.; Ma, Y.; Jiang, L. Facile Preparation of the Porous PDMS Oil-Absorbent for Oil/Water Separation. *Adv. Mater. Interfaces* **2017**, *4*, 1600862. [CrossRef]
19. Li, Q.; Duan, T.; Shao, J.; Yu, H. Fabrication Method for Structured Porous Polydimethylsiloxane (PDMS). *J. Mater. Sci.* **2018**, *53*, 11873–11882. [CrossRef]
20. Jung, S.; Kim, J.H.; Kim, J.; Choi, S.; Lee, J.; Park, I.; Hyeon, T.; Kim, D.-H. Reverse-Micelle-Induced Porous Pressure-Sensitive Rubber for Wearable Human-Machine Interfaces. *Adv. Mater.* **2014**, *26*, 4825–4830. [CrossRef]
21. Liu, W.; Chen, Z.; Zhou, G.; Sun, Y.; Lee, H.R.; Liu, C.; Yao, H.; Bao, Z.; Cui, Y. 3D Porous Sponge-Inspired Electrode for Stretchable Lithium-Ion Batteries. *Adv. Mater.* **2016**, *28*, 3578–3583. [CrossRef] [PubMed]
22. Liang, S.; Li, Y.; Yang, J.; Zhang, J.; He, C.; Liu, Y.; Zhou, X. 3D Stretchable, Compressible, and Highly Conductive Metal-Coated Polydimethylsiloxane Sponges. *Adv. Mater. Technol.* **2016**, *1*, 1600117. [CrossRef]
23. Han, J.-W.; Kim, B.; Li, J.; Meyyappan, M. Flexible, Compressible, Hydrophobic, Floatable, and Conductive Carbon Nanotube-Polymer Sponge. *Appl. Phys. Lett.* **2013**, *102*, 051903. [CrossRef]
24. Pedraza, E.; Brady, A.-C.; Fraker, C.A.; Molano, R.D.; Sukert, S.; Berman, D.M.; Kenyon, N.S.; Pileggi, A.; Ricordi, C.; Stabler, C.L. Macroporous Three-Dimensional PDMS Scaffolds for Extrahepatic Islet Transplantation. *Cell Transplant.* **2013**, *22*, 1123–1135. [CrossRef]
25. Zhao, X.; Li, L.; Li, B.; Zhang, J.; Wang, A. Durable Superhydrophobic/Superoleophilic PDMS Sponges and Their Applications in Selective Oil Absorption and in Plugging Oil Leakages. *J. Mater. Chem. A* **2014**, *2*, 18281–18287. [CrossRef]

26. Shi, J.; Zhang, H.; Jackson, J.; Shademani, A.; Chiao, M. A Robust and Refillable Magnetic Sponge Capsule for Remotely Triggered Drug Release. *J. Mater. Chem. B* **2016**, *4*, 7415–7422. [CrossRef]
27. Liang, S.; Li, Y.; Chen, Y.; Yang, J.; Zhu, T.; Zhu, D.; He, C.; Liu, Y.; Handschuh-Wang, S.; Zhou, X. Liquid Metal Sponges for Mechanically Durable, All-Soft, Electrical Conductors. *J. Mater. Chem. C* **2017**, *5*, 1586–1590. [CrossRef]
28. Bélanger, M.-C.; Marois, Y. Hemocompatibility, Biocompatibility, Inflammatory and in Vivo Studies of Primary Reference Materials Low-Density Polyethylene and Polydimethylsiloxane: A Review. *J. Biomed. Mater. Res.* **2001**, *58*, 467–477. [CrossRef]
29. Lötters, J.C.; Olthuis, W.; Veltink, P.H.; Bergveld, P. The Mechanical Properties of the Rubber Elastic Polymer Polydimethylsiloxane for Sensor Applications. *J. Micromech. Microeng.* **1997**, *7*, 145–147. [CrossRef]
30. Mata, A.; Fleischman, A.J.; Roy, S. Characterization of Polydimethylsiloxane (PDMS) Properties for Biomedical Micro/Nanosystems. *Biomed. Microdevices* **2005**, *7*, 281–293. [CrossRef]
31. Dong, C.-H.; He, L.; Xiao, Y.-F.; Gaddam, V.R.; Ozdemir, S.K.; Han, Z.-F.; Guo, G.-C.; Yang, L. Fabrication of High-Q Polydimethylsiloxane Optical Microspheres for Thermal Sensing. *Appl. Phys. Lett.* **2009**, *94*, 231119. [CrossRef]
32. Piruska, A.; Nikcevic, I.; Lee, S.H.; Ahn, C.; Heineman, W.R.; Limbach, P.A.; Seliskar, C.J. The Autofluorescence of Plastic Materials and Chips Measured under Laser Irradiation. *Lab. Chip.* **2005**, *5*, 1348. [CrossRef] [PubMed]
33. Yilgör, E.; Yilgör, I. Silicone Containing Copolymers: Synthesis, Properties and Applications. *Prog. Polym. Sci.* **2014**, *39*, 1165–1195. [CrossRef]
34. González-Rivera, J.; Iglio, R.; Barillaro, G.; Duce, C.; Tinè, M. Structural and Thermoanalytical Characterization of 3D Porous PDMS Foam Materials: The Effect of Impurities Derived from a Sugar Templating Process. *Polymers* **2018**, *10*, 616. [CrossRef] [PubMed]
35. Lee, J.N.; Park, C.; Whitesides, G.M. Solvent Compatibility of Poly(Dimethylsiloxane)-Based Microfluidic Devices. *Anal. Chem.* **2003**, *75*, 6544–6554. [CrossRef]
36. Phenix, A. The Swelling of Artists' Paints in Organic Solvents. Part 2, Comparative Swelling Powers of Selected Organic Solvents and Solvent Mixtures. *J. Am. Inst. Conserv.* **2002**, *41*, 61–90. [CrossRef]
37. Duncan, T.T.; Chan, E.P.; Beers, K.L. Quantifying the 'Press and Peel' Removal of Particulates Using Elastomers and Gels. *J. Cult. Herit.* **2021**, *48*, 236–243. [CrossRef]
38. Duncan, T.T.; Berrie, B.H.; Weiss, R.G. Soft, Peelable Organogels from Partially Hydrolyzed Poly(Vinyl Acetate) and Benzene-1,4-Diboronic Acid: Applications to Clean Works of Art. *ACS Appl. Mater. Interfaces* **2017**, *9*, 28069–28078. [CrossRef]
39. Duncan, T.T.; Vicenzi, E.P.; Lam, T.; Brogdon-Grantham, S.A. A Comparison of Materials for Dry Surface Cleaning Soot-Coated Papers of Varying Roughness: Assessing Efficacy, Physical Surface Changes, and Residue. *J. Am. Inst. Conserv.* **2023**, *62*, 152–167. [CrossRef]
40. Duncan, T.T.; Chan, E.P.; Beers, K.L. Maximizing Contact of Supersoft Bottlebrush Networks with Rough Surfaces To Promote Particulate Removal. *ACS Appl. Mater. Interfaces* **2019**, *11*, 45310–45318. [CrossRef]
41. Carretti, E.; Dei, L.; Macherelli, A.; Weiss, R.G. Rheoreversible Polymeric Organogels: The Art of Science for Art Conservation. *Langmuir* **2004**, *20*, 8414–8418. [CrossRef]
42. Carretti, E.; Dei, L.; Weiss, R.G.; Baglioni, P. A New Class of Gels for the Conservation of Painted Surfaces. *J. Cult. Herit.* **2008**, *9*, 386–393. [CrossRef]
43. Samorì, C.; Galletti, P.; Giorgini, L.; Mazzeo, R.; Mazzocchetti, L.; Prati, S.; Sciutto, G.; Volpi, F.; Tagliavini, E. The Green Attitude in Art Conservation: Polyhydroxybutyrate-Based Gels for the Cleaning of Oil Paintings. *ChemistrySelect* **2016**, *1*, 4502–4508. [CrossRef]
44. Yiming, J.; Sciutto, G.; Prati, S.; Catelli, E.; Galeotti, M.; Porcinai, S.; Mazzocchetti, L.; Samorì, C.; Galletti, P.; Giorgini, L.; et al. A New Bio-Based Organogel for the Removal of Wax Coating from Indoor Bronze Surfaces. *Herit. Sci.* **2019**, *7*, 34. [CrossRef]
45. Prati, S.; Volpi, F.; Fontana, R.; Galletti, P.; Giorgini, L.; Mazzeo, R.; Mazzocchetti, L.; Samorì, C.; Sciutto, G.; Tagliavini, E. Sustainability in Art Conservation: A Novel Bio-Based Organogel for the Cleaning of Water Sensitive Works of Art. *Pure Appl. Chem.* **2018**, *90*, 239–251. [CrossRef]
46. Çakmak, Y.; Çakmakçi, E.; Apohan, N.K.; Karadag, R. Isosorbide, Pyrogallol, and Limonene-Containing Thiol-Ene Photocured Bio-Based Organogels for the Cleaning of Artworks. *J. Cult. Herit.* **2022**, *55*, 391–398. [CrossRef]
47. Johnson, L.M.; Gao, L.; Shields IV, C.W.; Smith, M.; Efimenko, K.; Cushing, K.; Genzer, J.; López, G.P. Elastomeric Microparticles for Acoustic Mediated Bioseparations. *J. Nanobiotechnol.* **2013**, *11*, 22. [CrossRef] [PubMed]
48. Zhang, A.; Cheng, L.; Hong, S.; Yang, C.; Lin, Y. Preparation of Anti-Fouling Silicone Elastomers by Covalent Immobilization of Carboxybetaine. *RSC Adv.* **2015**, *5*, 88456–88463. [CrossRef]
49. Hörner, T.G.; Klüfers, P. The Species of Fehling's Solution. *Eur. J. Inorg. Chem.* **2016**, *2016*, 1798–1807. [CrossRef]
50. Almdal, K.; Dyre, J.; Hvidt, S.; Kramer, O. Towards a Phenomenological Definition of the Term 'Gel'. *Polym. Gels Netw.* **1993**, *1*, 5–17. [CrossRef]
51. Horie, V.C. *Materials for Conservation*, 1st ed.; Architectural Press: Oxford, UK, 1998.
52. Sandu, I.C.A.; Candeias, A.; van den Berg, K.J.; Sandbakken, E.G.; Tveit, E.S.; van Keulen, H. Multi Technique and Multiscale Approaches to the Study of Ancient and Modern Art Objects on Wooden and Canvas Support. *Phys. Sci. Rev.* **2019**, *4*, 20180016. [CrossRef]
53. Stulik, D.; Miller, D.; Khanjian, H.; Khandekar, N.; Richard, W.; Carlson, J.; Petersen, W.C. *Solvent Gels for the Cleaning of Works of Art: The Residue Question*; Dorge, V., Ed.; Getty Publications: Los Angeles, CA, USA, 2004.

54. Gupta, N.S.; Lee, K.-S.; Labouriau, A. Tuning Thermal and Mechanical Properties of Polydimethylsiloxane with Carbon Fibers. *Polymers* **2021**, *13*, 1141. [CrossRef] [PubMed]
55. Xia, Y.; Whitesides, G.M. Soft lithography. *Annu. Rev. Mater. Sci.* **1998**, *28*, 153–184. [CrossRef]
56. Lisensky, G.C.; Campbell, D.J.; Beckman, K.J.; Calderon, C.E.; Doolan, P.W.; Ottosen, R.M.; Ellis, A.B. Replication and Compression of Surface Structures with Polydimethylsiloxane Elastomer. *J. Chem. Educ.* **1999**, *76*, 537. [CrossRef]
57. Salvadori, B.; Errico, V.; Mauro, M.; Melnik, E.; Dei, L. Evaluation of Gypsum and Calcium Oxalates in Deteriorated Mural Paintings by Quantitative FTIR Spectroscopy. *Spectrosc. Lett.* **2003**, *36*, 501–513. [CrossRef]

Disclaimer/Publisher's Note: The statements, opinions and data contained in all publications are solely those of the individual author(s) and contributor(s) and not of MDPI and/or the editor(s). MDPI and/or the editor(s) disclaim responsibility for any injury to people or property resulting from any ideas, methods, instructions or products referred to in the content.

Article

Linear Polyethyleneimine-Based and Metal Organic Frameworks (DUT-67) Composite Hydrogels as Efficient Sorbents for the Removal of Methyl Orange, Copper Ions, and Penicillin V

Luis M. Araque [1,2], Roberto Fernández de Luis [3], Arkaitz Fidalgo-Marijuan [3,4], Antonia Infantes-Molina [5], Enrique Rodríguez-Castellón [5], Claudio J. Pérez [6], Guillermo J. Copello [1,2] and Juan M. Lázaro-Martínez [1,2,*]

[1] Departamento de Ciencias Químicas, Facultad de Farmacia y Bioquímica, Universidad de Buenos Aires, Buenos Aires 1113, Argentina; lmaraque@conicet.gov.ar (L.M.A.); gcopello@ffyb.uba.ar (G.J.C.)
[2] Consejo Nacional de Investigaciones Científicas y Técnicas (CONICET), Instituto de Química y Metabolismo del Fármaco (IQUIMEFA-UBA-CONICET), Buenos Aires 1113, Argentina
[3] BCMaterials, Basque Center for Materials, Applications and Nanostructures, UPV/EHU Science Park, 48940 Leioa, Spain; roberto.fernandez@bcmaterials.net (R.F.d.L.); arkaitz.fidalgo@ehu.eus (A.F.-M.)
[4] Departamento de Química Orgánica e Inorgánica, Facultad de Ciencia y Tecnología, University of the Basque Country (UPV/EHU), 48940 Leioa, Spain
[5] Departamento de Química Inorgánica, Cristalografía y Mineralogía, Facultad de Ciencias, Universidad de Málaga, 29010 Malaga, Spain; ainfantes@uma.es (A.I.-M.); castellon@uma.es (E.R.-C.)
[6] Consejo Nacional de Investigaciones Científicas y Técnicas (CONICET), Instituto de Investigaciones en Ciencia y Tecnología de Materiales (INTEMA), Facultad de Ingeniería, Universidad de Mar del Plata, Mar del Plata 7600, Argentina; cjperez@fi.mdp.edu.ar
* Correspondence: lazarojm@ffyb.uba.ar; Tel.: +54-11-5287-4323

Citation: Araque, L.M.; Fernández de Luis, R.; Fidalgo-Marijuan, A.; Infantes-Molina, A.; Rodríguez-Castellón, E.; Pérez, C.J.; Copello, G.J.; Lázaro-Martínez, J.M. Linear Polyethyleneimine-Based and Metal Organic Frameworks (DUT-67) Composite Hydrogels as Efficient Sorbents for the Removal of Methyl Orange, Copper Ions, and Penicillin V. *Gels* **2023**, *9*, 909. https://doi.org/10.3390/gels9110909

Academic Editor: Shiyang Li

Received: 27 October 2023
Revised: 8 November 2023
Accepted: 14 November 2023
Published: 16 November 2023

Copyright: © 2023 by the authors. Licensee MDPI, Basel, Switzerland. This article is an open access article distributed under the terms and conditions of the Creative Commons Attribution (CC BY) license (https://creativecommons.org/licenses/by/4.0/).

Abstract: This research explores the integration of DUT-67 metal organic frameworks into polyethyleneimine-based hydrogels to assemble a composite system with enough mechanical strength, pore structure and chemical affinity to work as a sorbent for water remediation. By varying the solvent-to-modulator ratio in a water-based synthesis path, the particle size of DUT-67 was successfully modulated from 1 μm to 200 nm. Once DUT-67 particles were integrated into the polymeric hydrogel, the composite hydrogel exhibited enhanced mechanical properties after the incorporation of the MOF filler. XPS, NMR, TGA, FTIR, and FT Raman studies confirmed the presence and interaction of the DUT-67 particles with the polymeric chains within the hydrogel network. Adsorption studies of methyl orange, copper(II) ions, and penicillin V on the composite hydrogel revealed a rapid adsorption kinetics and monolayer adsorption according to the Langmuir's model. The composite hydrogel demonstrated higher adsorption capacities, as compared to the pristine hydrogel, showcasing a synergistic effect, with maximum adsorption capacities of 473 ± 21 mg L^{-1}, 86 ± 6 mg L^{-1}, and 127 ± 4 mg L^{-1}, for methyl orange, copper(II) ions, and penicillin V, respectively. This study highlights the potential of MOF-based composite hydrogels as efficient adsorbents for environmental pollutants and pharmaceuticals.

Keywords: hydrogels; DUT-67; adsorption; multifunctional composites; emerging pollutants

1. Introduction

The quality and availability of water for human consumption are affected by industrial and urban activities, mainly due to the improper discharge of chemical and biological pollutants into water bodies. Particularly, dyes, heavy metal ions and emergent pollutants scape in different degrees from current wastewater treatment technologies, posing a risk to human health and the ecosystem equilibrium in general [1]. There is no regulation that restricts or determines a limit for the concentration of dyes in industrial effluents, as what is regulated is their use in products such as toys or textiles [2]. In the case of copper ions (Cu^{2+}), the World Health Organization establishes a Cu^{2+} limit of 2 mg L^{-1} in drinking water [3,4]. Drugs such as penicillin V are considered emerging contaminants, and there is still not enough information on their effects on health or regulations for their disposal or presence in aquifers [4]. This worldwide issue is accentuated in isolated areas lacking adequate sanitation facilities. Therefore, it is imperative to develop materials and technologies to treat polluted water sources. One promising alternative is to create filter systems based on adsorbents. Adsorption is a low-cost and easy-to-implement technique that requires minimal energy. It has been widely and effectively used for the removal of pollutants from water [5]. Hydrogels, which production is quite affordable, are known to have high specific surface areas and adsorption capacities, as well as chemical versatility to modify their adsorption affinity. For these reasons, they have emerged as promising adsorbents [6]. Among them, polyethyleneimine-based gels have revealed outstanding capacities to retain different adsorbates, but most of them still exhibit poor mechanical resistance to deformation.

In a previous research work, our group developed a pH- and ionic strength-responsive hydrogel based on linear polyethyleneimine hydrochloride (L-PEI·HCl) [7]. Although the material had high swelling and adsorption capacities, it exhibited poor resistance to deformation, which hindered its manipulation, affecting its applicability. This hydrogel has a high affinity for transition metal ions, which allows us to evaluate the release of Cu^{2+} ions from montmorillonite materials with paramagnetic metal complexes [8].

The development of composite materials is a promising strategy to improve the mechanical properties of hydrogels. Composite materials feature an interface between the hydrogel polymer chains and the particulate filler that distributes the deformation applied to the material through this interface. The effect of the filler depends on three factors, i.e., particle size, the type of interaction between the filler and the polymeric matrix, and the filler dispersion and distribution degree along the matrix [9]. In this sense, Rao et al. developed a composite hydrogel by crosslinking gelatin with genipin, employing amino-functionalized microfibrillated cellulose as the filler [10]. The composite hydrogel exhibited a compressive strength of 1.52 MPa, which was 41.2 times higher than that of the hydrogel without the filler. In another study, a composite hydrogel was prepared by 3D printing gelatin methacrylate with nano-attapulgite as the filler, increasing the compressive strength and modulus 4.3 and 16.4 times, respectively [11]. Additionally, Chen et al. developed composite hydrogels based on polyvinyl alcohol with acidified carbon nanotubes as a filler [12]. By adding 2% of the filler, the Young's modulus, maximum tensile stress, and toughness increased from 0.07 MPa, 108 kPa, and 83,819.2 J m^3 to 0.185 MPa, 234 kPa, and 163,697.2 J m^3, respectively [12–14].

Considering the potential of PEI hydrogels for water environmental remediation, the incorporation of sorbent-like metal organic framework (MOF) fillers into the PEI hydrogel matrix offers a dual advantage. This integration enhances both the mechanical properties and adsorption capacity of the hybridized system. MOFs are a class of porous materials that have gained popularity in recent years due to their application for water remediation purposes. Their very high specific surface areas, together with the chemically versatility to encode their frameworks and pore structures, has opened the way to their application for the removal and degradation via photocatalysis of myriad organic and inorganic pollutants [13–16]. The activity in this research area has been further intensified since the discovery of water-stable MOFs, including the DUT-67 one employed in this study as the filler of the PEI hydrogels. Further, the green synthesis protocols developed for trivalent and tetravalent metal carboxylate MOFs during the last years have facilitated the scale-up of these MOFs' production with a high degree of control over their particle size. This is the case for the low-temperature water synthesis path applied to the crystallization of DUT-67 particles applied in this study [17–19]. As many other Zr-MOFs, the structure of DUT-67 is built up from the archetypal hexanuclear $Zr_6(OH)_4O_4(CO_2^-)_{12}$ clusters connected via eight 2,5-thiophenedicarboxilate linkers to form a micro- to mesoporous structure with two pores of different characteristics (i.e., diameter, pore window, and surface chemistry).

All in all, the presence of thiophene fragments and uncoordinated positions occupied by modulator molecules within the zirconium cluster makes DUT-67 a versatile sorbent both for inorganics and organics with varied characteristics [20]. Few studies have assessed the use of fillers to enhance the mechanical properties of PEI-based hydrogels. There are no studies on the use of DUT-67 as a filler for PEI-based hydrogels or any other polymer. Furthermore, works on PEI-based hydrogels have predominantly focused on the use of the ramified architecture rather than on the linear one [17,21,22].

In this study, a high-yield low-temperature green synthesis of DUT-67 samples has been proposed. The modulation of DUT-67 crystallization by acetic acid has enabled a fine control over the particle size of the final material. Subsequently, DUT-67 nanoparticles were employed as a filler in the L-PEI·HCl hydrogels, which were obtained via a crosslinking with ethylene glycol diglycidyl ether (EGDE). The DUT-67 content in the composite was varied, and the physicochemical, thermal, and textural properties of the system were duly studied. The adsorptive properties of the composite gels for organic pollutants such as methyl orange (MO) azo dye, penicillin V (PEN) antibiotic, and copper ions (Cu^{2+}) were investigated and correlated with the chemistry and porosity of the hydrogels before and after the incorporation of DUT-67 nanoparticles as an active filler in the system.

2. Results and Discussion

2.1. DUT-67 Synthesis and Characterization

The water-based green synthesis paths of MOFs have received important attention during the last years, including the ones reported for Zr-MOFs and, more specifically, for DUT-67. Considering the usual parameter spaces explored in these synthesis paths, we adapted the water-based crystallization of DUT-67 while modulating its particle size by controlling the concentration of the acetic acid modulator in the media [23]. As revealed by scanning electron microscopy (SEM) (Figure 1A–C), as the concentration of the modulator was lowered, the particle sizes of the DUT-67 samples (acetic acid) decreased from 1.01 ± 17 μm to 202 ± 24 nm. It is interesting to note that L54 exhibited a quite homogenous particle size shape in comparison to the ones obtained under higher concentrations of acetic acid [24,25].

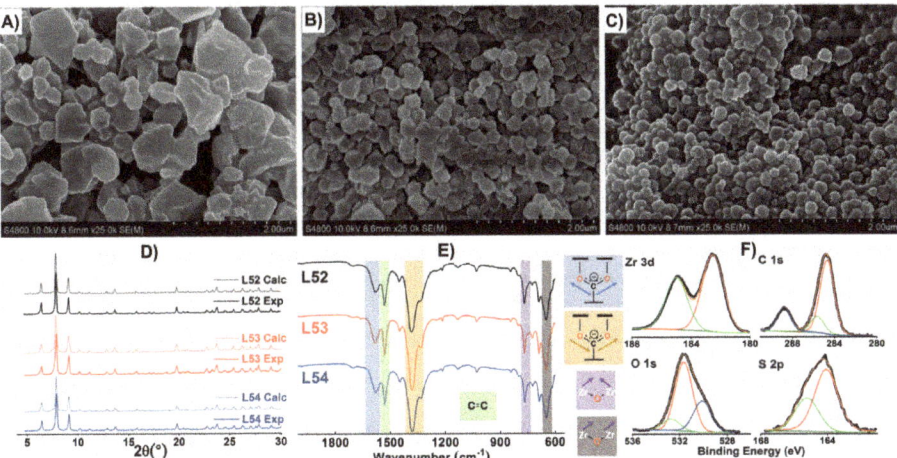

Figure 1. (**A–C**) SEM images, (**D**) PXRD patterns, (**E**) ATR—FTIR spectra of L52, L53, and L54; and (**F**) XPS spectra of L54.

The samples after the synthesis, washing, and activation process were characterized by powder X-ray diffraction (PXRD), attenuated total reflectance-Fourier-transform infrared spectroscopy (ATR-FTIR), FT Raman, and thermogravimetric analysis (TGA). First, the

PXRD patterns for the three DUT samples were very similar to the calculated pattern obtained from the structural model reported by Drache et al. [26]. Second, a slight broadening of the diffraction maxima was observed as the particles size of the DUT-67 samples decreased (Figure 1D). The FTIR and Raman spectroscopies data (Figure 1E and S1) also confirmed that the acetic acid modulation did not cause significant changes in the chemistry of the DUT-67 samples. In detail, the FTIR spectra exhibited the characteristic absorption maxima ascribed to the bending vibrations of the Zr-O-Zr bonds (649 and 772 cm^{-1}), the stretching vibrations of the O=C–O bonds (1574 (asymmetric) and 1393 (symmetric) cm^{-1}), and the vibration of the C=C bond (1531 cm^{-1}) [27]. The Raman spectra exhibited a characteristic peak at 1474 cm^{-1}, which corresponded to the vibration of the C=C bond in the aromatic ring [28]. Although it is well known that the monocarboxylate modulators can induce significant linker defects on the crystal structure of Zr-MOFs, the thermogravimetry curves of the three DUT-67 samples were quite similar (Figure S2). Thus, there was expected to be a similar linker-defectiveness in the three of them. As expected, thermal degradation occurred in three steps, the first related to the dehydration occurring between 30 °C and 100 °C, the second ascribed to the acetic acid modulator release and the zirconium hexanuclear cluster dehydration from 100 to 350 °C, and the last one to the organic linker pyrolysis (350–550 °C).

High-pressure absorption isotherms acquired at 0 °C (Figure S3) showed 2 (L52), 2 (L53), and 5 mmol g^{-1} (L54) capacity to adsorb CO_2 at 30 bar that were dependent on the modulator added to the synthesis. The lower the acetic acid concentration, the higher the capacity of the BET samples to capture CO_2, and, in parallel, the higher the value of the surface area value obtained (L52: 841 m^2 g^{-1}, L53: 1083 m^2 g^{-1}, and L54: 1231 m^2 g^{-1}) after the fitting of the low-pressure range data. In view of the textural and morphological characteristics, L54 was selected as the sample for further characterization and integration into the polymeric PEI matrix. The surface chemistry of this sample was studied by means of X-ray photoelectron spectroscopy (XPS) (Figure 1F and Table S2). The Zr 3d core level spectrum showed the doublet ascribed to the spin orbit splitting of the Zr $3d_{5/2}$ and $3d_{3/2}$ levels located at 182.5 and 184.9 eV, respectively, in line with the values reported previously for DUT-67 [29]. For C 1s, three noticeable contributions were observed at 284.7, 285.6, and 288.8 eV, which corresponded to adventitious carbon layers containing aliphatic C-C and aromatic C=C bonds at 284.7 eV, C–OH and C-O-C at 285.6 eV, and carboxylate in the ligand at 288.8 eV. The O 1s core level spectrum was fitted with three contributions at 530.1, 531.6, and 532.7 eV, corresponding to the lattice oxygen atoms in the cluster, oxygen atoms of the ligand, and oxygen atoms from water molecules adsorbed into the material, respectively. Finally, the S 2p signal exhibited two contributions at 164.1 and 165.3 eV that were attributed to the S $2p_{3/2}$ and $2p_{1/2}$ doublet, respectively, of the S atoms in the thiophene ring [30,31].

Solid-state nuclear magnetic resonance spectroscopy (*ss*-NMR) experiments were also done to further investigate the bulk structure of the material. Figure 2A,B show the ^{1}H magic angle spinning (MAS) and ^{13}C cross-polarization and magic angle spinning (CP-MAS) NMR spectra of L54, respectively. In the ^{1}H spectrum, signal c was assigned to the proton in the thiophene ring, while the other four signals did not belong to the DUT-67 components. Considering the shape and the chemical shift (~4.2 ppm), signal b was assigned to protons from free water molecules trapped within the MOF pores, even when the sample was oven-dried before the experiments. The three signals marked as a were assigned to methyl protons from three different acetic acid populations (modulator) that occupied the linker defective position of the zirconium hexanuclear clusters [32]. The signal widths confirmed, as expected, that the modulator populations are part of the structure and not molecules simply trapped within the material pores. The ^{13}C *ss*-NMR spectrum exhibited the signals arising from two carbons (2 and 3) of the aromatic ring and an additional one belonging to the carboxylate groups (4) of the ligand. Two populations of linkers were found where carbon 2 had a distinct chemical arrangement and environment within the MOF structure. Additionally, the methyl carbon signals (1) revealed three populations of the acetic acid modulator. Moreover, two signals corresponding to the carboxylate carbons (5) were observed. One of the signals was very broad, suggesting the existence of at least two overlapping signals and, hence, and the presence of three acetate populations in the DUT-67structure. Finally, 2D ss-NMR experiments were performed to unravel the carbon hydrogen interactions within the DUT-67 framework. Figure 3 shows the 2D ^{1}H-^{13}C heteronuclear correlation (HETCOR) and ^{1}H-^{1}H single quantum/double quantum (SQ/DQ) results.

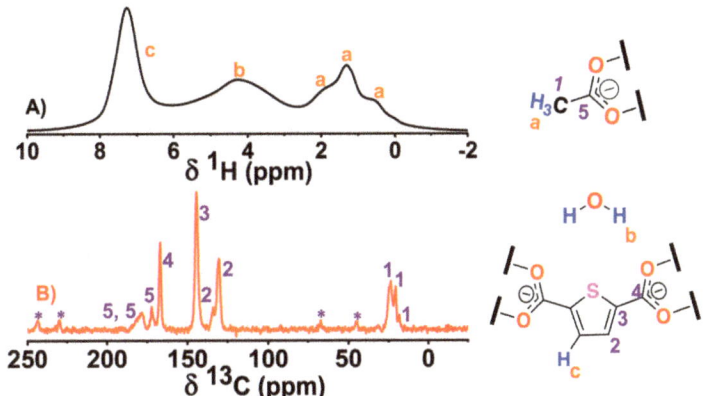

Figure 2. (**A**) ^{1}H MAS (@32 kHz) and (**B**) ^{13}C CP—MAS NMR (@15 kHz) spectra of DUT—67.

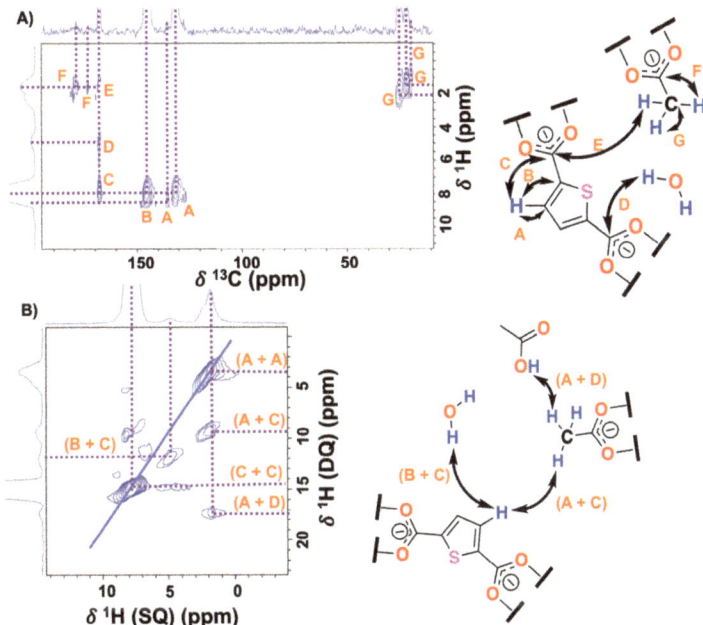

Figure 3. (**A**) The 2D ^{1}H—^{13}C HETCOR (@15 kHz), and (**B**) 2D ^{1}H—^{1}H SQ/DQ (@32 kHz) ss—NMR spectra of DUT—67.

In the 2D ^{1}H-^{13}C HETCOR spectrum (Figure 3A), short- and long-range correlations were observed between the ligand hydrogen and the different carbons of the thiophene ring (A and B) and the carbon of the carboxylate groups (C), respectively. Interestingly, correlations were also observed between the protons of water molecules and the carbons of the carboxylate groups belonging to the ligand (D). Thus, these interactions would participate in the maintenance of the stability of the supramolecular structure. Furthermore, different chemical environments for the methyl (G correlations) and carboxylate (F correlations) groups of the modulator were evidenced through three different ^{1}H and ^{13}C populations. In addition, the correlation between the ligand and modulator (E) was clearly differentiated. In the 2D ^{1}H-^{1}H SQ/DQ spectrum (Figure 3B), correlations were observed between the ligand protons with free water molecules (B and C) and those of the modulator (A and C). Additionally, a correlation was detected between the modulator protons and an acidic hydrogen at ~12 ppm (A and D), which was not observed in the ^{1}H-MAS experiments at 32 kHz. This correlation was assigned to carboxylic acid protons from modulator molecules found within the pores, as, in the structure, they are coordinated as carboxylates. These observations provide further evidence for the presence of free water molecules inside the pores, while the modulator is found both in the pores and as part of the MOF structure.

2.2. Composite Hydrogels Synthesis and Characterization

Composite hydrogels were synthesized via a click reaction in water between the amine groups of L-PEI·HCl and the epoxy ring of the EGDE component under mild conditions. Apart from the conditions implemented in the filler synthesis, the environmentally friendly synthesis conditions of click chemistry were adopted to minimize the ecological and economic impacts of the materials employed. The gelation consisted of a nucleophilic attack of the nitrogen from the polymer to the electrophilic carbon from a previously protonated epoxy ring of the crosslinker. Overall, a covalent bond between the two species was generated, giving rise to the reaction of both extremes of the EGDE molecules, the crosslinking of the different polymer chains, and finally, forming a 3D network structure. The same crosslinking reaction was effective when dispersed with the DUF-67 nanoparticles in the gelation media. Indeed, composites with 5, 10, and 15 wt% of the MOF with respect to the polymeric matrix were assembled by applying this strategy. In the last step, an acidic treatment was carried out to ensure the hydrolysis of any remaining epoxy rings to form 1,2-diol groups. The three composite hydrogels were named 5%D, 10%D, and 15%D.

The influence of DUT-67 integration into the structure, chemistry, and mechanical properties of the composite hydrogels was characterized before their application for water remediation ends. First, the texture of the composite hydrogels hindered the detection of DUT-67 fingerprints by PXRD. Complementarily, the ^{1}H high-resolution magic angle spinning (HRMAS) NMR data displayed two distinct zones with ^{1}H chemical shifts ranging from 3.5 to 4.3 and 2.2 to 3.2 ppm, corresponding to the hydrogen atoms of the crosslinker and the polymer, respectively (Figure 4A). No resonance signals related to the filler were detected, which was expected due to the sample size and the low filler load (0.37 mg of the 2.5 mg employed to perform the experiment). Nonetheless, the resonance signals for the crosslinker and the polymer were affected by the MOF particles, which induced changes in the chemical environment and the structure of the hydrogel. These results also indicated that the distribution of the filler might be homogenous throughout the hydrogel structure.

The XPS spectra of 15%D corroborated the presence of zirconium ions with a similar Zr $3d_{5/2}$ and $3d_{3/2}$ binding energies (182.4 and 184.9 eV) compared to the ones of DUT-67 fillers (Figure 4B and Table S2). In the high-resolution C 1s spectrum, the three contributions observed at 284.7, 286.0, and 288.6 eV were ascribed to C–C and C=C bonds in the adventitious carbon and hydrogel chains (284.8 eV), to C–N and C–O bonds of the hydrogel chains (286.0 eV), and to carboxylate groups (288.6 eV). The contributions of C 1s in the composite hydrogels were consistent with those of P1.5E (Table S4). The contributions of carbon atoms from DUT-67 overlapped with those of the hydrogel components. A similar scenario was found for the XPS signals ascribed to O 1s (532.0 and 533.2 eV), which were almost identical to those of the pristine hydrogel (Table S4). Overall, the O 1s peak was ascribed to the sum of the contributions from the C–O bonds within the hydrogel structure and water molecules adsorbed on the material surface, respectively. Last, the bimodal N 1s spectrum (399.1 (C–N) and 401.5 (C–N$^+$) eV) was related to neutral and protonated amine groups of the hydrogel [30,31].

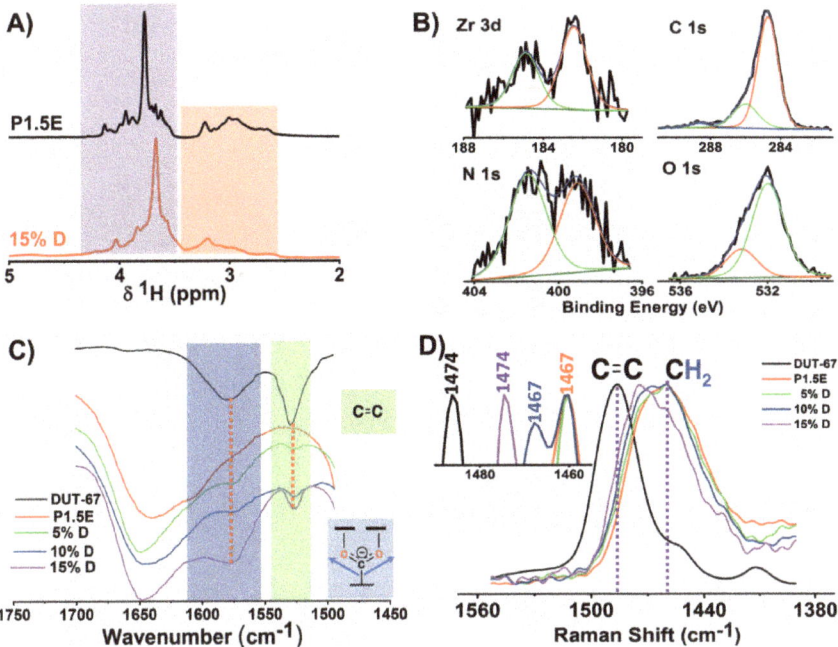

Figure 4. (**A**) ^{1}H HRMAS NMR spectra of P1.5E and 15%D materials. (**B**) XPS spectra of 15%D. (**C**) ATR–FTIR and (**D**) FT Raman of DUT—67, pristine hydrogel, and composite hydrogels.

ATR-FTIR and FT Raman spectroscopies can be used to confirm the presence of the inorganic and organic components of DUT-67 in composite hydrogels. If the MOF structure was altered during the hydrogel synthesis, the ligand would be in the form of carboxylic acid instead of carboxylate. Each form of the ligand absorbs IR radiation at different wavelengths. Figure S4 compares the FTIR spectra for DUT-67, pristine, and composite hydrogels. At first, there were no apparent differences among the spectra, which showed the characteristic signals for the hydrogel dominating the experimental contribution. These included the stretching vibration of the O–H and N–H bonds at 3000 cm^{-1}, the vibration of the C–O–C (from the EGDE crosslinker) and C–N (from the polymer and the bond formed between the polymer and the crosslinker bridges) at 1070 cm^{-1}, the bands corresponding to the vibrations of the aliphatic chains in both hydrogel components at 1454 and 2850 cm^{-1}, and a low-intensity band at 1640 cm^{-1} from the bending vibration of the N–H bond of the secondary amines. However, after amplifying and normalizing the curves in the 1400–1700 cm^{-1} range (Figure 4C), the symmetric stretching vibration from the carboxylate (1574 cm^{-1}) and the vibration of the C=C bonds from the aromatic ring (1531 cm^{-1}) of the 2,5-thiophenedicarboxylate linker were detected [27]. Moreover, there was a direct correlation between the DUT-67 content in the composite hydrogels and the intensity of the signals. Similarly, the normalization and amplification of the RAMAN spectra were necessary to extract the signature associated with DUT-67 from the ones of the polymeric hydrogel matrix (Figure S5). The characteristic vibrational modes of C–H (2800 and 3000 cm^{-1}) and the bending vibration of -CH$_2$- (1467 cm^{-1}) dominated the Raman spectra. The characteristic vibration of the C=C bond from 2,5-thiophenedicarboxylate (1474 cm^{-1}) was very close to the signals arising from the -CH$_2$- group of the hydrogel [33].

To complete the structural characterization of the materials, the morphology of the samples was studied (Figure 5) by SEM, after being swollen in water and then lyophilized. Even though the structure of the gels collapsed after lyophilization, an interconnected porous structure could still be observed. The inclusion of DUT-67 particles did not affect the pore size of the hydrogels. Additionally, a partial agglomeration of filler particles was observed. Nevertheless, these agglomerated along with smaller DUT-67 particles and were distributed throughout the hydrogel matrix.

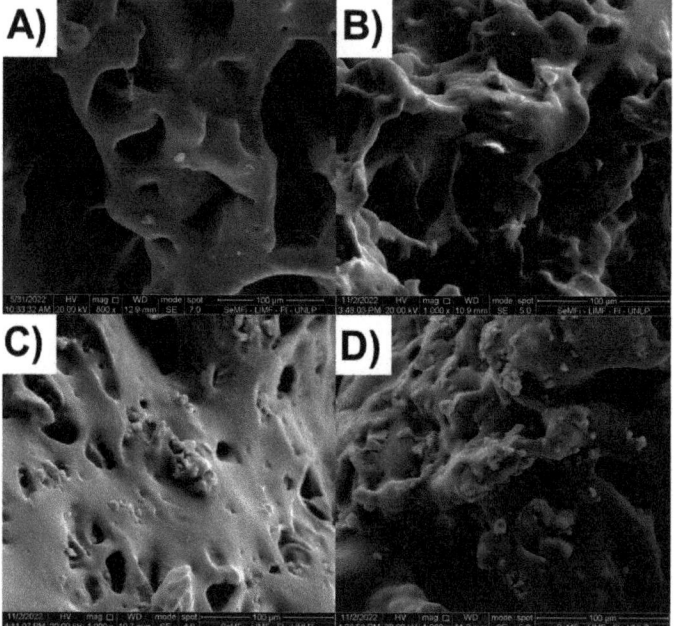

Figure 5. SEM images of (**A**) P1.5E, (**B**) 5%D, (**C**) 10%D, and (**D**) 15%D hydrogels.

2.3. Thermal, Swelling, and Mechanical Properties of L-PEI·HCl@DUT-67 Hydrogels

The thermal behavior of the samples was studied through TGA (Figure S6). For all the samples a two-stage thermal evolution was observed. The first step occurred between 10 and 200 °C and accounted for the dehydration of the hydrogels, with a weight loss of approximately 8% of the initial mass. At temperatures above 230 °C, the pyrolysis of the hydrogel structure began, ending at approximately 570 °C. Overall, the weight loss ascribed to this second stage was approximately of 70% of the initial one. In agreement with the presence of DUT-67, the weigh percentage of the residue observed at 800 °C increased as the MOF loading did in the composite hydrogels, because the final thermal degradation product of DUT-67 was a ZrO$_2$ residue. Although there were subtle differences on the TGA curve between the different hydrogels, we could not find a correlation between them and the degradation profile of DUT-67 sample.

Next, the swelling of the hydrogels was investigated, as illustrated in Figure 6A. The hydrogels rehydrated rapidly (~5 min), hindering the study of the kinetics. The swelling tended to decrease with the MOF loading. This tendency could be attributed to three factors. First, the filler particles restricted the freedom of the polymer chains to accommodate and support more water molecules in their network structure. Second, the interaction among polymer chains and the MOF particles (as indicated by the ^{1}H HRMAS NMR experiments) reduced the number of available water adsorption sites for water adsorption into the hydrogel network. Third, the interactions between the two components leads to structural changes in the material, which affected its water retention capacity. Additionally, it is worth noting that DUT-67 is a micro-porous material that does not exhibit breathing-triggered morphological changes associated with water adsorption due to its rigid framework, and thus, it does not contribute to the overall swelling of the composites.

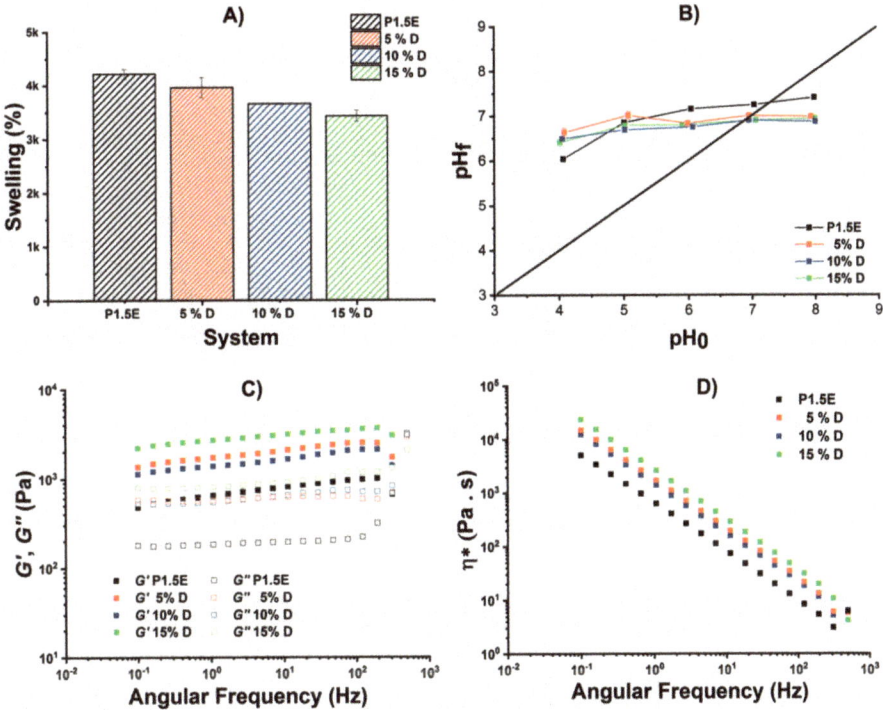

Figure 6. (**A**) Swelling capacities, (**B**) pH$_{pzc}$, (**C**) G' and G'', and (**D**) η^* of P1.5E, 5%D, 10%D, and 15%D hydrogels.

The point of zero charge (pH$_{pzc}$), which is an important factor in evaluating the effect of pH in adsorption studies [34], was evaluated as well in our system, as shown in Figure 6B. It should be noted that the amine groups in the polymer were the most sensitive points to pH changes in the hydrogel. The pK_a of PEI was approximately 10, indicating its basic nature [35]. During the hydrogel formation, the secondary amines were crosslinked and became tertiary, which were less basic, resulting in a decrease in the pH with a pH$_{PZC}$ = 7.3 (pristine hydrogel). Moreover, DUT-67 softened the basicity of the material about 0.3 units, resulting in a pH$_{PZC}$ = 7.0 (composite hydrogel), which was related to its interaction with the free electron pair of the amines of the hydrogel skeleton. Therefore, this result demonstrated that there was not only physical support of the DUT-67 into the hydrogel but a chemical interaction between the filler and the nitrogen atoms of the polymeric matrix.

The deformation response of the hydrogels studies in this work was studied by measuring several rheological parameters such as the storage (G') and loss (G'') moduli and complex viscosity (η^*). Figure 6C shows the G' and G'' of the pristine hydrogel and the composite hydrogels. Within the studied frequency range (0.1–500 Hz), all four materials exhibited a linear viscoelastic response. For the four hydrogels, G' (which was related to the elastic response) was higher than G'' (which was related to the viscous response).

This behavior indicated that the three-dimensional network of the hydrogels presented a solid-like and elastic behavior [12]. On the other hand, the G' and G'' values increased concomitantly with the percentage of DUT-67, pointing out that MOF-particles increased the material resistance to the deformation of the hydrogel. This phenomenon could be predicted, since the results of ^{1}H HRMAS and pH_{pzc} allowed inferring that the polymer chains and the filler interacted with each other. Additionally, SEM images revealed that DUT-67 particles were distributed throughout the matrix. The complex viscosity (η^*) is related to the stability of the hydrogel structure [36]. Figure 6D shows η^* values as a function of the angular frequency for the pristine hydrogel and the composite hydrogels. For all materials, η^* decreased linearly as the frequency increased. This finding indicates that the materials behaved as non-Newtonian fluids with a pseudo-plastic character [37,38]. The 15%D hydrogel presented the highest η^* values, which decreased and as the filler content decreased, with the pristine hydrogel presenting the lowest values. This finding further supports the hypothesis that DUT-67 is acting as a mechanical reinforcement of the three-dimensional network of the hydrogels obtained, since it interacts adequately with the polymer chains and is distributed throughout the entire matrix, allowing for a greater stress-bearing capacity of the materials.

2.4. Functional Assessment of PEI@DUT-67 Hydrogels as Sorbents for Water Remediation

Due to its mechanical and chemical properties, 15%D was selected to assess its capacity and kinetics for the adsorption of MO, PEN antibiotic, and Cu^{2+} metal ions. To this end, experimental adsorption kinetics and isotherms data were obtained, and these data were fitted to mathematical models described in Experimental Section S1. Since the fittings were performed with non-linearized data, chi-square (χ^2) and the residual sum of squares (RSS) were taken as the criteria to evaluate their degree of matching. The kinetics models used to fit the data were the pseudo-first order, pseudo-second order, and the Freundlich modified models. The isotherm models used to fit the data were the Langmuir and Freundlich models.

The experimental adsorption rates of MO at pH = 6, along with the modeled kinetic data, are presented in Figure 7A and Table S5, respectively. The adsorption process for both hydrogels exhibited rapid kinetics, reaching equilibrium after approximately 120 min of contact. The experimental data were better fitted by the Freundlich modified model for the 15%D hydrogel, while the Elovich model provided a better fit for P1.5E. In the case of P1.5E, the suitability of the Elovich model suggested the presence of heterogeneous adsorption sites with chemisorption as the rate-controlling step of the process [39]. Since MO is an anionic pollutant and the hydrogel has a positive charge density, electrostatic attraction likely played a role in the adsorption process [40]. Conversely, for the 15%D hydrogel, the kinetics of MO adsorption was not controlled by the intra-particle diffusion process, as indicated by the value of $m > 2$. In general, adding DUT-67 particles altered the response of the material improving its kinetics of adsorption but also the adsorption capacity for a fixed MO concentration.

The experimental and modeled adsorption kinetics of P1.5E and 15%D for the capture of Cu^{2+} at pH = 4 are presented in Figure 7B and Table S6, respectively. Like the adsorption of MO, the adsorption of Cu^{2+} was rapid, even faster than that of MO. In this case, equilibrium was achieved after approximately 60 min. The interaction between Cu^{2+} ions and nitrogen atoms from secondary amines became stronger as stable coordinated complexes were formed. The experimental data were better fitted by the pseudo-second order model for the 15%D hydrogel, and by the pseudo-first model for P1.5E. However, for both hydrogels, the χ^2 and RSS values (Table S6) of the pseudo-first and pseudo-second order models were found to be similar. Thus, the adsorption kinetics of Cu^{2+} by both hydrogels could be adequately described by both models, indicating that the adsorption of Cu^{2+} involved physical and chemical adsorption processes [41–45].

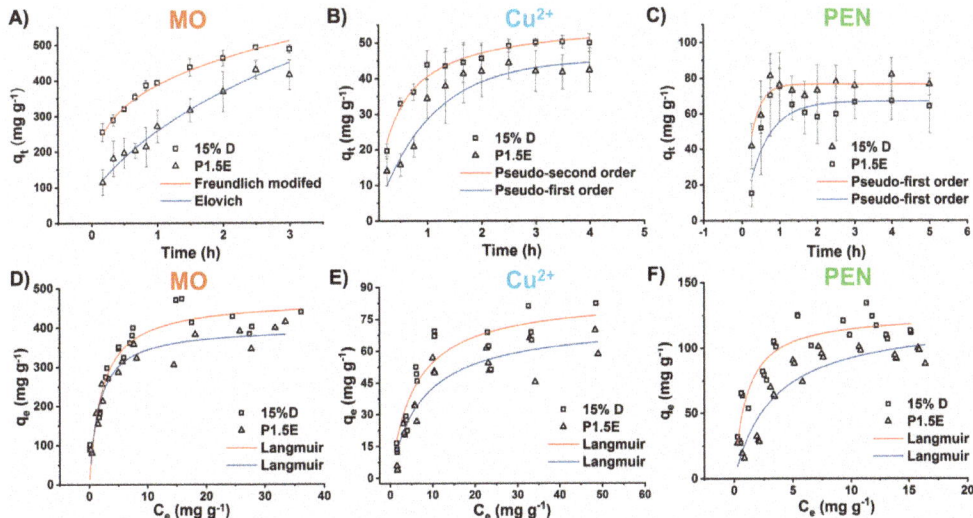

Figure 7. Adsorption kinetics of (**A**) MO, (**B**) Cu^{2+}, and (**C**) PEN by the hydrogels and kinetics models fittings. Adsorption isotherms of (**D**) MO, (**E**) Cu^{2+}, and (**F**) PEN and fittings to the Langmuir model.

The adsorption rates of PEN at pH = 4 and the modeled kinetic data are presented in Figure 7C and Table S7, respectively. The adsorption of PEN was the most rapid of the three pollutants studied, reaching equilibrium after approximately 45 min. At pH = 4, the hydrogel amine groups were protonated (positively charged), while most carboxylic acid molecules were deprotonated (negatively charged), suggesting an electrostatic interaction between the material surface and the PEN molecules. For both P1.5E and the 15%D hydrogel, the experimental data were better fitted by the pseudo-first model. However, the values of the evaluation criteria for the pseudo-first and pseudo-second order models exhibited only a slight difference, suggesting that the controlling step of PEN adsorption involved physical and chemical processes.

Adsorption isotherms describe the medium conditions under which the equilibrium is reached between the adsorbate, that interacts with the host, and the dissolved species in the liquid phase, at a constant temperature. By fitting the experimental data obtained from the adsorption isotherm experiments to mathematical models, the mechanism of the adsorption and the maximum capacity (q_e) to host the adsorbate by the adsorbent can be determined. That is, it delivers information about the number of adsorption sites and the strength/affinity of the interaction between the host and the guest. In this study, the adsorption isotherms of MO, Cu^{2+}, and PEN on the hydrogels at different initial concentrations and at 25 °C were determined. The fitted parameters are shown in Figure 7D–F, and Tables S8, S9 and S10, respectively. For all three pollutants (MO, Cu^{2+}, and PEN), the experimental isotherms data for P1.5E and the 15%D hydrogel were better fitted by the Langmuir model, indicating that MO, PEN, and Cu^{2+} could form a monolayer on the hydrogel surface, favoring the chemical adsorption process.

The experimental values of the maximum adsorption capacities (q_{max}) for MO were 381 ± 24 mg g^{-1} and 420 ± 14 mg g^{-1}, for P1.5E and 15%D, respectively, while the calculated values were 402 ± 14 mg g^{-1} and 473 ± 21 mg g^{-1}. For Cu^{2+}, the experimental q_{max} values were 59 ± 9 mg g^{-1} and 71 ± 9 mg g^{-1}, with calculated values of 74 ± 6 mg g^{-1} and 86 ± 6 mg g^{-1}, for P1.5E and 15%D, respectively. Finally, the experimental q_{max} values for PEN were 98 ± 9 mg g^{-1} and 115 ± 4 mg g^{-1}, with calculated values of 123 ± 9 mg g^{-1} and 127 ± 4 mg g^{-1}, for P1.5E and 15%D, respectively. All experimental and calculated values exhibited differences within a range of 4–11%, which is considered acceptable. Additionally, the synergistic effect between both components of the composite hydrogel was evident, as the adsorption capacity for 15%D was consistently higher than that of P1.5E for the three pollutants. Complementarily, the adsorption of pollutants on the composite hydrogels was studied by using ATR-FTIR spectroscopy (Figures S7–S9). The adsorption process was evident for MO, where absorption bands were added to those of hydrogels due to the high content in relation to the adsorbent. However, no significant changes were shown for Cu^{2+} ions and PEN due to the lower uptake by the adsorbents compared to MO. In general, it is difficult to visualize the adsorption process with this technique, because the matrix masks the absorption signals of the adsorbed molecules.

Considering the results from the functional assessment of our hydrogels, the adsorption kinetics and capacity of composite 15%D were always faster and higher than the observed ones for P1.5E, indicating a synergistic effect between both porous components, the hydrogel network structure and the micropore space of the DUT-67 nanoparticles. Similar synergic effects have been reported for polymer/MOF composite systems, where the kinetics and capacity of the systems overcame the average of their separate components [46,47].

2.5. Effect of the pH on the Adsorption Capacity

The pH value plays a crucial role in the adsorption process as it can influence the speciation and the charge of both the adsorbent and the adsorbate. To investigate the effect of the pH on the adsorption capacities of P1.5E and 15%D, adsorption experiments with MO and PEN at pH = 4, 7, and 10 were conducted, and the results are presented in Table 1. For Cu^{2+} ions, the effect of the pH on the adsorption process could not be studied, as Cu^{2+} precipitates as copper hydroxide at pH $\geq$ 5.2 [48]. Notably, for both materials, as the pH values increased, a decrease in the adsorption capacity was observed. This outcome was anticipated due to the nature of the hydrogels and pollutants involved. At pH $\geq$ 4, most of the MO (pK_a = 4.3) and PEN (pK_a = 4.3) molecules were negatively charged. Consequently, at pH = 4, the secondary and tertiary amine groups on the polymer were protonated, enabling the material to interact effectively with the negatively charged MO and PEN. As the pH increased, the amine groups in the polymer became deprotonated, leading to a reduced capacity for the hydrogels to interact electrostatically with the pollutant molecules [7].

Table 1. Effect of the pH on the adsorption capacities of MO and PEN by P1.5E and 15%D hydrogels.

System	pH 4		pH 7		pH 10	
	MO (mg g^{-1})	PEN (mg g^{-1})	MO (mg g^{-1})	PEN (mg g^{-1})	MO (mg g^{-1})	PEN (mg g^{-1})
P1.5E	696 ± 4	98 ± 9	462 ± 9	24 ± 3	78 ± 5	1.6 ± 1.0
15%D	799 ± 32	115 ± 4	584 ± 20	29 ± 3	92 ± 11	2.3 ± 0.6

2.6. Competitive Adsorption

Competitive adsorption experiments on a binary mixture were conducted to assess the effect of multiple pollutants on the adsorption process. The binary mixture consisted of Cu^{2+} and PEN, as each interacted differently with the materials. The experiments employed the same initial molar concentration, and the results are presented in Table 2.

Table 2. Competitive adsorption of Cu^{2+} and PEN by the hydrogels.

System	Cu^{2+} (mmol g^{-1})	PEN (mmol g^{-1})	Cu^{2+} Mixed (mmol g^{-1})	PEN Mixed (mmol g^{-1})
P1.5E	0.20 ± 0.01	0.32 ± 0.02	0.34 ± 0.06	0.22 ± 0.01
15%D	0.38 ± 0.02	0.36 ± 0.01	0.59 ± 0.02	0.29 ± 0.01

Once again, the synergistic effect between both components in 15%D was observed, with its adsorption capacity being higher than that of P1.5E in all cases. In the case of the mixture, both P1.5E and 15%D showed a preference for Cu^{2+} ions, resulting in an increase in the Cu^{2+} adsorption capacity, while decreasing the PEN adsorption capacity. The opposite was expected, since the association (k_a) constant from the Langmuir model was higher for PEN than for Cu^{2+} (Tables S9 and S10, respectively). This result can be attributed to the chemical nature of PEN: a β-lactam compound prompt to coordinate metal ions [49,50]. After being adsorbed, PEN molecules can coordinate Cu^{2+} ions with the nitrogen atoms of the PEI structure, as illustrated in Figure 8, thereby increasing the capacity of the material to uptake Cu^{2+} from the solution.

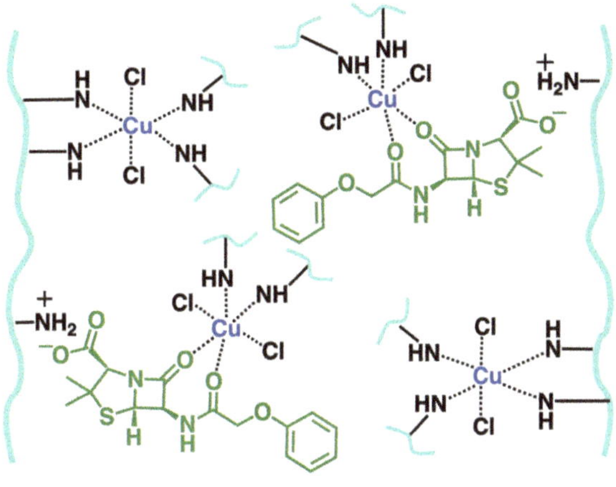

Figure 8. Diagram of the interaction between Cu^{2+} and PEN within the material. The secondary amine groups are simplified as —NH for clarity.

3. Conclusions

The modulation of the particle size of DUT-67 was achieved in a green-synthesis path by controlling the water-to-modulator ratios. Through experiments using ss-NMR spectroscopy, the presence of (i) modulator molecules likely coordinated to the zirconium hexanuclear clusters of the DUT-67 structure and (ii) water molecules adsorbed into the pores were identified. XPS, NMR, FTIR, and FT Raman studies confirmed the presence and interaction of the DUT-67 particles with the polymeric chains within the hydrogel network. The composite hydrogel with 15% filler content exhibited the highest resistance to deformation, indicating a reinforcement of the hydrogel network structure by the porous MOF filler. Additionally, 15%D showed an improved adsorption capacity and kinetics as compared to the pristine hydrogel. The adsorption of MO, Cu^{2+} ions, and PEN on the hydrogels involved both physical and chemical adsorption processes. The Langmuir model provided the best fit for the adsorption isotherms, suggesting a monolayer-type adsorption of the pollutants on the hydrogel surface. The adsorption capacity decreased as the pH increased, as the deprotonation of amine groups in the polymer reduced the interaction with the negatively charged pollutants. In the binary mixture experiments, the hydrogels showed a preference for Cu^{2+} ions over PEN, resulting in an increased Cu^{2+} adsorption capacity and a decreased PEN adsorption capacity.

4. Materials and Methods

4.1. Materials

L-PEI·HCl (~87 kDa) was synthesized as described previously [51]. Analytical grade EGDE (TCI—Portland, OR, USA), and 2,5-thiophenedicarboxylic acid (H_2TDC), acetic acid, zirconium dichloride oxide ($ZrOCl_2·8H_2O$), sodium acetate, hydrochloric acid, sodium hydroxide, MO, PEN, phosphate buffer, sodium chloride, copper(II) sulphate, and D_2O (Merck—Rahway, NJ, USA) were of analytic grade and were used without further purification.

4.2. DUT-67 Synthesis

The DUT-67 samples were prepared by a high-yield water-based reflux-synthesis described previously by Reinsch et al. [52]. with some modifications. Briefly, 6.44 g (20 mmol) of $ZrOCl_2 \cdot 8H_2O$ were dissolved in 100 mL of water and acetic acid solution, and then, 2.29 g (13.3 mmol) of H_2TDC were added to the reactor under stirring conditions. The concentration of acetic acid in the water solution was employed as the tool to control the particle size of the final DUT-67 material, as detailed in Table S1. The mixture was then heated at 95 °C and kept under vigorous stirring for 1 h. It is important to note that a white precipitate was generated after the addition of the organic linker to the $ZrOCl_2 \cdot 8H_2O$ solution. The DUT-67 particles were recovered by centrifugation (19,000 rpm, 30 min), redispersed twice in aqueous sodium acetate solution (15 mL, 0.1 M) and water (15 mL), and finally dried at 100 °C overnight. After drying, the solids were milled and named L52 (3.70 g), L53 (3.95 g), and L54 (3.87 g).

4.3. MOF@L-PEI·HCl Composite Hydrogels Synthesis

The L54 sample was employed as the filler to assemble the MOF@L-PEI·HCl composite hydrogels. First, DUT-67 200 nm particles were dispersed in 6.4 mL of H_2O and sonicated for 1 h. The concentration of the MOF particles in the dispersion was controlled to obtain composite hydrogels with 5, 10, and 15 wt% of the filler that were added to 6.4 mL of H_2O. The mixture was sonicated for 1 h. Then, 320 mg (~1.7 mmol of N) of L-PEI·HCl and 0.8 mL (2.6 mmol) of EGDE were added to the MOF-dispersions. The solutions were heated at 90 °C for 180 min while stirring. After the synthesis, the hydrogels were swelled with distilled water, and further washes were performed to remove the unreacted species. The samples were sieved through a 1000 μm mesh to homogenize the particle size. Residual epoxide groups were hydrolyzed with 6 M HCl for 1 h. Finally, the samples were washed three times with 10 mM phosphate buffer to equilibrate to pH = 6 and then with distilled water. The initial L-PEI·HCl hydrogel and the composite ones obtained after the integration of the DUT-67 particles were named P1.5E, 5%D, 10%D, and 15%D, respectively. The reaction yields were calculated considering the initial mass of both precursors and the filler.

4.4. Characterization Techniques

Details related to the characterization techniques can be found in the Supplementary Materials (Experimental S1).

The morphology and hydrogel cross-section microstructure of the DUT-67 samples were studied by SEM. PXRD patterns of the DUT-67 samples were obtained using a Panalytical X 'pert CuKα diffractometer. ATR-FTIR and FT Raman spectra were recorded on a Nicolet iS50 spectrometer (Thermo Scientific—Waltham, MA, USA) using a one-reflection diamond crystal. HRMAS NMR and ss-NMR spectra were acquired with a Bruker Avance-III HD spectrometer equipped with a 14.1 T narrow bore magnet. The TGA was recorded with a TGA-50 Shimadzu. XPS analysis was carried out with a Physical Electronics (Versa-Pro II) operating with a monochromatic X-ray source Al (k-alpha) of photons at 1486 eV under ultra-high vacuum using a pressure of 10^{-6} Pa. CO_2 high-pressure adsorption isotherms were acquired using ISorb-1 equipment from 0 to 30 bars and after activating the sample at 120 °C for 4 h. The BET surface areas were calculated following the protocol described by Kim et al. [53] and considering the fitting of the data between 0.05 and 0.35 bars. The viscoelastic behavior was determined with an Anton Paar rotational rheometer (MCR-301).

4.5. Kinetic and Adsorption Capacity Assessment

Adsorption tests were conducted in batches at room temperature (25 °C) and with constant stirring (100 rpm). Adsorption isotherms data were determined using 5 mg of dried hydrogel added to an aqueous solution (100 mL) of MO, Cu^{2+}, or PEN with concentrations ranging from 4 to 45 mg L^{-1}, 6 to 60 mg L^{-1}, and 2 to 18 mg L^{-1}, respectively. The concentration difference after reaching the equilibrium allowed for estimating the adsorption capacity of each of the studied materials at a given concentration of the pollutant. In parallel, the kinetics of the adsorption were determined by measuring the sorbate concentration decay in the solution supernatant during the adsorption process. To this end, MO, Cu^{2+}, and PEN solutions with 60, 30, and 12 mg L^{-1} concentrations were employed to perform the experiments.

The concentrations of MO, Cu^{2+}, and PEN before, during, and after adsorption were determined by UV Vis spectroscopy. MO presented a characteristic peak at 468 nm. For Cu^{2+}, the chromogenic complexing agent 1-(2-pyridylazo)-2-naphtol (PAN) [54] was used, obtaining a Cu^{2+} complex that presented a characteristic absorption peak at 548 nm. PEN presented a characteristic peak at 220 nm. For the three target pollutants studied in this work, calibration curves were constructed. The experimental data obtained from the adsorption experiments were adjusted to kinetics and isotherm models (Experimental S1).

Supplementary Materials: The following supporting information can be downloaded at: https://www.mdpi.com/article/10.3390/gels9110909/s1, Experimental S1. Characterization techniques. Experimental S2. Kinetic and Isotherm models. Figure S1. FT Raman spectra f L52, L53, and L54; Figure S2. TGA curves of L52 and L53; Figure S3. High pressure CO_2 adsorption isotherms at 0 °C of L52, L53, and L54; Figure S4. ATR-FTIR spectra of L54, P1.5E, 5%D, 10%D, and 15%D; Figure S5. FT RAMAN spectra of L54, P1.5E, 5%D, 10%D, and 15%D; Figure S6. TGA curves of L54, P1.5E, 5%D, 10%D, and 15%D; Figure S7. ATR-FTIR spectra of 15% D after MO adsorption; Figure S8. ATR-FTIR spectra of 15% D after Cu^{2+} adsorption; Figure S9. ATR-FTIR spectra of 15% D after PEN adsorption. Table S1. Solvent (water): modulator (acetic acid) proportions for DUT-67 synthesis; Table S2. Fitting parameters from XPS data of L54; Table S3. Fitting parameters from XPS data of 15%D; Table S4. Fitting parameters from XPS data of P1.5E; Table S5. Kinetic parameters for MO adsorption; Table S6. Kinetic parameters for Cu^{2+} adsorption; Table S7. Kinetic parameters for PEN adsorption; Table S8. Isotherm model parameters for MO adsorption; Table S9. Isotherm model parameters for Cu^{2+} adsorption; Table S10. Isotherm model parameters for PEN adsorption. References [55–66].

Author Contributions: Conceptualization, L.M.A., G.J.C. and J.M.L.-M.; methodology, L.M.A., G.J.C. and J.M.L.-M.; validation, L.M.A., G.J.C. and J.M.L.-M.; formal analysis, L.M.A. and J.M.L.-M.; investigation, L.M.A., G.J.C., R.F.d.L., C.J.P., A.F.-M., A.I.-M., E.R.-C. and J.M.L.-M.; resources, G.J.C., R.F.d.L., A.F.-M., A.I.-M., E.R.-C. and J.M.L.-M.; data curation, L.M.A., G.J.C. and J.M.L.-M.; writing—original draft preparation, L.M.A. and J.M.L.-M.; writing—review and editing, L.M.A., G.J.C. and J.M.L.-M.; visualization, L.M.A. and J.M.L.-M.; supervision, J.M.L.-M.; project administration, J.M.L.-M.; funding acquisition, J.M.L.-M. All authors have read and agreed to the published version of the manuscript.

Funding: This research was funded by the Agencia Nacional de Promoción de la Investigación, el Desarrollo Tecnológico y la Innovación (ANPCYT, Argentina), project PICT 2019-00845; Universidad de Buenos Aires, project UBACyT 2020-2024/11BA and Ministry of Science and Innovation (Spain), project PID2021-126235OB-C32 funded by MCIN/AEI/10.13039/501100011033 and FEDER and project TED2021-130756B-C31 funded by MCIN/AEI/10.13039/501100011033 and by "ERDF A way of making Europe" by the European Union NextGenerationEU/PRTR. The MSCA-RISE-2017 (No 778412) INDESMOF, which received funding from the European Union's Horizon 2020 research and innovation program, is also acknowledged.

Institutional Review Board Statement: Not applicable.

Informed Consent Statement: Not applicable.

Data Availability Statement: The data presented in this study are available in the Supplementary Material.

Acknowledgments: L.M.A. is grateful for his doctoral fellowship granted by AGENCIA & CONICET. We would like to thank A. Lucena Serrano (SCAI, UMA) for the NMR experimental setup, and G. Nuñez-Taquía for the English grammar revision.

Conflicts of Interest: The authors declare no conflict of interest.

References

1. Foong, C.Y.; Wirzal, M.D.H.; Bustam, M.A. A Review on Nanofibers Membrane with Amino-Based Ionic Liquid for Heavy Metal Removal. *J. Mol. Liq.* **2020**, *297*, 111793. [CrossRef]
2. Alzain, H.; Kalimugogo, V.; Hussein, K. A Review of Environmental Impact of Azo Dyes. *Int. J. Res. Rev.* **2023**, *10*, 64–689. [CrossRef]
3. Pal, A.; Jayamani, J.; Prasad, R. An Urgent Need to Reassess the Safe Levels of Copper in the Drinking Water: Lessons from Studies on Healthy Animals Harboring No Genetic Deficits. *Neurotoxicology* **2014**, *44*, 58–60. [CrossRef]
4. Peña-Guzmán, C.; Ulloa-Sánchez, S.; Mora, K.; Helena-Bustos, R.; Lopez-Barrera, E.; Alvarez, J.; Rodriguez-Pinzón, M. Emerging Pollutants in the Urban Water Cycle in Latin America: A Review of the Current Literature. *J. Environ. Manag.* **2019**, *237*, 408–423. [CrossRef] [PubMed]
5. Sellaoui, L.; Gómez-Avilés, A.; Dhaouadi, F.; Bedia, J.; Bonilla-Petriciolet, A.; Rtimi, S.; Belver, C. Adsorption of Emerging Pollutants on Lignin-Based Activated Carbon: Analysis of Adsorption Mechanism via Characterization, Kinetics and Equilibrium Studies. *Chem. Eng. J.* **2023**, *452*, 139399. [CrossRef]
6. Zhu, H.; Chen, S.; Duan, H.; He, J.; Luo, Y. Removal of Anionic and Cationic Dyes Using Porous Chitosan/Carboxymethyl Cellulose-PEG Hydrogels: Optimization, Adsorption Kinetics, Isotherm and Thermodynamics Studies. *Int. J. Biol. Macromol.* **2023**, *231*, 123213. [CrossRef]
7. Araque, L.M.; Pérez, C.J.; Infantes-Molina, A.; Rodríguez-Castellón, E.; Copello, G.J.; Lázaro-Martínez, J.M. Linear PEI-based Responsive Hydrogels: Synthesis and Characterization. *J. Appl. Polym. Sci.* **2023**, *140*, e54047. [CrossRef]
8. Crespi, A.F.; Zomero, P.N.; Pérez, A.L.; Brondino, C.D.; Molina, A.I.; Linck, Y.G.; Monti, G.A.; Fernández, M.A.; Rodríguez-Castellón, E.; Lázaro-Martínez, J.M. Montmorillonite Materials with Paramagnetic Metal Complexes: Structural Studies and Catalytic Degradation of Emerging Pollutants. *J. Environ. Chem. Eng.* **2023**, *11*, 111420. [CrossRef]
9. Sugawara, A.; Asoh, T.A.; Takashima, Y.; Harada, A.; Uyama, H. Composite Hydrogels Reinforced by Cellulose-Based Supramolecular Filler. *Polym. Degrad. Stab.* **2020**, *177*, 109157. [CrossRef]
10. Rao, Z.; Dong, Y.; Liu, J.; Zheng, X.; Pei, Y.; Tang, K. Genipin-Crosslinked Gelatin-Based Composite Hydrogels Reinforced with Amino-Functionalized Microfibrillated Cellulose. *Int. J. Biol. Macromol.* **2022**, *222*, 3155–3167. [CrossRef]
11. Liu, C.; Dai, T.; Wu, X.; Ma, J.; Liu, J.; Wu, S.; Yang, L.; Zhao, H. 3D Bioprinting of Cell-Laden Nano-Attapulgite/Gelatin Methacrylate Composite Hydrogel Scaffolds for Bone Tissue Repair. *J. Mater. Sci. Technol.* **2023**, *135*, 111–125. [CrossRef]
12. Chen, L.; Wu, Y.; Xing, W.; Su, Q.; Tang, L.; Xue, H.; Gao, J. Mechanically Robust Composite Hydrogels for High Performance Solar Driven Interface Evaporation. *Chem. Eng. Sci.* **2023**, *267*, 118330. [CrossRef]
13. Rehman Shah, H.U.; Ahmad, K.; Naseem, H.A.; Parveen, S.; Ashfaq, M.; Rauf, A.; Aziz, T. Water Stable Graphene Oxide Metal-Organic Frameworks Composite (ZIF-67@GO) for Efficient Removal of Malachite Green from Water. *Food Chem. Toxicol.* **2021**, *154*, 112312. [CrossRef]
14. Ahmad, K.; Shah, H.U.R.; Parveen, S.; Aziz, T.; Naseem, H.A.; Ashfaq, M.; Rauf, A. Metal Organic Framework (KIUB-MOF-1) as Efficient Adsorbent for Cationic and Anionic Dyes from Brackish Water. *J. Mol. Struct.* **2021**, *1242*, 130898. [CrossRef]
15. Ahmad, K.; Nazir, M.A.; Qureshi, A.K.; Hussain, E.; Najam, T.; Javed, M.S.; Shah, S.S.A.; Tufail, M.K.; Hussain, S.; Khan, N.A. Engineering of Zirconium Based Metal-Organic Frameworks (Zr-MOFs) as Efficient Adsorbents. *Mater. Sci. Eng. B* **2020**, *262*, 114766. [CrossRef]
16. Ahmad, K.; Naseem, K.; Shah, H.-R.; Riaz, N.N.; Alhadhrami, A.; Majeed, H.; Ahmad, M.M.; Afzal Awan, M.M.; Ahmad, S.; Ashfaq, M. Towards Sustainable Water Purification: MOFs as a Promising Solution to Eliminate Toxic Water Pollutant Resorcinol. *Z. Phys. Chem.* **2023**, *237*, 1669–1689. [CrossRef]
17. Chen, Z.; Zhang, Z.-B.; Zeng, J.; Zhang, Z.J.; Ma, S.; Tang, C.-M.; Xu, J.-Q. Preparation of Polyethyleneimine-Modified Chitosan/Ce-UIO-66 Composite Hydrogel for the Adsorption of Methyl Orange. *Carbohydr. Polym.* **2023**, *299*, 120079. [CrossRef]
18. Jin, Y.; Li, Y.; Du, Q.; Chen, B.; Chen, K.; Zhang, Y.; Wang, M.; Sun, Y.; Zhao, S.; Jing, Z.; et al. Efficient Adsorption of Congo Red by MIL-53(Fe)/Chitosan Composite Hydrogel Spheres. *Microporous Mesoporous Mater.* **2023**, *348*, 112404. [CrossRef]
19. Drache, F.; Bon, V.; Senkovska, I.; Marschelke, C.; Synytska, A.; Kaskel, S. Postsynthetic Inner-Surface Functionalization of the Highly Stable Zirconium-Based Metal-Organic Framework DUT-67. *Inorg. Chem.* **2016**, *55*, 7206–7213. [CrossRef]
20. Zhuang, G.; Bai, J.; Tan, L.; Huang, H.; Gao, Y.; Zhong, X.; Zhong, C.; Wang, J. Preparation and Catalytic Properties of Pd Nanoparticles Supported on Micro-Crystal DUT-67 MOFs. *RSC Adv.* **2015**, *5*, 32714–32719. [CrossRef]
21. Liu, D.; Qiu, J.; Xu, R.; Liu, J.; Feng, J.; Ouyang, L.; Qian, S.; Qiao, Y.; Liu, X. β-CD/PEI/PVA Composite Hydrogels with Superior Self-Healing Ability and Antibacterial Activity for Wound Healing. *Compos. B Eng.* **2022**, *238*, 109921. [CrossRef]
22. Godiya, C.B.; Revadekar, C.; Kim, J.; Park, B.J. Amine-Bilayer-Functionalized Cellulose-Chitosan Composite Hydrogel for the Efficient Uptake of Hazardous Metal Cations and Catalysis in Polluted Water. *J. Hazard. Mater.* **2022**, *436*, 129112. [CrossRef] [PubMed]
23. Mguni, L.L.; Yao, Y.; Ren, J.; Liu, X.; Hildebrandt, D. Modulated Synthesized Ni-Based MOF with Improved Adsorptive Desulfurization Activity. *J. Clean. Prod.* **2021**, *323*, 129196. [CrossRef]
24. Usman, K.A.S.; Maina, J.W.; Seyedin, S.; Conato, M.T.; Payawan Jr, L.M.; Dumée, L.F.; Razal, J.M. Downsizing Metal–Organic Frameworks by Bottom-up and Top-down Methods. *NPG Asia Mater.* **2020**, *12*, 58. [CrossRef]
25. Bigdeli, F.; Fetzer, M.N.A.; Nis, B.; Morsali, A.; Janiak, C. Coordination Modulation: A Way to Improve the Properties of Metal–Organic Frameworks. *J. Mater. Chem. A Mater.* **2023**, *11*, 22105–22131. [CrossRef]

26. Drache, F.; Bon, V.; Senkovska, I.; Getzschmann, J.; Kaskel, S. The Modulator Driven Polymorphism of Zr(IV) Based Metal–Organic Frameworks. *Philos. Trans. R. Soc. A Math. Phys. Eng. Sci.* **2017**, *375*, 20160027. [CrossRef]
27. Liu, C.; Liu, Y.; Shi, Y.; Wang, Z.; Guo, W.; Bi, J.; Wu, L. Au Nanoparticles-Anchored Defective Metal–Organic Frameworks for Photocatalytic Transformation of Amines to Imines under Visible Light. *J. Colloid Interface Sci.* **2023**, *631*, 154–163. [CrossRef]
28. Liu, H.-X.; Liu, T.-T.; Huang, T.; Fang, Z.-B.; Li, L.; Yin, Q.; Cao, R.; Gong, X.-Q.; Liu, T.-F. Trace of Molecular Doping in Metal–Organic Frameworks: Drastic Change in the Electronic Band Structure with a Preserved Topology and Porosity. *J. Mater. Chem. A Mater.* **2020**, *8*, 12370–12377. [CrossRef]
29. Liu, Y.; Zou, J.; Guo, B.; Ren, Y.; Wang, Z.; Song, Y.; Yu, Y.; Wu, L. Selective Photocatalytic Oxidation of Thioanisole on DUT-67(Zr) Mediated by Surface Coordination. *Langmuir* **2020**, *36*, 2199–2208. [CrossRef]
30. Xu, M.; Zhao, X.; Jiang, H.; Song, X.; Zhou, W.; Liu, X.; Liu, Z.; Wang, H.; Huo, P. Constructing Schottky Junctions via Pd Nanosheets on DUT-67 Surfaces to Accelerate Charge Transfer. *J. Colloid Interface Sci.* **2022**, *608*, 3022–3029. [CrossRef]
31. Zhao, X.; Xu, M.; Song, X.; Zhou, W.; Liu, X.; Wang, H.; Huo, P. Integration of 3D Macroscopic Reduced Graphene Oxide Aerogel with DUT-67 for Selective CO2 Photoreduction to CO in Gas-Solid Reaction. *Chem. Eng. J.* **2022**, *446*, 137034. [CrossRef]
32. Drache, F.; Cirujano, F.G.; Nguyen, K.D.; Bon, V.; Senkovska, I.; Llabrés i Xamena, F.X.; Kaskel, S. Anion Exchange and Catalytic Functionalization of the Zirconium-Based Metal–Organic Framework DUT-67. *Cryst. Growth Des.* **2018**, *18*, 5492–5500. [CrossRef]
33. Sangeetha Margreat, S.; Ramalingam, S.; Sebastian, S.; Xavier, S.; Periandy, S.; Daniel, J.C.; Maria Julie, M. DFT, Spectroscopic, DSC/TGA, Electronic, Biological and Molecular Docking Investigation of 2,5-Thiophenedicarboxylic Acid: A Promising Anticancer Agent. *J. Mol. Struct.* **2020**, *1200*, 127099. [CrossRef]
34. Borges, M.M.C.; Pires, B.C.; Vieira, S.S.; Borges, K.B.; de Lima Guimarães, L.G. Magnetic and PH Responsive Composite Hydrogel-Based on Poly(2-(Diethylamino)Ethyl Methacrylate)/Chitosan for Fipronil Removal from Aqueous Medium. *React. Funct. Polym.* **2021**, *168*, 105050. [CrossRef]
35. Curtis, K.A.; Miller, D.; Millard, P.; Basu, S.; Horkay, F.; Chandran, P.L. Unusual Salt and PH Induced Changes in Polyethylenimine Solutions. *PLoS ONE* **2016**, *11*, e0158147. [CrossRef]
36. Shu, X.; Wei, Y.; Luo, X.; Liu, J.; Mao, L.; Yuan, F.; Gao, Y. κ-Carrageenan/Konjac Glucomannan Composite Hydrogel Filled with Rhamnolipid-Stabilized Nanostructured Lipid Carrier: Improvement of Structure and Properties. *Food Hydrocoll.* **2023**, *134*, 108088. [CrossRef]
37. Zhang, Z.; Lucia, L. Toward Synergistic Reinforced Graphene Nanoplatelets Composite Hydrogels with Self-Healing and Multi-Stimuli Responses. *Polymer* **2021**, *234*, 124228. [CrossRef]
38. Bhat, M.A.; Rather, R.A.; Shalla, A.H. Texture and Rheological Features of Strain and PH Sensitive Chitosan-Imine Graphene-Oxide Composite Hydrogel with Fast Self-Healing Nature. *Int. J. Biol. Macromol.* **2022**, *222*, 3129–3141. [CrossRef]
39. Nnaji, N.J.N.; Sonde, C.U.; Nwanji, O.L.; Ezeh, G.C.; Onuigbo, A.U.; Ojukwu, A.M.; Mbah, P.C.; Adewumi, A.O.; Unoka, E.C.; Otedo, J.O.; et al. Dacryodes Edulis Leaf Derived Biochar for Methylene Blue Biosorption. *J. Environ. Chem. Eng.* **2023**, *11*, 109638. [CrossRef]
40. Aldahash, S.A.; Siddiqui, S.; Uddin, M.K. Eco-Friendly Synthesis of Copper Nanoparticles from Fiber of Trapa Natans L. Shells and Their Impregnation Onto Polyamide-12 for Environmental Applications. *J. Nat. Fibers* **2023**, *20*, 2224976. [CrossRef]
41. Yang, J.; Huang, Y.; Cheng, Y.; Wu, X.; Lu, J.; Wan, Q.; Feng, J.; Zeng, Q.; Zhao, S.; Yu, L.; et al. Long-Acting Removal of High-Toxic p-Nitrophenol in Wastewater via Peroxymonosulfate Activation by Cyclic Membrane Catalysis. *J. Clean. Prod.* **2023**, *401*, 136739. [CrossRef]
42. Asgari, P.; Mousavi, S.H.; Aghayan, H.; Ghasemi, H.; Yousefi, T. Nd-BTC Metal-Organic Framework (MOF); Synthesis, Characterization and Investigation on Its Adsorption Behavior toward Cesium and Strontium Ions. *Microchem. J.* **2019**, *150*, 104188. [CrossRef]
43. Rao, R.A.K.; Kashifuddin, M. Pottery Glaze-An Excellent Adsorbent for the Removal of Cu(II) from Aqueous Solution. *Chin. J. Geochem.* **2012**, *31*, 136–146. [CrossRef]
44. Semenova, A.; Giles, L.W.; Vidallon, M.L.P.; Follink, B.; Brown, P.L.; Tabor, R.F. Copper-Binding Properties of Polyethylenimine–Silica Nanocomposite Particles. *Langmuir* **2022**, *38*, 10585–10600. [CrossRef] [PubMed]
45. Valverde, A.; Tovar, G.I.; Rio-López, N.A.; Torres, D.; Rosales, M.; Wuttke, S.; Fidalgo-Marijuan, A.; Porro, J.M.; Jiménez-Ruiz, M.; García Sakai, V.; et al. Designing Metal-Chelator-like Traps by Encoding Amino Acids in Zirconium-Based Metal–Organic Frameworks. *Chem. Mater.* **2022**, *34*, 9666–9684. [CrossRef]
46. Tovar, G.I.; Valverde, A.; Mendes-Felipe, C.; Wuttke, S.; Fidalgo-Marijuan, A.; Larrea, E.S.; Lezama, L.; Zheng, F.; Reguera, J.; Lanceros-Méndez, S. Chitin/Metal-Organic Framework Composites as Wide-Range Adsorbent. *ChemSusChem* **2021**, *14*, 2892–2901. [CrossRef]
47. Valverde, A.; de Fernandez-de Luis, R.; Salazar, H.; Gonçalves, B.F.; King, S.; Almásy, L.; Kriechbaum, M.; Laza, J.M.; Vilas-Vilela, J.L.; Martins, P.M. On The Multiscale Structure and Morphology of Pvdf-Hfp@ Mof Membranes in The Scope of Water Remediation Applications. *Adv. Mater. Interfaces* **2023**, *10*, 2300424. [CrossRef]
48. Zhong, J.; Wang, B.; Sun, K.; Duan, J. Hyperbranched Polyethylenimine–Based Polymeric Nanoparticles: Synthesis, Properties, and an Application in Selective Response to Copper Ion. *Colloid Polym. Sci.* **2021**, *299*, 1577–1586. [CrossRef]
49. Sher, A.; Veber, M.; Marolt-Gomišček, M. Spectroscopic and Polarographic Investigations: Copper(II)-Penicillin Derivatives. *Int. J. Pharm.* **1997**, *148*, 191–199. [CrossRef]

50. van Krimpen, P.C.; van Bennekom, W.P.; Bult, A. A Study of the Metal Complexation Behaviour of Some Penicillins, Cephalosporins and Their Derivatives. *Pharm. Weekbl.* **1988**, *10*, 259–266. [CrossRef]
51. Lázaro-Martínez, J.M.; Rodríguez-Castellón, E.; Vega, D.; Monti, G.A.; Chattah, A.K. Solid-State Studies of the Crystalline/Amorphous Character in Linear Poly (Ethylenimine Hydrochloride) (PEI·HCl) Polymers and Their Copper Complexes. *Macromolecules* **2015**, *48*, 1115–1125. [CrossRef]
52. Reinsch, H.; Waitschat, S.; Chavan, S.M.; Lillerud, K.P.; Stock, N. A Facile "Green" Route for Scalable Batch Production and Continuous Synthesis of Zirconium MOFs. *Eur. J. Inorg. Chem.* **2016**, *2016*, 4490–4498. [CrossRef]
53. Kim, K.C.; Yoon, T.-U.; Bae, Y.-S. Applicability of Using CO_2 Adsorption Isotherms to Determine BET Surface Areas of Microporous Materials. *Microporous Mesoporous Mater.* **2016**, *224*, 294–301. [CrossRef]
54. Cheng, K.L.; Bray, R.H. 1-(2-Pyridylazo)-2-Naphthol as Possible Analytical Reagent. *Anal. Chem.* **1955**, *27*, 782–785. [CrossRef]
55. Hartmann, S.R.; Hahn, E.L. Nuclear Double Resonance in the Rotating Frame. *Phys. Rev.* **1962**, *128*, 2042–2053. [CrossRef]
56. Fung, B.M.; Khitrin, A.K.; Ermolaev, K. An Improved Broadband Decoupling Sequence for Liquid Crystals and Solids. *J. Magn. Reson.* **2000**, *142*, 97–101. [CrossRef]
57. van Rossum, B.-J.; Förster, H.; de Groot, H.J.M. High-Field and High-Speed CP-MAS 13C NMR Heteronuclear Dipolar-Correlation Spectroscopy of Solids with Frequency-Switched Lee–Goldburg Homonuclear Decoupling. *J. Magn. Reson.* **1997**, *124*, 516–519. [CrossRef]
58. Feike, M.; Demco, D.E.; Graf, R.; Gottwald, J.; Hafner, S.; Spiess, H.W. Broadband Multiple-Quantum NMR Spectroscopy. *J. Magn. Reson. A* **1996**, *122*, 214–221. [CrossRef]
59. Sun, H.; Zhan, J.; Chen, L.; Zhao, Y. Preparation of CTS/PAMAM/SA/Ca^{2+} Hydrogel and Its Adsorption Performance for Heavy Metal Ions. *Appl. Surf. Sci.* **2023**, *607*. [CrossRef]
60. Lopez-Ramon, M.V.; Stoeckli, F.; Moreno-Castilla, C.; Carrasco-Marin, F. On the Characterization of Acidic and Basic Surface Sites on Carbons by Various Techniques. *Carbon* **1999**, *37*, 1215–1221. [CrossRef]
61. Ho, Y. The Kinetics of Sorption of Divalent Metal Ions onto Sphagnum Moss Peat. *Water Res.* **2000**, *34*, 735–742. [CrossRef]
62. Anirudhan, T.S.; Rejeena, S.R.; Tharun, A.R. Preparation, Characterization and Adsorption Behavior of Tannin-Modified Poly(Glycidylmethacrylate)-Grafted Zirconium Oxide-Densified Cellulose for the Selective Separation of Bovine Serum Albumin. *Colloids Surf. B Biointerfaces* **2012**, *93*, 49–58. [CrossRef] [PubMed]
63. Cheung, C.W.; Porter, J.F.; Mckay, G. Sorption Kinetic Analysis for the Removal of Cadmium Ions from Effluents Using Bone Char. *Water Res.* **2001**, *35*, 605–612. [CrossRef] [PubMed]
64. Teng, H.; Hsieh, C.-T. Activation Energy for Oxygen Chemisorption on Carbon at Low Temperatures. *Ind. Eng. Chem. Res.* **1999**, *38*, 292–297. [CrossRef]
65. Kuo, S.; Lotse, E.G. Kinetics of Phosphate Adsorption and Desorption by Hematite and Gibbsite. *Soil. Sci.* **1973**, *116*, 400–406. [CrossRef]
66. Suen, S.-Y. A Comparison of Isotherm and Kinetic Models for Binary-Solute Adsorption to Affinity Membranes. *J. Chem. Technol. Biotechnol.* **1996**, *65*, 249–257. [CrossRef]

Disclaimer/Publisher's Note: The statements, opinions and data contained in all publications are solely those of the individual author(s) and contributor(s) and not of MDPI and/or the editor(s). MDPI and/or the editor(s) disclaim responsibility for any injury to people or property resulting from any ideas, methods, instructions or products referred to in the content.

MDPI AG
Grosspeteranlage 5
4052 Basel
Switzerland
Tel.: +41 61 683 77 34

Gels Editorial Office
E-mail: gels@mdpi.com
www.mdpi.com/journal/gels

Disclaimer/Publisher's Note: The title and front matter of this reprint are at the discretion of the . The publisher is not responsible for their content or any associated concerns. The statements, opinions and data contained in all individual articles are solely those of the individual Editors and contributors and not of MDPI. MDPI disclaims responsibility for any injury to people or property resulting from any ideas, methods, instructions or products referred to in the content.

www.ingramcontent.com/pod-product-compliance
Lightning Source LLC
LaVergne TN
LVHW070631100526
838202LV00012B/783